Problems and Methods of Optimal Structural Design

MATHEMATICAL CONCEPTS AND METHODS IN SCIENCE AND ENGINEERING

Series Editor: Angelo Miele
Mechanical Engineering and Mathematical Sciences
Rice University

Recent volumes in this series:

11 **INTEGRAL TRANSFORMS IN SCIENCE AND ENGINEERING** • *Kurt Bernado Wolf*

12 **APPLIED MATHEMATICS: An Intellectual Orientation** • *Francis J. Murray*

14 **PRINCIPLES AND PROCEDURES OF NUMERICAL ANALYSIS** • *Ferenc Szidarovszky and Sidney Yakowitz*

16 **MATHEMATICAL PRINCIPLES IN MECHANICS AND ELECTROMAGNETISM Part A: Analytical and Continuum Mechanics** • *C. C. Wang*

17 **MATHEMATICAL PRINCIPLES IN MECHANICS AND ELECTROMAGNETISM, Part B: Electromagnetism and Gravitation** • *C. C. Wang*

18 **SOLUTION METHODS FOR INTEGRAL EQUATIONS: Theory and Applications** • *Edited by Michael A. Golberg*

19 **DYNAMIC OPTIMIZATION AND MATHEMATICAL ECONOMICS** • *Edited by Pan-Tai Liu*

20 **DYNAMICAL SYSTEMS AND EVOLUTION EQUATIONS: Theory and Applications** • *J. A. Walker*

21 **ADVANCES IN GEOMETRIC PROGRAMMING** • *Edited by Mordecai Avriel*

22 **APPLICATIONS OF FUNCTIONAL ANALYSIS IN ENGINEERING** • *J. L. Nowinski*

23 **APPLIED PROBABILITY** • *Frank A. Haight*

24 **THE CALCULUS OF VARIATIONS AND OPTIMAL CONTROL: An Introduction** • *George Leitmann*

25 **CONTROL, IDENTIFICATION, AND INPUT OPTIMIZATION** • *Robert Kalaba and Karl Spingarn*

26 **PROBLEMS AND METHODS OF OPTIMAL STRUCTURAL DESIGN** • *N. V. Banichuk*

A Continuation Order Plan is available for this series. A continuation order will bring delivery of each new volume immediately upon publication. Volumes are billed only upon actual shipment. For further information please contact the publisher.

Problems and Methods of Optimal Structural Design

N. V. Banichuk
Institute for Problems of Mechanics
Academy of Sciences of the USSR
Moscow, USSR

Translated by
Vadim Komkov
Winthrop College

Translation edited by
Edward J. Haug
University of Iowa

Plenum Press · New York and London

Library of Congress Cataloging in Publication Data

Banichuk, Nikolai Vladimirovich.
Problems and methods of optimal structural design.

(Mathematical concepts and methods in science and engineering; vol. 26)
Translation of: Optimizatsiia form uprugikh tel.
Includes bibliographies and index.
1. Structural design—Mathematical models. 2. Elastic analysis (Theory of structures)
I. Haug, Edward J. II. Title. III. Series.
TA658.2.B3413 1983 624.1′771 83-8103
ISBN 0-306-41284-5

ОПТИМИЗАЦИЯ ФОРМ УПРУГИХ ТЕЛ
OPTIMIZATSIYA FORM UPRUGIKH TEL

A Division of Plenum Publishing Corporation
233 Spring Street, New York, N.Y. 10013

Printed in the United States of America

Preface to the English Edition

The author offers a systematic and careful development of many aspects of structural optimization, particularly for beams and plates. Some of the results are new and some have appeared only in specialized Soviet journals, or as proceedings of conferences, and are not easily accessible to Western engineers and mathematicians. Some aspects of the theory presented here, such as optimization of anisotropic properties of elastic structural elements, have not been considered to any extent by Western research engineers.

The author's treatment is "classical", i.e., employing classical analysis. Classical calculus of variations, the complex variables approach, and the Kolosov–Muskhelishvili theory are the basic techniques used. He derives many results that are of interest to practical structural engineers, such as optimum designs of structural elements submerged in a flowing fluid (which is of obvious interest in aircraft design, in ship building, in designing turbines, etc.). Optimization with incomplete information concerning the loads (which is the case in a great majority of practical design considerations) is treated thoroughly. For example, one can only estimate the weight of the traffic on a bridge, the wind load, the additional loads if a river floods, or possible earthquake loads.

It is of interest to compare this monograph with some comprehensive literature reviews and related books, such as: (a) McIntosh, S. C., "Structural Optimization via Optimal Control Techniques; A Review," in *Structural Optimization Symposium*, American Society of Mechanical Engineers, New York, N.Y. (1974), pp. 49–64; (b) Pierson, B. L., "A Survey of Optimal Structural Design under Dynamic Constraints," *International Journal for Numerical Methods in Engineering*, Vol. 4 (1972), pp. 491–499; (c) Haug, E. J., "A Review of Distributed Parameter Structural Optimization Literature," *Optimization of Distributed Parameter Structures* (E. J. Haug and J. Cea, eds.), Sijthoff and Noordhoff, Alphen aan den Rijn, Netherlands (1981), pp. 3–74; (d) Haug, E. J., and Arora,

J. S., *Applied Optimal Design*, Wiley-Interscience, New York (1979); (e) Shield, R. T., *Optimum Design of Structures through Variational Principles*, Lecture Notes in Physics, Vol. 21, No. 1 (1973).

This monograph contains material that is not discussed in any of the above. It truly complements many developments in Western countries. However, one should mention other points of view. Developments in design sensitivity analysis and direct applications of functional analysis to optimal design theory have rapidly advanced in the years 1980–82 in the United States, France, Great Britain, and Denmark. In particular, three recent conferences contain a number of papers shedding new light on optimization techniques in general, and on optimization of structural stability in particular. The NATO-NSF sponsored conference held in Iowa City, Iowa in May of 1980 witnessed a number of presentations in which authors approached problems of structural optimization from the point of view of modern functional analysis. Papers by Haug and Arora,[346] Haug and Komkov,[352] Zolesio,[529, 530] Cea,[286, 287] and Rousselet[484] contained this approach, and problems closely related to this point of view were given by Olhoff and Niordson[445] and Masur.[196] Some authors (Olhoff, Masur, Haug) made specific comments concerning the appearance of singular design and corresponding occurrence of multiple eigenvalues in optimization problems concerning elastic stability and elastic vibrations. It can be shown that optimization against a fundamental eigenmode produces, in some cases, a decrease in the values of higher eigenvalues. Eventually, both the fundamental and the higher eigenmodes "cross over" with disasterous effect on the so-called optimum solution. This discovery led to a new surge of activity in the investigation of multiple eigenvalue solutions to problems of optimization for elastic bodies.

The 17th Midwestern Mechanics Conference held in Ann Arbor, Michigan in May of 1981 witnessed further research developments along these lines. Specifically, the papers of Masur[418] and Komkov[390] pursued related topics. Komkov and Haug observed that designs that are close to singular may be very sensitive to effects of terms that are normally neglected, because of "smallness," and reexamined the commonly used engineering hypothesis in the context of optimal design theory.

These new research directions were again emphasized in a special session of the American Mathematical Society held in Pittsburgh, Pennsylvania in May of 1981. Functional analytic techniques were emphasized by Miersemann,[427] Knightly and Sather,[386] and Payne.[452] The paper by Payne offers a beginning in the direction of postulating, in functional analytic terms, the stabilizability of an elastic or viscoelastic system.

A list of recent advances would not be complete without mentioning some new numerical techniques that offer insight into difficult optimization problems. We complement the author's references by an additional bibliography of mainly Western research articles and collections. This bibliography is appended to the author's list.

EDWARD J. HAUG
VADIM KOMKOV

Foreword

This monograph is devoted to the exposition of new ways of formulating problems of structural optimization and techniques of solution. We recall here some research results concerning the optimum shape and structural properties of elastic bodies subjected to external tractions. We study problems of optimal design with incomplete information, accounting for the interaction between the structure and its environment. This study is devoted to overcoming some basic mathematical difficulties caused by the two-dimensional nature of our problems, by the existence of local functionals, by unknown boundaries, and by incompleteness of information.

This monograph contains results of research that was completed in the Department of Mechanics of Controlled Systems, in the Institute for Problems in Mechanics of the Academy of Sciences of the USSR. It is based mainly on results established by the author. Many results that are described in this book were derived by the author in cooperation with V. M. Kartvelishvili, A. A. Mironov, and A. P. Seiranjan.

Professor F. L. Chernous'ko, who has shown a consistent interest in our work, directed the author's attention to the current interest in some problems considered in this book. In his research, the author had the support of A. Ju. Ishlinskii and A. I. Lur'e. We have included in this book advice and remarks that were given in our study of specific problems by N. H. Arutjunjan, V. I. Birjuk, V. V. Bolotin, L. A. Galin, V. M. Entov, L. M. Kurshin, Ju. P. Lepik, K. A. Lur'e, F. Niordson, G. K. Pozharitskii, V. I. Feodos'ev, V. M. Frolov, and N. N. Janenko. To all the above the author expresses his sincere gratitude.

N. V. BANICHUK

Introduction

Problems of optimization of structures are attracting considerable attention at the present time. A considerable number of papers have been devoted to these problems, mainly published during the last 15 to 20 years. Interest in research in the area of optimal design has grown in connection with the rapid development of aeronautical and space technologies, shipbuilding, and design of precision machinery.

Using optimal design, one can lower considerably the weight of aeronautical structures and equipment and improve mechanical properties of structures. Optimization problems also arise in engineering structural design. Hence, research in this field has definite and valuable applications.

Problems of optimal design also have definite theoretical importance. Introduction and investigation of new types of mathematical problems are interesting in themselves. As we consider optimal design of various physical properties, we also develop effective techniques of optimization, utilizing specific properties of the problem that is investigated. The search for the optimum shape and composition of an elastic body generally encounters very serious mathematical difficulties. Many cases of optimal design can be reduced to the solution of a variational problem with an unknown boundary. Game-theoretic optimization problems also arise, for which there do not exist any systematic techniques of investigation. Known difficulties are related to the fact that optimization of elastic bodies is closely connected with nonlinear problems of mechanics. These nonlinearities are caused by the nonlinear nature of optimality conditions.

Complexity of problems of optimization explains the fact that up to the mid-Sixties, research in this area was limited to a few one-dimensional problems. Commencement of research of a more general nature became possible during the succeeding period of time, due to the growth of mathematical techniques of optimization (such as the calculus of variations, the theory of optimal processes, nonlinear programming, etc.) and the availability of powerful computer technology.

Without any pretense regarding completeness, we shall introduce here a review of some research concerning optimization of elastic structures. We mention only certain classical research results that directly effect the problems considered in this book.

In 1638 Galileo Galilei[173] introduced the concept of uniform strength and determined the shape of a uniformly strong beam. He considered the bending of a cantilevered beam (having a rectangular cross section of constant width, but variable height) subjected to a concentrated load applied at the free end (see Fig. 1). He proved that the condition of uniform strength is satisfied if the variation of the height h satisfies a parabolic law. Subsequently, it turned out that the problem of finding the shape of a beam having minimal weight, with the constraints that normal stress σ_x does not exceed some magnitude σ_0, i.e., $\sigma_x \leqslant \sigma_0$, can be reduced to the problem solved by Galileo. Hence, the cantilevered, uniform strength beam also happens to be the beam of minimum weight.

Other examples were discovered in which the condition of uniform strength insures the minimum weight of a structure. It is this relationship that is generally responsible for the interest in structures of uniform strength. However, further research into the bending of beams with more complex hypotheses made it clear that the concepts of uniform strength and optimality of design are not the same. Various problems of finding an optimum or uniform strength shape of a beam, or of a structure that is composed of beams (while taking into account the specific weight, torsion, and other factors), were considered in Refs. 1, 62, 113-115, 138, 139, 168, 169, 191, 193, 195, 197, 199, 200, 214, and 232.

Real progress in the theory of optimal design came about in connection with research into the shape of a compressed beam (a column) that has minimum weight and can withstand a given load, without the loss of stability (see Fig. 2). This problem was posed by J. Lagrange,[188] but his solution turned out to be erroneous. The optimal shape of a compressed beam was derived by

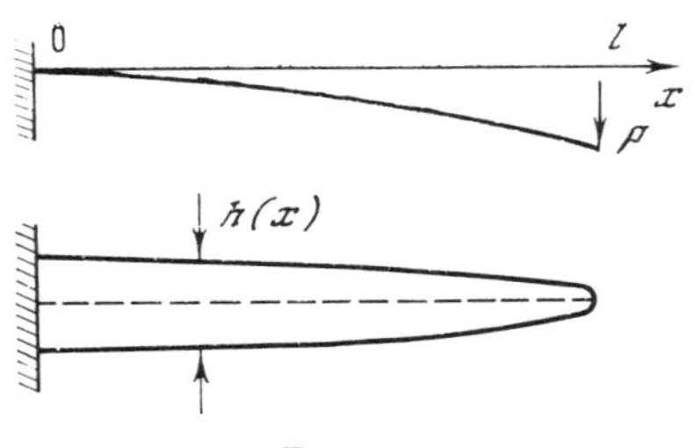

Figure 1

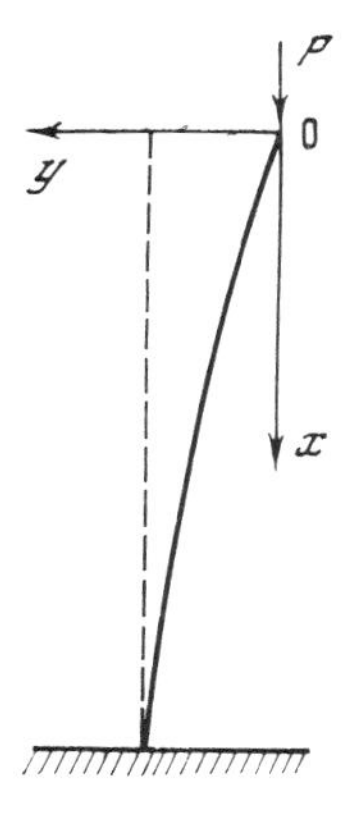

Figure 2

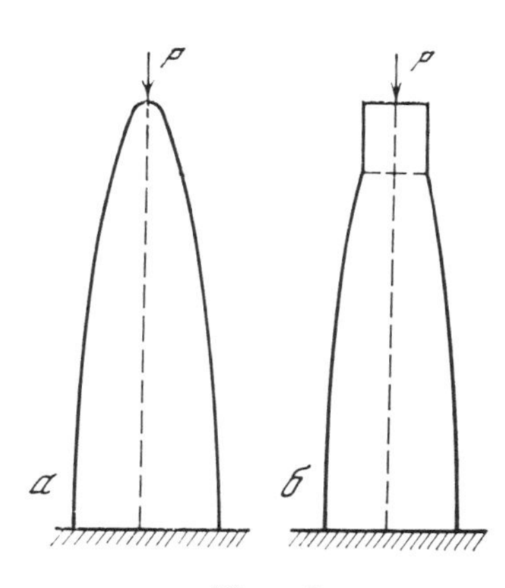

Figure 3

T. Clausen[171] (Fig. 3*a*). As the distance from the free end of the beam approaches zero, the thickness of this optimum beam also approaches zero and the compressive stress increases unboundedly. To prevent this feature, E. L. Nikolai[106,107] introduced an additional constraint concerning the magnitude of permissible stress. The distribution of thickness obtained in this case is shown in Fig. 3*b*.

Detailed investigations of the column problem were carried out for various types of beams and for different support conditions in subsequent research (see Refs. 11, 142, 184, 185, 186, 198, 212, 233, 235). Problems considered in these papers concern minimization of weight of a beam for a fixed magnitude of the load that causes loss of stability. Also considered is the dual problem of maximizing the critical load, for a given volume of the beam. In particular, it was shown in Ref. 184 that if we consider convex cross-sectional areas, the optimum beam will have a cross section that is an equilateral triangle. In comparison with a cylindrical beam with circular cross section, an optimally shaped triangular beam showed a gain in the magnitude of the critical force of 61.2%. A rigorous mathematical justification of optimality of these previously discovered shapes was offered in Ref. 233. Certain problems in optimization of elastic arches and circular plates were considered in Refs. 170, 172, and 246.

In the papers cited above, the authors considered static problems of bending and stability, so the effects of changes of shape on inertial properties were ignored. Dynamic problems of optimal design were first considered in the papers of M. G. Krein[73] and F. Niordson.[207] Krein considers the problem of finding the mass density distribution $\rho(x)$ in a string while he optimizes its natural frequency of vibration, with an additional constraint $\rho_1 \leqslant \rho(x) \leqslant \rho_2$ (where ρ_1 and ρ_2 are given constants). He assumes that the total volume of the

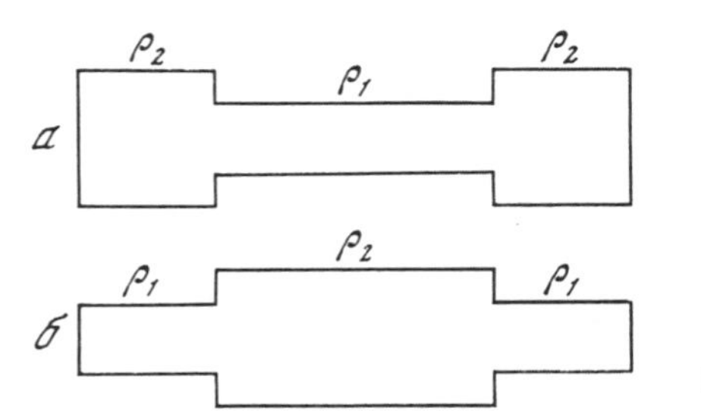

Figure 4

string is given. Since in modelling behavior of a string its strength in bending is assumed to be negligible, this class of problems reduces to finding the best distribution of inertial properties. The distribution $\rho(x)$ that was found to be optimum, where x is the coordinate varying with the length of the string, was steplike. Figure 4*a* shows the distribution of $\rho(x)$, corresponding to the maximum fundamental frequency, while Figure 4*b* corresponds to the minimum fundamental frequency.

In Ref. 207, the problem of finding the distribution of thickness of a beam that has the maximum fundamental frequency of transverse vibration is treated. In this problem, a change of shape of the beam causes not only a change in the inertial properties, as was the case with a string, but also causes a change in the strength of the beam. The distribution of thickness found by a numerical technique is illustrated in Fig. 5. Subsequently, dynamical problems arising in the optimal design of beams and plates have been considered in papers by several authors (see Refs. 31-33, 52, 54, 56, 123, 133, 134, 158, 160, 182, 183, 202, 209-211, 216, 223, 228, 234, 236, 237, and 244).

Problems discussed in the papers cited above, concerning optimal design of elastic structural elements, all have the following property: the "control" variable that describes the distribution of the thickness enters into the coefficients of the differential equations of equilibrium. This circumstance explains the approximate nature of the equations of state used in the analysis, whereby only the average value of one of the space variables (namely, the thickness) is used. In the general case of determining the optimum shape of elastic bodies, it is necessary to determine the shape of the region in which we pose the equa-

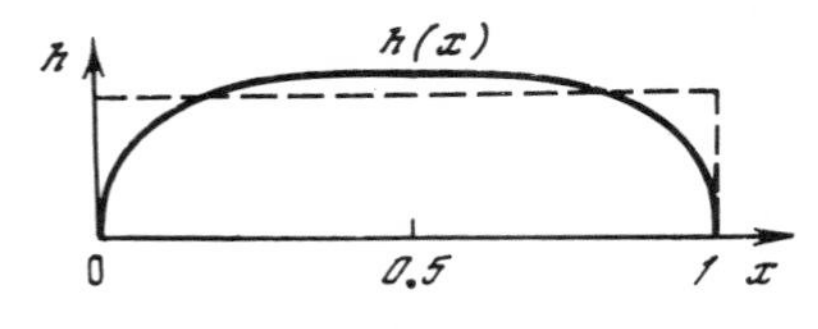

Figure 5

tions of equilibrium. These are problems with an unknown boundary. It appears that the first problem of this kind was posed by B. Saint Venant,[126] in connection with the search for the shape of the cross section of an elastic cylindrical beam that has the greatest torsional stiffness. He conjectures (supporting the conjecture by some computations and comparisons) that among all possible beams having the same cross-sectional area, the beam with a circular cross section has the greatest torsional stiffness. A series of research papers were devoted to the proof of this hypothesis.[110, 218] A rigorous proof of this conjecture, based on a symmetrization theorem, was given by G. Polya and G. Szegö.[110] Perturbation techniques and the highly developed techniques of the theory of complex variables permit a more general development of the theory of optimum shapes. Many results in the theory of optimization of unknown boundaries were obtained by the use of such techniques (see Refs. 14, 19, 20, 47, 78, 79, 95, 112, 144, 174, 192, and 245).

In connection with rather wide applications in technology and in construction of composite materials, engineers concerned with optimal design began to investigate optimization of the internal composition of elastic bodies. Many research investigations have been completed, containing results concerning the optimal design of structures utilizing inhomogeneous materials and problems of optimization of anisotropic elastic bodies (see Refs. 3, 21-24, 41, 43, 53, 103, 105, 108, 109, 186, and 187).

It is quite common in the theory of optimal design of structures to assume that external forces, conditions of support, and properties of materials are exactly known. The optimization problem is then posed as finding a shape and internal composition that allows a given quality criterion to attain a minimum (or a maximum). An additional difficulty is encountered in this type of problem if the external forces and properties of materials are known only approximately. There are many ways of mathematically modelling such problems. In the probabilistic approach, applied loads and material properties are assumed to be random quantities and we minimize the mathematical expectation of the weight or of some other quality criterion of design. Using the minimax (or the guaranteed) approach, which is a typical approach of game theory, we regard the set containing all possible realizations of the external loads, boundary conditions, and material properties that will be used in the design as known. We seek the shape and the internal material composition of the design that optimizes the quality criterion, while satisfying the strength and geometric constraints for all possible choices of loads and other factors mentioned above.[9,10,16] Problems of optimization with incomplete information have been considered in the context of control of dynamical systems by N. N. Krasovskii, A. B. Kurzanskii, F. L. Chernous'ko, and others. A detailed exposition of this theory and of specific problems solved with its help is contained in the monographs of Refs. 71, 77, and 150.

A sequence of research articles on the theory of optimal design is devoted to the problems of multicriteria optimization, in particular to optimization of structures subjected to moving loads. The problem of optimization for a dynamic load was considered within the framework of plastic design by M. Save and W. Prager,[226] while the problem considering a load of short duration was considered by R. T. Shield.[230] In Ref. 222, W. Prager and R. T. Shield solve the simplest problem of multicriteria optimization for an elastic, trilayer beam that is alternately subjected to bending and tensile loads.

In the research articles cited above, it is assumed that changes in the position and magnitude of loads occur in a quasi-static manner. Therefore, dynamic effects are ignored in the subsequent investigation. It is not hard to see that problems of multicriteria optimization and of optimal design with variable loads are equivalent to certain problems of optimal design with incomplete information concerning the applied loads, considered within the framework of the minimax approach. The solution of W. Prager and R. T. Shield obtained in Ref. 222 can be derived by considering the problem of optimal design of a beam in which we do not know which of the two possible loads will be applied and each is permitted to exist and must be considered. An analogous case arises when we consider the problem of a moving point load.[226] The design solution does not change if the load applied to the beam is regarded as a stationary point load, whose point of application is not known.

Further developments in techniques and in solution of specific problems of multicriteria optimization of elastic bodies were given in Refs. 12, 56, 121, 123, 124, 125, 164, and 194. For discussions of the general theoretical problems in the theory of the optimal design of structures, such as a general theoretical discussion of criteria of optimality, existence and uniqueness of an optimum solution, and cataloguing of the different types of problems that permit solutions by the use of standard techniques, we cite Refs. 14, 96, 112, 165, 178, 196, 219, 223, 225, 227, 231, 240, 241, and 242.

A substantial body of research in the area of optimization of elastic structures has been accomplished by the use of classical techniques of the calculus of variations. In parallel with the use of such techniques, a research direction has developed, based on application of theoretical ideas of the theory of optimal control. Beginning with the well-known work of A. I. Lur'e,[93] the techniques of optimal control theory, in particular the maximum principle of L. S. Pontryagin[111] (see also Refs. 37, 83, 84, 159, 189–191, and 216), have been applied in structural optimization.

Applications of control theory have also been developed for systems with distributed parameters.[4, 44, 69, 88, 89, 90, 95] We have already observed that the solution of optimal shape problems for elastic bodies can be reduced to solving nonlinear boundary-value problems of a system of differential equations. The nonlinear character of these problems precludes the possibility of applying known analytical techniques. Among the analytical techniques that can be applied to nonlinear problems, the most general and widely used is the perturbation technique. Application of this technique holds great promise for research in optimal design. The perturbation technique permits us to obtain simple approximate formulas and to analyze the dependence of the solution on parameters. The effectiveness of this technique increases in the solution of multiple parameter problems, where difficulties arise in numerical determination of dependence of the solutions on the parameters. The wide use of perturbation techniques in optimal control theory is illustrated in an article of F. L. Chernous'ko.[146] Applications of this technique in solving one- and two-dimensional problems in optimization of design are given in Refs. 14, 162, and 246.

At the present time, a large volume of research into optimization of elastic bodies is accomplished by using powerful computers. For this reason, in many articles authors derive computational algorithms intended for solving specific problems of optimization of design. Fundamental concepts used for construction of computational algorithms are given in works on optimal control theory,[42, 71, 72, 111, 149] nonlinear programming, calculus of variations, and numerical methods of optimization.[99, 100, 148, 157]

Additionally writings of an expository nature, concerning problems of optimization of elastic bodies, are contained in the books and articles of Refs. 4, 36, 45, 51, 62, 66, 83, 108, 109, 116, 129, 151, 159, 166, 208, 215, 220, 229, and 243.

N. V. Banichuk

Contents

1

Formulation of Problems and Research Techniques in Structural Optimization

In Sections 1.1 through 1.3 of this chapter we consider the ingredients of formulating problems of optimal design, such as the choice of a model of the state functions and control variables, and the assignment of constraints. We identify functionals that characterize the behavior of elastic bodies. Problems concerning application of variational principles in the formulation of and solution for optimal designs are explained in Section 1.4. In Section 1.5 we describe a technique that permits reduction of problems with local functionals to a much simpler class of problems with integral quality criteria. In Sections 1.6–1.8 we introduce optimality conditions for variational problems with nonadditive functionals and with perturbed boundaries in the presence of differential constraints. In Section 1.9 we consider some dual problems with homogeneous functionals and introduce some formulas that relate the solutions of these problems to each other. In Section 1.10 we present computational algorithms for iterative optimization, which are utilized in this monograph to solve problems of optimal design.

1.1. Formulation of Some Optimal Design Problems

Problems in the theory of optimal design that are considered here consist of determining the shape, the interior properties, and the working conditions of a structure that provide an extremal value (a minimum or a maximum) for a chosen criterion, in the presence of certain constraints. The broad multitude of formulations typifies the theory of optimal design. This can be explained by the fact that the equations determining the load and deformation of the structure, and our demands concerning its mechanical behavior, are fundamentally different for different types of structures (beams, columns, curved beams, plates, shells), different rheological properties (elasticity, plasticity, creep), external interactions (surface and body forces, static and dynamic loads, "dead" loads and depending on the design of the structure, thermal loads), the class of control variables (the shape of the structure, the distribution of physical properties inside the structure), and assumptions on completeness of the information and working conditions (for example, problems with incomplete information concerning the external interactions and the manner of supporting the structure). Exactness of our modelling and related data influences the formulation of the problem.

A complete formulation of problems in design optimization includes the basic state equations, the functional that is to be optimized, and constraints on the state functions and the control variables that are to be determined. From a mathematical point of view, these problems may be classified by the types of equations and boundary conditions, by the types of functionals to be minimized, by the types of constraints, by the manner in which the control variables enter into the basic relations (such as control by varying the coefficients of the equations or by varying the boundary of the domain), by the degree of completeness of the information concerning the data (problems with complete or incomplete information), and by other criteria.

Fundamental ingredients in formulating problems of optimal design include the following: the choice of (i) a model; (ii) the control variables of the functionals that depend on the state variables (or phase variables) and control variables; (iii) a functional that is to be optimized; and (iv) a system of constraints that are imposed on the control variables, on the state variables, and on the functionals considered in the problem.

First, we choose state variables u and a system of state equations

$$L\,(h)u = q \tag{1.1}$$

that relate these variables to the physical and geometric parameters of the solid body and to the external actions. Here, $u = \{u_1(x), \ldots, u_m(x)\}$ is a vector function describing the state of the medium. The independent variable $x = \{x_1, \ldots, x_s\}$ takes its value in the region Ω occupied by the solid body. The symbol $L(h)$ in Eq. (1.1) denotes the differential operator whose action depends on the spatial coordinates x_i. This operator is assumed to be linear, since we consider only linearly elastic bodies. The coefficients of the operator L depend on the control vector function $h = \{h_1(x), \ldots, h_n(x)\}$. The positive integers m, n, and s are given. For given loads and design variables, this system of equations should be a nonsingular and should uniquely determine the state variables that characterize the stress and deformation state of the structure. Finding these state variables for given control variables will be called the direct problem.

If the equations determining the state of the structure reflect the laws of physics, then the choice of control variables, of the functionals that we have to investigate—including the payoff functional (the quality criterion)—and of the system of constraints are dictated by the working conditions of the structure and by the technological feasibility of building such a structure. The control variables $h_i(x)$ determine the shape and physical properties of the deformable body. We can choose, for example, $h_i(x)$ as (i) the distribution of thickness or cross-sectional area of the body; (ii) the location of the neutral surface for a curvilinear beam or shell; (iii) the distribution of a reinforcing material; or (iv) the angles that determine the anisotropic properties (the axes of anisotropy) at each point in an elastic body.

In addition to the state and control variables, we have to consider the problem of optimal design with both integral and local constraints on characteristics of the structure, written in the form

$$J_i = \int_{\Omega} f_i(x, u, h)\, dx, \qquad i = 1, \ldots, r_1 \tag{1.2}$$

$$J_j = \max_x f_j(x, u(x), h(x)), \quad j = r_1 + 1, \ldots, r_1 + r_2 \tag{1.3}$$

where f_i denote expressions that involve the variables u and h, while r_1 and r_2 are nonnegative integers such that $r_1 + r_2 = r$. We can express in the form of an integral, or of a combination of integrals of the form of Eq. (1.2), such properties of the structure as weight, energy of elastic deformation (compliance), natural frequency of vibration, and critical load that causes the structure to lose its stability. Local characteristics include the maximum deflection or the magnitude of stresses, which may be expressed as in Eq. (1.3).

Demands that the structure must satisfy can be reduced to a system of constraints involving the control and the state variables, written in the form

$$F_i(J_1, \ldots, J_r) \leqslant 0, \quad i = 1, 2, \ldots, k \tag{1.4}$$

where F_i and k are given. In specific problems various types of inequalities may be expressed as inequality constraints of the form of Eq. (1.4). Certain types of inequalities are associated with strength conditions that are reduced to constraints on stresses. For example, the conditions $\max_x |\sigma_{ij}| - \sigma_{ij}^0 < 0$ could be investigated as strength conditions (here σ_{ij} are the components of the stress tensor, and σ_{ij}^0 are given positive constants). In this case we put a separate bound on the permissible value of each stress component. Alternately, we can impose the condition $\max_x g(\sigma_{ij}) - k^2 \leqslant 0$, which represents the criterion of transition to the plastic state of material (k is the constant of plasticity and $g(\sigma_{ij})$ is the second stress deviation invariant corresponding to the Von Mises theory). Constraints on elastic displacements that arise from either geometric or strength requirements are of a different type. As an example, we consider the condition $\max_x |u| - e \leqslant 0$, where u is the displacement of the elastic medium and e is a given positive number. We can include in the class of local constraints the two-sided inequality $\delta_1 \leqslant h(x) \leqslant \delta_2$ (where δ_1 and δ_2 are given constants). This inequality provides bounds on the thickness distribution in certain problems of optimal design of beams and plates. It can be written either in the form of Eq. (1.4) or in the form of a system of inequalities $\max_x h - \delta_2 \leqslant 0$ and $\delta_1 - \min_x h \leqslant 0$, or as a single inequality $\max_x (h - \delta_1)(h - \delta_2) \leqslant 0$.

As examples of constraints on integral functionals (i.e., constraints of the integral type), we can use the isoperimetric condition of constant volume of a plate $\int h\, dx = V$ and a constraint on compliance of the structure, $\int wq\, dx - c \leqslant 0$. Here, q denotes the transverse load, w is the deflection function, and V and c are given positive constants. Integration is performed over the volume occupied by the plate. We can take as the functional to be optimized one of the functionals considered of the form of Eqs. (1.2) or (1.3), or a function depending on these functionals, i.e.,

$$J = F(J_1, \ldots, J_r) \tag{1.5}$$

The optimization problem posed by Eqs. (1.1)-(1.5) consists in finding the function $h(x)$ that provides the functional of Eq. (1.5) with a minimum (or a maximum) and satisfies Eqs. (1.1)-(1.4).

It should be noted that the number of functionals considered and the constraints assigned to the problem, which are supposed to be noncontradictory, may theoretically be arbitrarily large. A single functional to be optimized (i.e., the quality criterion) must be defined for each specific case. For example, in the bending of a beam having a variable thickness, we can pose the problem of minimizing the weight of the beam with constraints on deflection or we can minimize the maximum deflection with a constant weight of the beam. However, the problem of simultaneously minimizing the weight and the maximal deflection, i.e., minimizing two functionals, does not make much sense.

In the choice of a model and formulation of the problem, an important part is played by a priori information concerning the properties of the solution. Information about the model and knowledge about the basic properties of the solution permit us to recognize genuine constraints and reject some constraints of "minor rank" in the formulation of an optimization problem, thereby reducing the problem to a formulation that may be solved by numerical or even analytic techniques. Hence, a considerable body of results in optimal design is related to models that have been thoroughly studied. However, it is frequently very difficult to guess in advance some properties of the required optimal solution. It sometimes turns out that after the optimization problem is formulated, the solution violates the hypotheses made in setting up the model. In many optimization problems concerning the shapes of plates that have been solved, the variation of the thickness reveals the existence of gradients that violate the assumptions made in formulating Kirchhoff's theory of thin plates. Other well-known singular solutions reveal the appearance of zero or infinite thickness in the optimal design. Once such a departure from the accepted model is discovered, or if we find violations in other conditions that were not accounted for in setting up the theoretical problem, it is necessary to introduce additional constraints. For example, in the plate bending problem, we may need additional constraints concerning its thickness. In this manner, the formulation of the problem is intertwined with the solution process. This interaction is not always clearly stated in the theory of optimal design. The *iterative* nature of research discussed here is not hard to discover in the great majority of problems that have been solved.

Considering what has been discussed above, it appears to be a useful suggestion that while we formulate and solve an optimization problem, we should also seek alternatives of sharpening the statements and conditions of the problem.

1.2. Basic Functionals

The choice of functionals that are considered in the process of optimal design comprises an essential part of formulating an optimization problem. This choice is affected by many circumstances: the basic purpose of the structure, its working conditions, technological feasibility of constructing the structure, constraints regarding its stability, properties of the models that are adapted for describing mechanical behavior of the structure, and a priori knowledge of properties of the optimal problem. In what follows, we shall study some typical functionals that are most frequently studied in problems involving optimization of elastic bodies.

1. Weight is a basic property of any structure. For this reason it is considered in the majority of articles on optimal design, either as the functional to be optimized, i.e., as the quality criterion, or included among the constraints.

The weight of a structure is descriptive of the amount of material that is necessary for its construction and also characterizes some of its working conditions. For example, an increase in the weight of the structure of an aircraft does not only increase the quantity of materials that must be prepared for the manufacture of this structure, but it also increases the fuel consumption during flight and adversely affects a number of other flight properties.

The weight is an integral characteristic of a structure. For continuous homogeneous bodies it is directly proportional to the volume occupied by the body

$$J = \gamma \int_{\Omega} dx \tag{1.6}$$

where γ is the specific weight of the material. In this case, to change the weight of a structure we need to change (or perturb) the domain of integration. For thin-walled structures made of homogeneous materials, the weight can be represented by an integral over the thickness distribution function h

$$J = \gamma \int_{\Omega} f(h)\, dx \tag{1.7}$$

For example, for a composite plate that lies in the $x_1 - x_2$ plane (x_1 and x_2 are Cartesian coordinates), which is fastened along the perimeter Γ of the region Ω occupied by the plate, $f = h(x_1, x_2)$. In this case the reduction of weight may be accomplished by either varying h for a fixed domain Ω, or by a simultaneous change of both the thickness and of the shape of the domain. In problems of structural optimization, the material may be classified as the *design material*, whose quantity and distribution in the structure is to be determined, and *nondesign material*, whose quantity and position is given *a priori*. Such is the case in the design of trilayer plates, where we usually wish to determine the optimum distribution of the thickness of the outer reinforcing layers, with a fixed middle layer. The weight functional J can then be represented by the sum of integrals $J_a + J_c$ (depending on the thickness of the reinforcing layers and the middle layer of the plate). Minimization of J is reduced to the problem of minimizing J_a, which is the weight of the outer layers. This is the essence of the additive property of the weight functional.

In cases of optimization of inhomogeneous bodies, the weight functional depends on the material structure; for example, on the concentration of the binding agent and on reinforcing additions for composite materials. If we denote by h_a, γ_a, h_c, γ_c the concentration and specific weight of the reinforcing and binding components respectively, then

$$J = \int_{\Omega} (\gamma_a h_a + \gamma_c h_c)\, d\Omega$$

where Ω is the domain occupied by the structure.

2. In many papers on the theory of optimal design, the magnitude of work performed by external forces during quasi-static loading of the body is regarded as a measure of the rigidity of the structure. This functional is called the compliance of the structure. Let an elastic body be supported on a part of the boundary Γ_u, while the loads q are applied to another part Γ_q; then the compliance is given by an integral over Γ_q of the inner product of the elastic displacements and external forces,

$$J = \frac{1}{2} \int_{\Gamma_q} qu\, d\Gamma_q \tag{1.8}$$

In a more general definition of the compliance functional, one can accept as q and u, respectively, the generalized forces and generalized displacements. For example, moments applied to a beam may be regarded as the distribution of generalized forces, while the angles of rotation of the cross sections at corresponding points may be regarded as the generalized displacements.

One of the reasons for the wide use of functionals of the type in Eq. (1.8) the theory of optimal design is the relative simplicity of obtaining conditions of optimality and solving problems. It should be noted that in the general case this functional does not characterize rigidity of the structure. Indeed, because the absolute value of the integral J is small, it does not follow that displacements are small, at some selected points within the elastic body. In some specific problems of optimal design, however, the functional of Eq. (1.8) may be designated as the quality criterion for rigidity of the structure. Let the load q causing the bending of a beam consist of a point force P applied at the point x_0, i.e., $q = P\delta(x - x_0)$, where δ denotes the Dirac delta function. Then the compliance of the beam is given by

$$J = \frac{1}{2} Pu(x_0)$$

Decreasing the compliance for a given value of P means decreasing the deflection of the beam at the point of application of the force.

If the bending load acting on a cantilever beam consists of a point moment M applied to the free end, causing the cross section at the free end to rotate by an angle ϕ, then compliance is given by

$$J = \frac{1}{2} M\varphi$$

In this case minimization of compliance (maximization of rigidity) consists in minimizing the angle of rotation of the cross section at the free end.

Let us consider an elastic body supported on a part Γ_u of its boundary and in contact with a perfectly rigid punch over a part Γ_q (the region of contact of its boundary). Let the displacement of the punch be denoted by u and let P be the resultant force applied to the punch. Then

$$J = \frac{1}{2} u \int_{\Gamma_q} q \, d\Gamma_q = \frac{1}{2} uP$$

Decreasing the compliance J for a given magnitude of P corresponds to reduction of the depth of penetration of the punch.

For a pipe having a circular cross section that is subjected to internal pressure q, the compliance J is proportional to the radial displacement of points on the boundary

$$J = \frac{1}{2} \int_0^{2\pi} qur \, d\varphi = 2\pi rqu$$

where r is the interior radius of the boundary.

3. The natural frequency of vibration, which can be represented by Rayleigh's quotient

$$\omega^2 = \frac{\int_\Omega \Pi(h, u) \, d\Omega}{\int_\Omega \mathrm{T}(h, u) \, d\Omega} \tag{1.9}$$

represents an important property of a structure. Here u denotes the amplitude function of the elastic displacements, h is the control variable, and Π and T are the potential and kinetic energy density functions, respectively.

In the case of longitudinal vibrations of a beam with a variable cross-sectional area $h = h(x)$, we obtain $\Pi = hu_x^2$ and $T = hu^2$. Here and in what follows we shall use dimensionless variables. For transverse vibrations of plates with variable thickness $h = h(x, y)$,

$$\Pi = h^3 [(u_{xx} + u_{yy})^2 + 2(1 - \nu)(u_{xx}u_{yy} - u_{xy}^2)], \qquad T = hu^2$$

where $u = u(x, y)$ is the amplitude function for the plate displacement.

The natural frequencies $\omega_i (i = 0, 1, 2, \ldots)$ that correspond to different eigenfunctions $u^i(x)$, form the spectrum $0 \leqslant \omega_0 \leqslant \omega_1 \leqslant \omega_2 \leqslant \cdots$. If the frequencies of the externally applied harmonic loads lie in the interval $0 < \omega < \omega_0$ or in an arbitrary interval $\omega_k < \omega < \omega_{k+1}$, then undesirable resonance phenomena do not occur in that structure. In applications, it is frequently necessary to widen the resonance-free frequency interval, i.e., to maximize ω_0, or if we consider the interval $\omega_k < \omega < \omega_{k+1}$ to maximize the difference $\Delta\omega_k = \omega_{k+1} - \omega_k$. Hence, in a certain class of dynamic problems of optimal design, either the fundamental frequency or a combination of natural frequencies appears as the functional to be optimized. If some other property is selected as the quality criterion, a typical constraint assigned to frequencies is of the type $\omega_0 \geqslant \mu$, where μ is a given number.

4. In the theory of optimal design of thin-walled structures that are compressed by applied forces and conserve energy, we can consider functionals of Rayleigh's type defining critical values for load parameters that cause the state of the structure to become unstable. We denote by p a load parameter, by pT the work performed in bringing a unit of volume of the structure into a critical state, and by Π the potential energy density function for the energy density function for the elastic deformation that is attained immediately after the loss of stability. We arrive then at the following critical value of p:

$$p = \frac{\int_\Omega \Pi(h, u)\, d\Omega}{\int_\Omega T(h, u)\, d\Omega} \tag{1.10}$$

For example, if we consider compression of a rectangular beam of variable cross-sectional area $h = h(x)$, then $\Pi = h^2 u_{xx}^2$ and $T = u_x^2$. The function $u = u(x)$ describes the deflection of the curved beam (after it has lost its stability). If an elastic plate of variable thickness $h = h(x, y)$ is compressed, then the expressions for Π and T are of the form

$$\Pi = h^3 [(u_{xx} + u_{yy})^2 + 2(1-\nu)(u_{xx}u_{yy} - u_{xy}^2)]$$

$$T = h [\sigma_x^0 u_x^2 + 2\tau_{xy}^0 u_x u_y + \sigma_y^0 u_y^2]$$

where $u(x,y)$ is the deflection function, and $\sigma_x^0(x,y)$, $\sigma_y^0(x,y)$ and $\tau_{xy}^0(x,y)$ are the stresses acting on the neutral surface of the plate, with $p = 1$.

Most common among problems of optimization of structures working in compression are problems of maximizing the critical value p_0 (p_0 is the smallest eigenvalue) for a given weight of the structure, or of minimizing the weight of the structure subject to the constraint $p \geqslant \mu$, where μ is a given number. This class of problems of optimal design, in which the stability criterion considers only the smallest eigenvalue, differs substantially from problems of optimal design for which constraints are given on the fundamental frequency and on higher natural frequencies.

5. The functionals considered above were of the integral type. Analysis of these functionals in optimization problems is carried out using techniques of the calculus of variations. However, many physically important problems lead to the formulation of functionals that depend on values of the state function at certain points that are not known in advance. Some basic strength and deflection criteria display such local properties. We shall offer some explanations concerning the introduction of such functionals into the problems of optimal design.

Let us consider the stresses in an elastic body that is subjected to external forces. Let the body occupy the domain Ω whose boundary is Γ. At each point $x \in \Omega + \Gamma$, we characterize the stressed state of the medium by the value of a function f of the stress tensor invariants I_1, I_2 and I_3, i.e., $f = f(I_1, I_2, I_3)$. It is understood that f has the following property: If at some point $x \in \Omega + \Gamma$, f attains a previously assigned value k^2 (the constant k is a material property) then at that point the material is in a transition state.

The material behaves elastically if the inequality $f < k^2$ is satisfied in the given domain. Various mechanical theories of materials credit the violation of this inequality with the appearance of a zone of plastic flow, with appearance of a region in which the deformations are not elastic, with loss of cohesion in the material, and with various other effects. In what follows we shall regard the condition $f = k^2$ as the plasticity criterion. The expression for f in terms of the components of the stress tensor is assumed to be a homogeneous function with the index of homogeneity equal to two.

For a given control variable (the shape of the body, the distribution of inhomogeneity or anisotropic properties), we solve the equilibrium problem for the elastic body and thereby find the stress components $\sigma_{ij}^0(x)$. It is then possible to determine the set of points $\Omega_0 (\Omega_0 \subset \Omega + \Gamma)$ on which the function f attains its maximum, which we denote by J,

$$J = (f)_{\Omega_0} = \max_x f \tag{1.11}$$

If we increase the loads by a factor of p, then the stresses also increase by the factor p. Here we utilize the hypothesis that the material is linearly elastic. We repeat that the assumption of linear elasticity will be used in every part of this monograph. When the value $p_0 = k/\sqrt{J}$ is attained by our parameter at some point $x \in \Omega_0$, this corresponds to the onset of plasticity. Clearly, the smaller the value of J, the higher are the loads (and the values of p_0) at which plastic deformation begins to occur within the body. To broaden the range of loads, such that the corresponding deformations of the body remain in the elastic range and creep does not occur, we can proceed to minimize the magnitude of J, i.e., if we attempt to choose the control function h satisfying the condition

$$J_* = \min_h J = \min_h \max_{x \in \Omega} f$$

6. Various properties characterizing the rigidity of a structure, such as the maximum pointwise displacement of the elastic medium, are described by local functionals. Rigidity of beams or plates subjected to bending may be evaluated by the magnitude of the greatest deflection

$$J = \max_{x \in \Omega} u(x)$$

and the problem of optimizing the rigidity of a structure through choice of the control variable h is naturally formulated as the problem of minimizing the maximal deflection ($J_* = \min_h J = \min_h \max_{x \in \Omega} u(x)$).

We need to explain in what sense plates can be optimized with respect to rigidity. Let q be a load applied to a plate that is proportional to the parameter p, i.e., $q = pq^0(x)$, where q^0 is a given function of the spatial coordinates, but that does not depend on the load parameter. The working conditions for a plate require fulfillment of the inequality $u < e$ (where e is a given number). In the linear theory considered here, where only small deflections are permitted, $u = pu^0$, where u^0 is the deflection corresponding to the load q^0, and $J = \max_x u^0$. Let us allow the loading parameter p to change within the range $0 \leqslant p \leqslant e/J$. Consequently, a plate which has the smallest magnitude of the maximum deflection can sustain the maximum load $p = e/J_*$ without violating the geometric constraint $u < e$.

1.3. Principal and Auxiliary Control Functions

The purpose of a substantial body of research into optimal design is to disclose the most effective technique of optimization. In the process of optimal design of a structure we usually have a large choice of control variables, whose change (variation) does affect the magnitude of the quality criterion. For example, a reduction in the weight of the structure may be attained by either a rational redistribution of its thickness, by changing the anisotropy of its material, by changing the reinforcement, by prestressing, etc. It is important to know which techniques of optimization, or which combination of such techniques, lead to the greatest gain in the value of the quality functional. Even in cases of cost limitation, or difficulties of a technological nature that prevent the construction of an optimal structure, research into an optimal design may be of great value, since it provides a theoretical judgement regarding the quality of traditional nonoptimal structures. Many structures that have been constructed using practical experience may turn out to be close to optimal, and further improvements may not be justified on purely economic grounds, but this can be clarified only if an optimal design investigation is carried out in each specific case.

We shall now consider certain problems related to the choice of control functions in some specific cases.

1. *Dimension.* By the dimension of a vector function we do not mean the number of components of the vector. We mean the number of independent variables that enter into the scalar state functions. The dimension of the state depends on many factors: the shape of the structure, material properties, the nature of external interactions, support conditions, etc. In some specific cases, reduction of the dimension of the state function from three to two, or even to one, may be possible, using conditions of symmetry or by averaging with respect to one or two spatial coordinates (such as the case of thin walled structures).

In contrast to properties of the state function, the dimension of a control function can be quite arbitrary. We can often use this property in the formulation of problems. In fact, in the optimal design of noncircular plates of variable thickness, the displacement function (which is the state variable) depends on two independent coordinates x and y lying in the plane of this plate. At the same time the distribution of thickness (which is the control function) can be regarded either as a function of the variables x and y, $h(x,y)$, or as a function of a single independent variable, say $h(x)$. We note that considering the "control" of a lower dimension, which is not effective in the sense of improving the value of the quality functional, nevertheless is of definite interest because of the simpler nature of practically fabricating an optimum design. Other examples of application of two-dimensional problems with one-dimensional control occur in the design of optimum rods or ribs reinforcing plates or shells.

2. *Beam and plate designs.* In optimal design of a beam, we must frequently consider the following problem: How does the cross section vary along the length of the beam? Various possibilities exist for choosing the shape of the cross-sectional area and the manner in which we select the change of parameters that define the cross section along the span of the beam. Let us consider, for example, beams of circular cross section and vary the radius along the length of the beam, i.e., $r = r(x)$. As an alternative, consider a beam of rectangular cross section, having height $h_1(x)$ and the width $h_2(x)$. We can choose as the control function (which will be denoted by h) either a single function or a collection of functions whose choice completely determines the shape of the beam, i.e., the shape of the cross-sectional area and the law of change for the geometric dimensions (parameters) of the cross-sectional areas as a function of distance measured along the length of the beam. For beams of circular cross section, assignment of the function $r = r(x)$ completely determines the shape of the beam, while for rectangular beams we need to choose two functions $h_1(x)$ and $h_2(x)$ to determine the shape of the beam.

A basic property of elastic beams is their rigidity, which turns up in the differential equations of equilibrium and in boundary value problems is the quantity EI, where E is Young's modulus and I is the moment of inertia of the cross-sectional area with respect to an axis perpendicular to the plane of bending and passing through the neutral axis of the beam. By means of this quantity, we can describe the effect of the distribution of "thicknesses" on the deflection function.

We shall limit our study to relations of the type

$$EI\,(x) = A_\alpha h^\alpha\,(x) \tag{1.12}$$

where A_α is a constant that depends on the shape of the cross-sectional area and on Young's modulus. For beams with constant rectangular cross-sectional area, $\alpha = 0$ and $A_0 = Eh_2 h_1^3/12$ (h_2 is the width and h_1 the height of the cross section).

For beams having a rectangular cross section with a variable height h_1 but having a constant width h_2, we have $\alpha = 3$, $h = h_1$, and $A_3 = Eh_2/12$. If only the width is allowed to vary, i.e., $h_2 = h_2(x)$ and the height remains constant, $\alpha = 1$, $h = h_2$, and $A_1 = Eh_1^3/12$. If we allow the ratio of height to the width to remain constant along the entire length ($h_1/h_2 = \mu$), all cross sections form similar rectangles, and $\alpha = 4$, $h = h1$, and $A_4 = E/12\mu$. For the case of circular beams of variable radius, $\alpha = 4$, $h = r$, and $A_4 = E\pi/4$.

For three-layered beams with external reinforcing layers having a variable height $\frac{1}{2} h_1$ and the middle layer having a constant height H [where $H \gg \max_x h(x)$], we have $\alpha = 1$, $h = h_1$, and $A_1 = Eh_2 H^2/4$, where h_2 is the width of the cross section.

In optimal design of beams, it is frequently convenient to adopt the area of the cross section as the control function, $S = S(x)$. The relation between S and EI differs for different types of beams. It can be represented by the formula $EI(x) = C_\alpha S^\alpha(x)$, in analogy with Eq. (1.12). For beams with rectangular cross section having constant height but variable width, this relationship is of the form $EI(x) = Eh_1^2 S(x)/12$. However, if the width is constant (h_2 = const) and the height varies, then we have $EI = (E/12h_2^2)\, S^3(x)$. For circular beams with variable radius, $EI(x) = (E/4\pi)\, S^2(x)$.

Let EI denote the cylindrical modulus of elasticity for plates (D is the more common notation in western literature) and let h denote the thickness. In optimal design of elastic plates, we generally use a relationship of the type in Eq. (1.12). The cases $\alpha = 1$ and $\alpha = 3$ correspond to three-layered and homogeneous plates, respectively. When $\alpha = 3$ the variable h in Eq. (1.12) has the physical interpretation of plate thickness and $A_3 = E/[12(1 - \nu^2)]$ where ν is Poisson's ratio for the plate material. In the case $\alpha = 1$, $h/2$ stands for the thickness of the exterior reinforcing layer of the plate and $A_1 = EH^2/[4(1 - \nu^2)]$, where H is the constant thickness of the middle layer.

3. *Auxiliary controls.* When we seek optimal distribution of the thickness h of a beam, a plate, or a shell, the constraint $h \geqslant 0$ is required for physical meaning of the control function. It becomes very difficult to interpret this inequality in optimization problems for which the thickness assumes the zero value at internal points or on some curves in the interior of the domain of definition Ω.

An approach originated by Valentine, introducing auxiliary functions, may be very useful in such cases. It permits us to "automatically" satisfy the given conditions. For example, if we introduce a new control function ϕ, related to h by the equality $h = \phi^2$, then it is clear that for any real value of ϕ the inequality $h \geqslant 0$ is satisfied.

More general constraints imposed on the control variable (i.e., the distribution of thickness) have the form

$$h_{\min} \leqslant h(x, \ y) \leqslant h_{\max} \tag{1.13}$$

where $h_{\min} \leqslant h_{\max}$ are given constants. Taking into account the inequalities of Eq. (1.13) in solving optimization problems leads to a necessity of computing the shapes of curves for which the function $h(x, y)$ satisfies one or the other of these bounds, or is strictly between them, and of "gluing together" these solutions. This leads to difficulties that are well known in this technique of solution. For this reason, we may find the ideas developed in[239] useful. Let us introduce an auxiliary control variable ϕ, with

$$h = \alpha + \beta \sin \varphi$$
$$\alpha = \frac{1}{2}(h_{\max} + h_{\min}), \qquad \beta = \frac{1}{2}(h_{\max} - h_{\min}) \tag{1.14}$$

Introduction of the function ϕ permits us to omit the conditions in Eq. (1.13) from future considerations. Indeed, substituting h given by Eq. (1.14) into Eq. (1.13) leads to inequalities that are satisfied for arbitrary values of ϕ.

In addition to inequalities of Eq. (1.13), we frequently have to consider isoperimetric conditions of the type

$$\int_{\Omega} h \, d\Omega = 1 \tag{1.15}$$

This condition can also be omitted from future considerations if we introduce an auxiliary control function

$$h = \varphi \Big/ \int_{\Omega} \varphi \, d\Omega \tag{1.16}$$

Direct substitution of h_1 defined by Eq. (1.16) into Eq. (1.15) should convince us that the equality of Eq. (1.15) is satisfied for arbitrary values of ϕ. It should be noted that the way an auxiliary control function is introduced is not unique. This feature can be exploited in solving the problem numerically, to speed convergence of a numerical algorithm.

4. In optimal design we can choose as control functions the coefficients of the equations, the boundary conditions, constraints, or even the boundary of the region defining the domain of definition for these equations (such is the case in

problems with unknown boundaries). Problems of optimizing coefficients and of the shape of the boundary are usually considered separately, since the techniques used for solving them are quite different. However, this division frequently turns out to be artificial. In many cases it is possible to use a mapping of the unknown region into a given canonical domain. The unknown functions that determine the mapping between these two domains also turn out to be the unknown coefficients in the resulting equations and functionals that are defined on the canonical domain. We shall explain this comment by offering an example.

Let us consider the problem of an elastic strip in tension (see Fig. 1.1). The side B_1B_4 is rigidly held, i.e., on that part of the boundary the projections u_1, u_2 of the displacement vector u in the directions of the x and y axes, respectively, are equal to zero. On the part B_2B_3 of the boundary of the strip the displacement in the direction of y is also equal to zero, but $u_1 = U$, where U is a given constant. We assume that $q = 0$ on the curves B_1B_2 and B_4B_3 that constitute the part of the boundary denoted by Γ_q. The problem of maximizing the rigidity of the strip by varying Γ_q is equivalent (according to Section 1.4) to the maximization of the potential energy of elastic deformation

$$J_* = \max_{\Gamma_q} \min_u \Pi \tag{1.17}$$

where

$$\Pi = \frac{1}{2} \iint_{\Omega} \left\{ a\,(u_{1x}^2 + u_{2y}^2) + 2b u_{1x} u_{2y} + \frac{1}{2} c\,(u_{1y} + u_{2x})^2 \right\} dx\,dy$$

and a, b, and c are coefficients that depend on the values of Young's modulus and Poisson's ratio. Let us suppose that the curves B_1B_2 and B_4B_3 form the perturbed boundary Γ_q, which is symmetric with respect to the x axis and satisfies the equation $y = Y(x)$. The region Ω has the shape $\Omega = \{0 \leqslant x \leqslant x_0,\ |y| \leqslant Y(x)\}$.

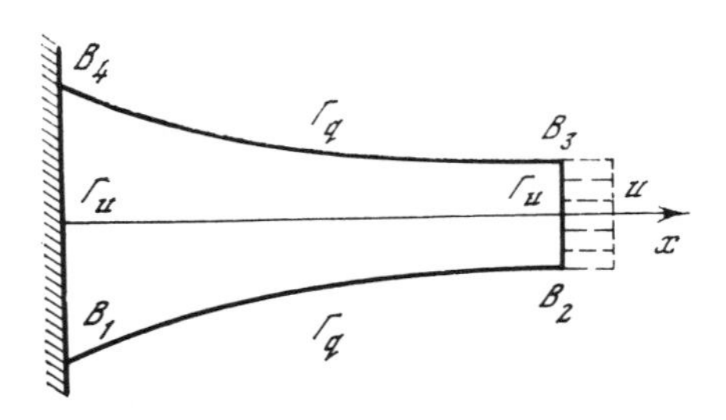

Figure 1.1

Let us transform the new independent variables according to the formulas

$$x' = x, \qquad y' = y/Y(x) \tag{1.18}$$

In the following discussion, we shall omit the primes. This transformation maps the region Ω into a rectangular region $\Omega_0 = \{0 \leqslant x \leqslant x_0,\ |y| \leqslant 1\}$ with the known boundary, while the functional Π is transformed into the form

$$\Pi = \frac{1}{2} \int_{\Omega} \left\{ a \left(u_{1x}^2 + \frac{1}{Y^2} u_{2y}^2 \right) + \frac{2b}{Y} u_{1x} u_{2y} + \frac{c}{2} \left(\frac{1}{Y} u_{1y} + u_{2x} \right)^2 \right\} Y \, dx \, dy \tag{1.19}$$

The original problem in Eq. (1.17) of finding an optimum boundary shape has thus been reduced with the help of Eq. (1.18) to the problem in Eq. (1.19) of optimizing coefficients.

1.4. Application of Variational Principles of the Theory of Elasticity to Eliminating Differential Relations

Variational principles are of great importance in the theory of elasticity. First, they allow us to formulate the boundary-value problems in the theory of elasticity in a compact form and in greater generality. The equilibrium equations for an elastic medium and certain types of boundary conditions (the natural conditions) can be derived as extremum conditions from the variational principles. Second, variational formulation of problems concerning the equilibrium of elastic bodies permits us to effectively apply direct methods of the calculus of variations to solve such problems. All this is true not only for boundary-value problems in the theory of elasticity, but for many other problems in mathematical physics.

In applications to problems of optimal design, variational principles permit us to bypass a study of differential relations and to omit the introduction of adjoint equations. Hence, the order of the general problem of optimization, related to the boundary-value problem, can be reduced, and the derivation of optimization conditions is greatly simplified. Aside from that, the variational principles and variational inequalities that can be derived with their help turn out to be very useful in analytical studies of optimization and in verifying optimal solutions. To be more concrete, we consider several examples.

1. Suppose that an elastic body occupies the region Ω, bounded by surfaces $\Gamma = \Gamma_u + \Gamma_q$. The body is rigidly fixed on Γ_u, while on Γ_q it is subjected to a traction q. We consider the problem of minimizing the compliance

$$J = \frac{1}{2} \int_{\Gamma_q} qu \, d\Gamma_q \to \min_{\Gamma_q} \tag{1.20}$$

by determining the shape of Γ_q. To state this problem in a "closed form," it is necessary to formulate the equations of state (equations of equilibrium), to define the constraints, and to introduce the adjoint variable that satisfies an auxiliary system of equations. In this approach, we generally encounter differential relations. However, taking into account the form of the functional of Eq. (1.20) and making use of a variational principle, we can reformulate the optimization problem so that it is no longer necessary to introduce adjoint variables. We shall offer a demonstration. According to a variational principle, the actual displacement field $u(x)$ in an elastic medium corresponds to a value of the functional $\Pi[u, \Gamma_q]$ (see Refs. 82 and 140)

$$\Pi(u, \Gamma_q) = \frac{1}{2} \int_{\Omega} \sigma_{ij} \varepsilon_{ij} \, d\Omega - \int_{\Gamma_q} uq \, d\Gamma_q \to \min_u \tag{1.21}$$

with the boundary condition $(u)_{\Gamma_u} = 0$. It has been assumed in Eq. (1.21) that the stresses σ_{ij} and strains ϵ_{ij} may be expressed in terms of displacements by using the kinematic conditions and Hooke's law. The actual displacement field $u(x)$ that minimizes the functional of Eq. (1.21) occurs when $J = -\Pi$. Using this equality and the variational principle of Eq. (1.21), it is possible to express compliance J by means of the functional Π

$$J = -\min_u \Pi \tag{1.22}$$

Consequently the problem of minimizing compliance can be reduced to performing successive operations of minimizing and maximizing the functional

$$J_* = \min_{\Gamma_q} J = \min_{\Gamma_q} (-\min_u \Pi) = -\max_{\Gamma_q} \min_u \Pi \tag{1.23}$$

In Eq. (1.23), the "interior" minimum with respect to u is computed taking into account the auxiliary condition $(u)_{\Gamma_u} = 0$, while the "exterior" maximum with respect to Γ_q may be subjected to additional conditions that are imposed on admissible variations of Γ_q, such as an isoperimetric condition that the volume of the domain Ω occupied by the elastic body remains constant.

There is no need to consider also the equilibrium equations written in terms of displacements, since these equations were already accounted for in formulating the problem of Eq. (1.23) as necessary conditions for extremizing Π with respect to u. The same remark applies to the boundary conditions assigned on Γ_q.

We observe that we can choose as control quantities not only the functions that determine the shape of the region Ω, but also variables h that determine the interior composition of the medium (distribution of the moduli of rigidity and anisotropy coefficients). All arguments remain valid, except that the "exterior" maximum in Eq. (1.23) is taken not only with respect Γ_q but also with respect to h.

2. Let us now assume that $q = 0$ on Γ_q and that nonzero displacements have been prescribed on Γ_u, i.e., the body is in a deformed state. To be specific, consider an elastic strip (see Fig. 1.1). On the lines $B_1 B_4$ and $B_3 B_2$ that form the boundary segment Γ_u, we specify displacements in the direction of the x axis: $(u_1)_{B_1 B_4} = 0$, $(u_1)_{B^2 B^3} = U(U > 0$ is a constant). The vertical displacements are all assumed to be equal to zero. We assume that $q = 0$ on the curves $B_1 B_2$ and $B_4 B_3$ that comprise Γ_q. We introduce the integral

$$J = \frac{1}{2} \int_{\Gamma_u} q u_1 \, d\Gamma_u = \frac{1}{2} U \int_{\Gamma_u} q \, d\Gamma_u = \frac{1}{2} UP \tag{1.24}$$

which helps us to estimate the rigidity of the strip when it is subjected to tension.

Assuming that U is given, we shall maximize the resultant force P, or equivalently the functional J, in optimizing the shape of the boundary segment Γ_q. We note that

$$\begin{aligned} J = \min_u \Pi, \qquad \Pi &= \frac{1}{2} \int_{\Omega} \sigma_{ij} \varepsilon_{ij} \, d\Omega \\ &= \frac{1}{2} \int_{\Omega} \left\{ a\,(u_{1x}^2 + u_{2y}^2) + 2b u_{1x} u_{2y} + \frac{c}{2} (u_{1y} + u_{2x})^2 \right\} dx\, dy \end{aligned} \tag{1.25}$$

and arrive at the formulation

$$J_* = \max_{\Gamma_q} J = \max_{\Gamma_q} \min_u \Pi \tag{1.26}$$

in analogy with Eq. (1.23). The constants a, b, and c are expressed in terms of Poisson's ratio ν and Young's modulus E, using the formulas $a = E(1 - \nu)/[(1 + \nu)(1 - 2\nu)]$, $b = \nu E/[(1 + \nu)(1 - 2\nu)]$, and $c = E/(1 + \nu)$, in the case of a plane deformation, and $a = E/(1 - \nu^2)$, $b = \nu E/(1 - \nu^2)$, and $c = E/(1 + \nu)$, in the case of plane stress.

3. Let the behavior of an elastic structure be described by a differential equation $L(h)\, u = q$ in a region Ω whose boundary is Γ, where L is a self-adjoint operator. The function h denotes the control variable, which defines the values of the coefficients of the operator L, and u is the state variable. The shape of the contour Γ is not given, but is to be determined during the optimization process. We define the energy of the elastic body as

$$\Pi = \frac{1}{2} \int_{\Omega} [uL(h)\, u - 2qu]\, d\Omega \tag{1.27}$$

Since the functional of Eq. (1.27) attains a minimum when the function u is substituted, the following equality is true:

$$J \equiv \frac{1}{2} \int_{\Omega} qu \, d\Omega = -\Pi \tag{1.28}$$

We can conclude, as in the first paragraph of this section, that minimization of the compliance functional J of Eq. (1.25), with a given differential connection, is equivalent to finding a minimum of the functional Π with respect to u and a maximum with respect to Γ and h

$$J_* = \min_\Gamma \min_h J = -\max_\Gamma \max_h \min_u \Pi \tag{1.29}$$

In particular, the problem of finding an optimum distribution of thickness and shape of a plate can be reduced to Eq. (1.29). In this case Ω is the region bounded by the plate contour, u is the distribution of displacements, and q is the normal load.

As an example we consider the case of pure torsion of an anisotropic cylindrical bar[86, 87] whose axis coincides with one of the coordinate of the Cartesian coordinate system xyz. The stress function $\phi(x, y)$ defined in the region Ω (whose boundary is Γ) is related to the stress components by the equalities $\tau_{xz} = \theta\phi_y$ and $\tau_{yz} = -\theta\phi_x$, where θ is the twist angle per unit length of the bar. It satisfies the following equations and boundary conditions:

$$L\varphi \equiv -(a\varphi_x - c\varphi_y)_x - (b\varphi_y - c\varphi_x)_y = 2 \tag{1.30}$$

$$(\varphi)_\Gamma = 0 \tag{1.31}$$

where a, b, and c denote the values of elastic constants of the anisotropic material satisfying the conditions $a > 0$ and $ab - c^2 > 0$.[87] We consider the problem of finding an optimum shape of the bar (i.e., of the shape of its cross-sectional area Ω) that maximizes torsional rigidity of the bar

$$K = 2 \int_\Omega \varphi \, dx \, dy \tag{1.32}$$

We note that for a given shape of the contour Γ and for given values of the coefficients a, b, and c, the solution of the boundary-value problem of Eqs. (1.30) and (1.31) minimizes the value of the functional[87]

$$\Pi = \int_\Omega (\varphi L\varphi - 4\varphi) \, dx \, dy = \int_\Omega (a\varphi_x^2 + b\varphi_y^2 - 2c\varphi_x\varphi_y - 4\varphi) \, dx \, dy \tag{1.33}$$

whose domain is the class of functions $\phi = \phi(x, y)$ satisfying the boundary condition of Eq. (1.31). The minimum value of the functional of Eq. (1.33) implies the equality $K = -\Pi$. Hence, the original problem of maximizing bar rigidity, i.e., the problem of Eqs. (1.30)-(1.32), can be reduced to the formulation

$$J_* = \max_\Gamma K = -\min_\Gamma \min_\varphi \Pi \tag{1.34}$$

The "interior" maximum of J_* in Eq. (1.34) with respect to ϕ is computed for a constant shape of the contour Γ and with the boundary condition of Eq. (1.31) satisfied. The exterior minimum with respect to Γ is computed with additional geometric constraints that must be satisfied by the contour Γ, for specific problems. For example, such a constraint may be the isoperimetric condition that the area of the cross section Ω must be constant.

4. A different type of a design optimization problem that permits us to omit the differential connection and utilize variational principles is related to the optimization of eigenvalues of self-adjoint boundary-value problems. As we have already observed in Section 1.2, the problem of maximizing (minimizing) an eigenvalue arises in optimization of stability or natural frequency of vibration of a conservative elastic system.

Let the state of an elastic system that occupies a region Ω with the boundary Γ be described by the equation

$$L(h)u - \lambda T(h)u = 0 \tag{1.35}$$

with homogeneous boundary conditions $N(h)u = 0$. We assume that $L(h)$ and $T(h)$ are positive and are self-adjoint differential operators (taking into account the boundary conditions). The coefficients of these operators depend on the control variable h. To compute the smallest eigenvalue and the corresponding eigenfunction, we can utilize Rayleigh's variational principle

$$\lambda_0 = \min_u \frac{J_1}{J_2}$$
$$J_1 = \int_\Omega uL(h)u\,d\Omega, \qquad J_2 = \int_\Omega uT(h)u\,d\Omega \tag{1.36}$$

Equation (1.35) is the Euler–Lagrange equation for the functional of Eq. (1.36). Consequently, the function u that realizes the minimum value of this functional automatically satisfies Eq. (1.35). Hence, the problem of choosing h and Γ to maximize the smallest eigenvalue can be reduced, after applying Rayleigh's principle, to finding a minimum with respect to u and a maximum with respect to h and Γ, i.e.,

$$\lambda_{0*} = \max_\Gamma \max_h \lambda_0 = \max_\Gamma \max_h \min_u \frac{J_1}{J_2} \tag{1.37}$$

If we consider the problem of optimizing the nth order eigenvalue (in the order of magnitude), instead of the smallest eigenvalues, then the minimum with respect to u in Eqs. (1.36) and (1.37) must be considered subject to auxiliary conditions that u must be orthogonal to the first $(n - 1)$ eigenfunctions.[67]

1.5. Reduction to Problems with Integral Functionals

Problems of optimal design can be classified, according to the the type of functional to be optimized and the form of constraints, into two basic groups. To the first group belong those optimization problems in which the quality criterion and the constraints are given in the form of integrals of the unknown functions. Thus, constraints appear as isoperimetric equalities and inequalities. Very often we find, in articles on optimal design, characteristic integral properties such as weight, energy of deformation (compliance), force causing loss of stability, and natural frequency of vibration (see Section 1.2). To the second group belong those problems of optimization in which the quality criterion and the constraints assigned to the problem are of the local (nonintegral) type. We shall also assign to this group problems of mixed type, in which we consider both integral and local properties of the structure. Typical local functionals that are minimized in the process of optimizing a structure are, for example, the maximum displacement of a deformed body or the maximum value of a stress component.

A large body of results obtained in the theory of optimal design is related to the first group of problems. This can be explained, first of all, by the fact that there exist well-known techniques for solving problems with integral functionals, such as the techniques of the classical calculus of variations and of nonlinear programming. Application of these techniques has resulted in successful completion of analytical and numerical research and has revealed a number of interesting general principles.(4,13,14,18,22,23,78,79,158,174,187,192,210) Certain types of problems with integral functionals can be substantially simplified because the equations of equilibrium turn out to be "natural" for these functionals and permit us to eliminate these differential equations.(14,234)

Much less research has been devoted to the investigation of two-dimensional optimization problems with local functionals. The reason for this lack of popularity is the absence of sufficiently broad, effective techniques for solving problems which belong to the second group. Typical difficulties associated with solving these problems are as follows. If, for example, we treat a problem of minimizing the maximum deflection of a plate by finding the optimum distribution of its thickness, then the derivation of necessary conditions for optimality and numerical realization of the statement of these conditions is complicated by the fact that the point of maximum deflection is not known a priori. The location of this point depends not only on the nature of the load, but also on the unknown distribution of thickness (i.e., on the control variable).

If we consider optimization problems with inequality constraints applied to the local properties, then identical difficulties arise in determining the location of points or regions in which a strict inequality is valid for these constraints. Considerable simplification in solving problems belonging to the second group can be attained when a priori information is either given, or is determined. For example, from symmetry of a problem, information concerning the point at which the desired local properties of the structure are located, or perhaps the location of the region containing the extremum of the quality functional, may be deduced.

We formulate here integral and local functionals, having primarily in mind two-dimensional problems of optimal design. In considering everything that has been discussed above, we should make clear our interest in deriving approximation techniques for reducing complex optimization problems with local functionals to problems with an integral quality criterion. In what follows we describe a technique of simplification, based on utilizing relations between norms assigned to spaces of continuous functions and norms in the spaces of p-integrable functions.

1. Consider first an optimization problem involving functionals of the forms of Eqs. (1.1)-(1.5), assuming that the constraints and the functional to be minimized are of the form

$$\int_{\Omega} f_i(x, u(x), u_x(x), h(x))\, dx \leqslant a_i, \qquad i = 1, \ldots, r_1 \tag{1.38}$$

$$\max_x |g_j(x, u(x), u_x(x), h(x))| \leqslant b_j, \qquad j = 1, \ldots, r_2 \tag{1.39}$$

$$J_* = \min_h J = \min_h \max_x g(x, u(x), u_x(x), h(x)) \tag{1.40}$$

where f_i, g_i, and g are known functions of x, u, u_x, and h; a_i and b_j are given constants; $u = \{u_1, \ldots, u_m\}$ and $h = \{h_1, \ldots, h_n\}$ are vector functions; $x = \{x_1, \ldots, x_s\}$ is the vector of independent variables; $u_x = \{u_{1x_1}, \ldots, u_{mx_3}\}$ is the vector of partial derivatives of the state variables; and m, n, s, r_1, and r_2 are given natural numbers.

The conditions of Eq. (1.39) may be replaced by inequalities

$$\begin{gathered} \max_x |g'_j(x, u(x), u_x(x), h(x))| \leqslant 1 \\ g'_j = g_j/b_j, \qquad j = 1, \ldots, r_2 \end{gathered} \tag{1.41}$$

From now on the primes will be omitted.

The functional to be minimized and the expressions given on the left-hand side of Eq. (1.41) represent the (L_∞) norms of the functions g and g_j in the space C of continuous functions, i.e., $J = \|g\|_C$ and $J_j = \|g_j\|_C$. Together with these norms given on the space C, we consider norms assigned to the L_p spaces of p-integrable functions. It is well-known in functional analysis that for an arbitrary $p > 0$ the following inequality is true $\|g\|_{L_p} \leqslant \|g\|_C$, and that

$$\lim_{p \to \infty} \|g\|_{L_p} = \|g\|_C \tag{1.42}$$

Relations analogous to Eq. (1.42) are also true for the functions g_j, $j = 1, 2, \ldots, r_2$. Since the C norm and the L_p norm differ by only a small amount for large values of p, we can replace the functionals $J = \|g\|_C$ and $J_j = \|g_j\|_C$ by

$$J_p = \left(\frac{1}{\mu(\Omega)} \int_\Omega |g|^p \, dx\right)^{1/p}, \qquad (J_j)_{p_j} = \left(\frac{1}{\mu(\Omega)} \int_\Omega |g_j|^{p_j} \, dx\right)^{1/p_j} \tag{1.43}$$

where $\mu(\Omega)$ is the measure of the set Ω. We make use of this fact in reducing a problem with local functionals to one with integral functionals. We use as an approximate solution the problem of Eqs. (1.1), (1.38), (1.40), and (1.41), the functions u and h satisfying Eq. (1.1), the constraint of Eq. (1.38), and the following conditions:

$$(J_j)_{p_j} \leqslant 1, \qquad j = 1, \ldots, r_2, \tag{1.44}$$

$$J_{p*} = \min_h J_p = \min_h \left(\frac{1}{\mu(\Omega)} \int_\Omega |g|^p \, dx\right)^{1/p} \tag{1.45}$$

Thus, the original problem given by Eqs. (1.1), (1.38), (1.40), and (1.41), is replaced by the problem given by Eqs. (1.1), (1.38), (1.43), (1.44), and (1.45) with integral functionals. In the study of the problem Eqs. (1.1), (1.38), (1.44), and (1.45), it is possible to utilize some well-known techniques of the calculus of variations by applying them to the integral functionals.

If the number of constraints of the type in Eqs. (1.39) and (1.41) is large (i.e., if r_2 is a larger integer), the presence of $r_2 + 1$ parameters p and p_j in the problem of Eqs. (1.1), (1.38), (1.44), and (1.45) causes considerable difficulties in estimating p and p_j to insure a given degree of accuracy. In such cases, it is possible to make use of a more "rigid" scheme of convergence with a single parameter p by assuming $p_1 = p_2 = \cdots = p_r = p$. It is also possible to simplify the system of constraints in Eq. (1.44) by considering the functions $g_j (j = 1, 2, \ldots, r_2)$ as components of a vector $g_v = \{g_1, g_2, \ldots, g_{r_2}\}$ and taking as the norm of that vector the quantity $\|g_j\|_C = \max_j \max_x |g_j[x, u(x), u_x(x), h(x)]|$. Thus, the system of inequalities of Eq. (1.41) can be reduced to a single inequality

$$\|g_v\|_C \leqslant 1 \tag{1.46}$$

Finally, replacing the C norm by the L_p norm leads to the inequality

$$\left\{\frac{1}{\mu(\Omega)} \int_\Omega \left[\frac{1}{r_2} \sum_{j=1}^{r_2} |g_j|^p\right] dx\right\}^{1/p} \leqslant 1 \tag{1.47}$$

We need to keep in mind the fact that $||g_v||_{L_p} \leqslant ||g_v||_C$ and that therefore the condition of Eq. (1.46) and the system of inequalities of Eq. (1.41) are only approximately satisfied. If Eq. (1.46) is not to be violated we may have to replace the inequality of Eq. (1.47) for the reduced problem by the following condition

$$\left\{\frac{1}{\mu(\Omega)}\int_{\Omega}\left[\frac{1}{r_2}\sum_{j=1}^{r_2}|g_j|^p\right]dx\right\}^{1/p} \leqslant 1-\varepsilon(p) \tag{1.48}$$

where $\epsilon(p)$ is a certain function of the parameter p. The above remarks are also applicable to the system of constraints of Eq. (1.44). The function $\epsilon(p)$ can be determined from estimates on the L_p norm.

2. We introduce an estimate of the error that has been caused by a problem with local functionals by one with integral functionals. Suppose that we are given two functionals that depend on the control function

$$J = \| f(x, u(x), h(x)) \|_C, \quad J = \| f(x, u(x), h(x)) \|_{L_p}$$

where $u(x)$ is a function that depends on the control function $h(x)$, because of the existence of a differential relation of the form of Eq. (1.1). We obtain an a priori estimate from above and below for the L_p norm using the C norm. We take into account the fact that the domain Ω is a subset of the s-dimensional Euclidean space: $\Omega \subset R^s$.

In what follows we shall use the notation $f(x) = f[x, u(x), h(x)]$. The estimate from above is obtained directly. Clearly, if we know the C-norm then $|f(x)| \leqslant \max_{x \in \Omega} |f(x)| \leqslant ||f||_C$. Therefore

$$\| f \|_{L_p} = \left(\frac{1}{\mu(\Omega)}\int_{\Omega}|f(x)|^p\,dx\right)^{1/p} < \| f \|_C$$

Now let us estimate the L_p norm from below, using the following concepts and symbols. Given ϵ, with $0 \leqslant \epsilon \leqslant 1$, let us define the set Ω_ϵ (see Fig. 1.2) for which the following inequality holds

$$| f(x) | \geqslant \| f \|_C (1 - \varepsilon) \tag{1.49}$$

We shall say that f belongs to the set $W_{H,\delta}$ if there exists an $\epsilon_0 > 0$ such that for every $0 \leqslant \epsilon \leqslant \epsilon_0$ the following inequality holds

$$\mu(\Omega_\varepsilon) \geqslant \mu(\Omega)\, H\varepsilon^\delta \tag{1.50}$$

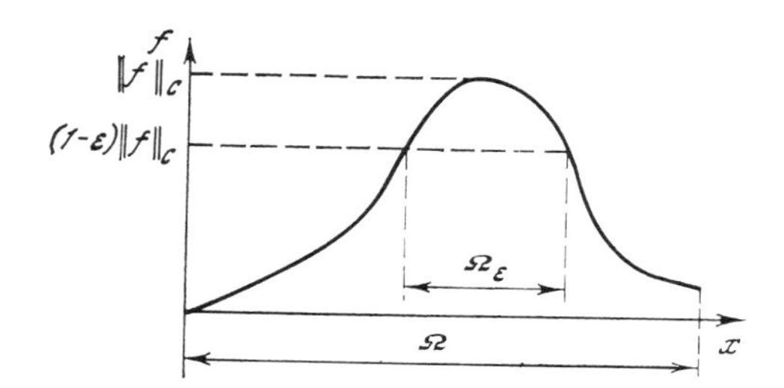

Figure 1.2

where $H > 0$ and $\delta > 0$ are given constants. We shall illustrate the meaning of our definition of the set $W_{H,\delta}$ by offering examples:

(a) Let $|f(x)| = \|f\|_C$ in the entire region Ω_0. Then $\Omega_0 \subset \Omega_\epsilon$ for any $0 \leqslant \epsilon \leqslant 1$, and consequently $f \in W_{H,0}$, $H = \mu(\Omega_0)/\mu(\Omega)$, for all ϵ_0.

(b) Suppose that there exist constants K and m such that

$$|f(x') - f(x)| \leqslant A \, \| x' - x \|^{\alpha} \tag{1.51}$$

where x' is some point for which $|f(x')| = \|f\|_C$, $x \in \Omega$ is an arbitrary point, and $\|x' - x\|$ is the Euclidean norm of $x' - x$ in R^s. If the inequality of Eq. (1.51) is satisfied, then the set Ω' defined by the inequality

$$A \, \| x' - x \|^{\alpha} \leqslant \varepsilon \, \| f \|_C \tag{1.52}$$

is contained in Ω_ϵ. In fact, for any $x \in \Omega'$, we have

$$|f(x') - f(x)| \leqslant A \, \| x' - x \|^{\alpha} \leqslant \varepsilon \, \| f \|_C$$

Since $|f(x') - f(x)| \geqslant \|f\|_C - |f(x)|$, we have $|f(x)| \geqslant \|f\|_C (1 - \epsilon)$. Therefore, $\Omega' \subseteq \Omega_\epsilon$ and consequently $\mu(\Omega_\epsilon) \geqslant \mu(\Omega')$. However, according to Eq. (1.52) the set Ω' is a ball. Therefore

$$\begin{gathered} \mu(\Omega_\varepsilon) \geqslant \mu(\Omega') = \mu(\Omega) H \varepsilon^{\delta} \\ H = \frac{V_s}{\mu(\Omega)} \left(\frac{\| f \|_C}{A} \right)^{\delta}, \qquad \delta = \frac{s}{\alpha} \end{gathered} \tag{1.53}$$

where V_s is the volume of a unit sphere in R_s. Hence, the function f is contained in $W_{H,\delta}$ for any ϵ_0, H, and δ that satisfy Eq. (1.53).

We continue by estimating the L_p norm from below, assuming that

$f \in W_{H,\delta}$. We have

$$\|f\|_{L_p} = \|f\|_C (1-\varepsilon)\left(\frac{1}{\mu(\Omega)}\int_\Omega \left|\frac{f(x)}{(1-\varepsilon)\|f\|_C}\right|^p dx\right)^{1/p}$$
$$\geqslant \|f\|_C (1-\varepsilon)\left(\frac{1}{\mu(\Omega)}\int_{\Omega_\varepsilon} \left|\frac{f(x)}{(1-\varepsilon)\|f\|_C}\right|^p dx\right)^{1/p}$$

We make use of the definition of Eq. (1.49) of the set Ω_ϵ and of Eq. (1.50) of the class of functions $W_{H,\delta}$. Assuming that $0 \leqslant \epsilon \leqslant \epsilon_0$, we obtain

$$\|f\|_{L_p} \geqslant \|f\|_C (1-\varepsilon)(H\varepsilon^\delta)^{1/p} \tag{1.54}$$

Applying the well-known inequality $a^\zeta \geqslant 1 + \zeta \ln a$, which is true for any ζ and any $a > 0$, we transform the inequality of Eq. (1.54) into the form

$$\|f\|_{L_p} \geqslant \|f\|_C \left[\frac{p+\ln H}{p} + \chi(\varepsilon) - \frac{\delta}{p}\varepsilon \ln \varepsilon\right]$$
$$\chi(\varepsilon) = \frac{\delta}{p}\ln\varepsilon - \varepsilon\left(\frac{p+\ln H}{p}\right) \tag{1.55}$$

Since $0 \leqslant \epsilon \leqslant 1$, it follows that $\epsilon\delta \ln \epsilon/p < 0$ and the last term in the inequality of Eq. (1.55) may be disregarded. The function $\chi(\epsilon)$ attains its maximum at the value $\epsilon_* = \delta/(p + \ln H)$. Therefore, we have the following bound

$$p \geqslant \frac{\delta}{\varepsilon_0} - \ln H \tag{1.56}$$

Substituting the above expression for ϵ_* into Eq. (1.55), it is possible to derive the following estimate of Eq. (1.57).

Using the estimate of Eq. (1.57) it is possible to obtain an estimate for the functional J of our original problem, in terms of the integral functional J_p:

$$\|f\|_C \psi(p) \leqslant \|f\|_{L_p} \leqslant \|f\|_C$$
$$\psi(p) = 1 + \frac{\delta}{p}\ln\left(\frac{\delta}{p+\ln H}\right) + \frac{1}{p}[\ln H - \delta] \tag{1.57}$$
$$J_p \leqslant J \leqslant J_p/\psi(p)$$

For a specific value h of the control variable we have

$$0 \leqslant J - J_p \leqslant J_p (1 - \psi(p))/\psi(p)$$

We thus obtain an error estimate for the functional that was derived in our transition from the original optimization problem to the auxiliary problem with an integral functional J_p. Let h_* be the design variable corresponding to the minimum of J and h_{p*} the one corresponding to the minimum of J_p. Hence, in finding the minimum value we have

$$\frac{J(h_*) - J_p(h_*)}{J(h_*)} \leqslant \frac{J(h_*) - J_p(h_{p*})}{J(h_*)} \leqslant \frac{J(h_{p*}) - J_p(h_{p*})}{J(h_{p*})}$$

Applying the inequality of Eq. (1.57) we finally derive the bound

$$0 \leqslant \frac{J(h_*) - J_p(h_{p*})}{J(h_*)} \leqslant 1 - \psi(p)$$

It should be noted that the error estimate Eq. (1.57) is based on *a priori* information concerning the function $f(x)[f \in W_{H,\delta}]$. In specific problems, information may be supplied by physical arguments.

1.6. Necessary Conditions for Optimality

1. We shall first describe a variational approach that is used to obtain optimality conditions and to reduce the optimization problem to a boundary-value problem with a system of differential equations. Let $u = \{u_1(x), \ldots, u_m(x)\}$ be a vector function that satisfies the differential equation

$$L(h)u = q \tag{1.58}$$

in the domain Ω, with the boundary conditions

$$(N(h)u)_\Gamma = 0 \tag{1.59}$$

on the boundary Γ of Ω. Here $h = \{h_1(x), \ldots, h_n(x)\}$ is the vector of the control variables. The differential operators $L(h)$ and $N(h)$ have coefficients depending on h. We denote by $J(u, h)$ and $J_i(u, h)$ $(i = 1, 2, \ldots, r)$ the integral functionals

$$J = \int_\Omega f(x, u, h)\, dx, \qquad J_i = \int_\Omega f_i(x, u, h)\, dx \tag{1.60}$$

where f and f_i are given functions of the variables x, u, and h. We consider here minimization of the functional J

$$J_* = \min_h J(u, h) \tag{1.61}$$

in the presence of integral equality constraints imposed on the control and state variables

$$J_i(u, h) - c_i = 0, \qquad i = 1, \ldots, r \tag{1.62}$$

Here c_i are constants, while f and f_i are given functions of the variables x, u, and h.

We shall now derive optimality conditions for the problem of Eqs. (1.58)-(1.62). To do that we first write formulas for the first variation of the integrals in Eq. (1.60) and equations relating variations as terms in u and h, which correspond to Eqs. (1.58) and (1.59)

$$\delta J = \int_\Omega \left(\frac{\partial f}{\partial u}\delta u + \frac{\partial f}{\partial h}\delta h\right) dx, \qquad \delta J_i = \int_\Omega \left(\frac{\partial f_i}{\partial u}\delta u + \frac{\partial f_i}{\partial h}\delta h\right) dx \tag{1.63}$$

$$L(h)\ \delta u + M(u, h)\, \delta h = 0 \tag{1.64}$$

$$N(h)\ \delta u + T(u, h)\, \delta h = 0 \tag{1.65}$$

where $\partial f/\partial u = \{\partial f/\partial u_1, \ldots, \partial f/\partial u_m\}$ and $\partial f_i/\partial u = \{\partial f_i/\partial u_1, \ldots, \partial f_i/\partial u_m\}$.

The variations δJ and δJ_i depend on variations in the state and control variables. Note that Eqs. (1.64) and (1.65) depend linearly on δu and δh. The variational equations of Eq. (1.64) and the boundary conditions of Eq. (1.65) are obtained from Eqs. (1.58) and (1.59) by substituting $u + \delta u$ and $h + \delta h$ for u and h, respectively, and by isolating terms that are linear in δu and δh. Here, $M(u, h)$, $T(u, h)$ denote operators acting on the vector δh.

We express the variation of the functional to be minimized in terms of the variation δh. To achieve that, we follow a classical technique of the calculus of variations and introduce an auxiliary vector function (the adjoint variable $v(x) = \{v_1(x), \ldots, v_m(x)\}$, which we determine from the condition that the variation of the functional to be minimized should not contain the variation of the state variable δu. To accomplish this objective, it is necessary that v satisfy a certain system of differential equations and boundary conditions, i.e., we must define v as the solution of a certain boundary-value problem. Let us form the scaler product of the left-hand side of Eq. (1.64) and the vector v. Taking the integral over the domain Ω as the definition of this scalar product, we have

$$\int_{\Omega} v\,[L(h)\,\delta u + M(u,\, h)\,\delta h]\,dx = 0$$

Next, we perform integration by parts, transforming this integral into the form

$$\int_{\Omega} v\,[L(h)\,\delta u + M(u,\, h)\,\delta h]\,dx = \int_{\Omega} [\delta u L^*(h)\,v + \delta h M^*(u,\, h)\,v]\,dx \tag{1.66}$$

Some boundary terms that arise in the integration by parts are equal to zero because of Eq. (1.65), which is a consequence of the boundary condition of Eq. (1.59) and relates the boundary values of the variations δu and δh. The remaining boundary terms that arise are set equal to zero, requiring that the function v obey the boundary condition $N^*(h)\,v = 0$. The coefficients of the operator L^* depend on h, but are independent of u. The coefficients of the operator M^* may depend on both h and u.

Considering Eqs. (1.63)-(1.66) and observing that Eq. (1.62) implies that $\delta J_i = 0$, we represent the variation δJ in the form

$$\delta J = \int_{\Omega} \left\{ \left[\frac{\partial f}{\partial h} + \sum_{i=1}^{r} \lambda_i \frac{\partial f_i}{\partial h} + M^*(u,\, h)\,v \right] \delta h + \left[L^*(h)\,v + \frac{\partial f}{\partial u} + \sum_{i=1}^{r} \lambda_i \frac{\partial f_i}{\partial u} \right] \delta u \right\} dx \tag{1.67}$$

where $\lambda = \{\lambda_1, \ldots, \lambda_r\}$ is the vector of Lagrange multipliers associated with the constraints of Eq. (1.62). In agreement with our previous remarks, we require the term multiplying δu must vanish, i.e., we let $L^*(h)\,v + \partial f/\partial u + \sum_{i=1}^{r} \lambda_i\,\partial f_i/\partial u = 0$. We have thus derived the required formula for the first variation of the functional to be optimized in terms of only δh

$$\delta J = \int_{\Omega} \left[M^*(u,\, h)\,v + \frac{\partial f}{\partial h} + \sum_{i=1}^{r} \lambda_i \frac{\partial f_i}{\partial h} \right] \delta h\,dx \tag{1.68}$$

In the derivation of Eq. (1.68), we have required that the function v satisfy the following differential equation and boundary condition:

$$L^*(h)v + \frac{\partial f}{\partial u} + \sum_{i=1}^{r} \lambda_i \frac{\partial f_i}{\partial u} = 0 \tag{1.69}$$

$$(N^*(h)v)_\Gamma = 0 \tag{1.70}$$

As is common usage in control theory, the function v shall be called the adjoint variable.

A necessary condition for the minimum of the functional J is given in the form $\delta J = 0$. Hence, we obtain the necessary condition of optimality

$$M^*(u, h)v + \frac{\partial f}{\partial h} + \sum_{i=1}^{r} \lambda_i \frac{\partial f_i}{\partial h} = 0 \tag{1.71}$$

In this manner the optimization problem of Eqs. (1.58)-(1.62) is reduced to the solution of the boundary-value problems of Eqs. (1.58), (1.59), (1.69), and (1.70), and to a determination of the control variable h from the optimality condition of Eq. (1.71). The constants λ_i are found by using the isoperimetric conditions of Eq. (1.62).

If we assume that f and f_i depend linearly on u, i.e., $f = ug(h)$ and $f_i = ug_i(h)$, where the functions g and g_i are independent of u, then u does not enter into the boundary-value problem of Eqs. (1.69) and (1.70) so it does not influence the adjoint variable v, which is the solution of

$$\begin{aligned} L^*(h)v + g + \sum_{i=1}^{r} \lambda_i g_i &= 0 \\ (N^*(h)\ v)_\Gamma &= 0 \end{aligned} \tag{1.72}$$

In this special case, the boundary-value problems of Eqs. (1.58), (1.59), and (1.72) are related to each other only through the unknown control variable h.

2. We now consider a special case in which the boundary-value problem of Eqs. (1.58) and (1.59) is self-adjoint, i.e., the linear operators L and N are self-adjoint. (This is[104] denoted by $L = L^*$ and $N = N^*$.) We also assume that $f = uq$, $f_i = g_i(h)$ $(i = 1, \ldots, r)$. In this case the boundary-value problem of Eqs. (1.69) and (1.70) becomes

$$L(h)\ v = -q \tag{1.73}$$

$$(N(h)\ v)_\Gamma = 0 \tag{1.74}$$

The boundary-value problem of Eqs. (1.73) and (1.74), up to the plus or minus sign of the right-hand side of Eq. (1.73), is identical with the boundary-value problem of Eqs. (1.58) and (1.59), so

$$v = -u \tag{1.75}$$

We can thus eliminate the function v from the optimality condition of Eq. (1.71), obtaining

$$M^*(u,\ h)u + \sum_{i=1}^{r} \lambda_i \frac{\partial g_i}{\partial h} = 0 \tag{1.76}$$

In this fashion, by using Eq. (1.75), we can eliminate the adjoint variable from further consideration in self-adjoint boundary-value problems, with the stated auxiliary assumptions. In this special case, the order of the boundary-value optimization problem is reduced by a factor of two.

Similarly, in the case for which $f = g(h)$, $f_i = g_i(h)$, $(i \neq k)$, and $f_k = uq$, we obtain

$$L(h)v + \lambda_k q = 0 \tag{1.77}$$

subject to boundary conditions of Eq. (1.74). Then $v = -\lambda_k u$ [see Eqs. (1.58), (1.59), (1.74), and (1.77)] and the optimality condition of Eq. (1.71) can be written as

$$-\lambda_k M^*(u,\ h)u + \frac{\partial g}{\partial h} + \sum_{i=1,\ i\neq k}^{r} \lambda_i \frac{\partial g_i}{\partial h} = 0 \tag{1.78}$$

3. *Examples.* We shall now derive optimality conditions for a beam of minimum weight, with constraints on compliance. Let the beam be freely supported at the left end ($x = 0$) and built in at the right end ($x = l$). We denote by q the transverse load acting on the beam and by w the deflection function. The displacement equation, boundary conditions, constraints concerning the compliance, and functional to be minimized are

$$
\begin{aligned}
&L(h)w \equiv (h^{\alpha}w_{xx})_{xx} = q \\
&w(0) = (h^{\alpha}w_{xx})_{x=0} = 0, \qquad w(l) = w_x(l) = 0 \\
&\int_0^l qw\,dx = c, \qquad \int_0^l h\,dx \to \min_h
\end{aligned}
\tag{1.79}
$$

Let us multiply all terms of the variational equation $(h^{\alpha}\delta w_{xx})_{xx} + \alpha(h^{\alpha-1}w_{xx}\delta h)_{xx} = 0$ [derived from Eq. (1.79)] by v and integrate the product, making use of integration by parts, to obtain

$$
\begin{aligned}
&\int_0^l v\,[(h^{\alpha}\delta w_{xx})_{xx} + \alpha(h^{\alpha-1}w_{xx}\delta h)_{xx}]\,dx \\
&\quad = \int_0^l [(h^{\alpha}v_{xx})_{xx}\delta w + \alpha(h^{\alpha-1}w_{xx}v_{xx})\,\delta h]\,dx + [v\,((h^{\alpha}\delta w_{xx})_x \\
&\quad + \alpha(h^{\alpha-1}w_{xx}\delta h)_x) - v_x(h^{\alpha}\delta w_{xx} + \alpha h^{\alpha-1}w_{xx}\delta h) \\
&\quad + \delta w_x(h^{\alpha}v_{xx}) - \delta w\,(h^{\alpha}v_{xx})_x]_{x=0}^{x=l}
\end{aligned}
\tag{1.80}
$$

In simplifying Eq. (1.80) by setting the boundary terms equal to zero, it suffices to set $v(0) = (h^{\alpha}v_{xx})_x = 0$ and $v(l) = v_x(l) = 0$. These conditions represent the boundary conditions to be satisfied by the adjoint variable. The basic equation for v and the optimality condition is written in the form

$$
(h^{\alpha}v_{xx})_{xx} + \lambda q = 0, \qquad \alpha h^{\alpha-1}w_{xx}v_{xx} + 1 = 0 \tag{1.81}
$$

where λ is a Lagrange multiplier that represents the isoperimetric condition of Eq. (1.79). From the formulation of the boundary-value problems for w and v, it follows that $v = -\lambda w$. We can thus eliminate the adjoint variable v from the optimality condition [the second equality in Eq. (1.81)] and from the Eq. (1.78), to derive

$$
h^{\alpha-1}w_{xx}^2 = \text{const} \tag{1.82}
$$

The optimality condition of Eq. (1.82) and the deflection equation with the boundary conditions of Eq. (1.79) form a boundary-value system of equations for determining the thickness $h(x)$ and deflection $w(x)$.

As a second example, consider the optimization problem for a bent elastic plate that is freely supported along the contour Γ

$$L(h)w \equiv [h^{\alpha}(w_{xx} + \nu w_{yy})]_{xx} + [h^{\alpha}(w_{yy} + \nu w_{xx})]_{yy} + 2(1-\nu)(h^{\alpha} w_{xy})_{xy} = q$$
$$(w)_{\Gamma} = 0, \quad \left(h^{\alpha}\left[\Delta w - \frac{1-\nu}{R}\frac{\partial w}{\partial n}\right]\right)_{\Gamma} = 0 \tag{1.83}$$
$$\int_{\Omega} \psi(w)\,dx\,dy = c, \quad \int_{\Omega} h\,dx\,dy \to \min_h$$

Here, $h(x, y)$ is plate thickness, $w(x, y)$ is plate deflection, $\psi(w)$ is a given function of w, and $\partial w/\partial n$, Δ, R, and Ω denote, respectively, the derivative in the direction of the outward normal to the contour Γ, the two-dimensional Laplacian in variables x and y, the radius of curvature, and the region bounded by Γ. In this example, the adjoint boundary-value problem and the optimality conditions assume the form

$$L(h)v + \frac{\partial\psi}{\partial w} = 0$$
$$(v)_{\Gamma} = 0, \quad \left(h^{\alpha}\left[\Delta v - \frac{1-\nu}{R}\frac{\partial v}{\partial n}\right]\right)_{\Gamma} = 0 \tag{1.84}$$
$$\alpha h^{\alpha-1}[\Delta w \Delta v - (1-\nu)(w_{xx}v_{yy} + w_{yy}v_{xx} - 2w_{xy}v_{xy})] = \text{const}$$

4. In the above arguments, we have assumed that the integrand functions in Eq. (1.60) depend only on x, u, and h, and are independent of derivatives of the state variables. Consider now the case in which $f = f(x, u, u_x, h)$ and $f_i = f_i(x, u, u_x, h)$. Then in Eq. (1.63) for the variation δJ, the following additional composite terms must be included: $(\partial f/\partial u)\,\delta u_x = (\partial f/\partial u_{1x})\,\delta u_{1x_1} + \cdots + (\partial f/\partial u_{mx_s})\,\delta u_{mx_s}$. Similar composite terms must be added to the expression for δJ_i

Equating the Lagrange functional to zero

$$\delta J + \sum_{i=1}^{r} \lambda_i \delta J_i + \int_{\Omega} v\,[L(h)\,\delta u + M(u, h)\,\delta h]\,dx$$

and carrying out some routine manipulations (integration by parts), we obtain

$$\int_{\Omega} \left\{\left[\frac{\partial f}{\partial h} + \sum_{i=1}^{r} \lambda_i \frac{\partial f_i}{\partial h} + M^*(u, h)\,v\right]\delta h + \left[L^*(h)\,v + \frac{\partial f}{\partial u} - \frac{\partial}{\partial x}\left(\frac{\partial f}{\partial u_x}\right) + \sum_{i=1}^{r} \lambda_i \left(\frac{\partial f_i}{\partial u} - \frac{\partial}{\partial x}\left(\frac{\partial f_i}{\partial u_x}\right)\right)\right]\delta u\right\} dx = 0$$
$$\frac{\partial}{\partial x}\frac{\partial f}{\partial u_x} \equiv \sum_{i,j=1}^{m,s} \frac{\partial}{\partial x_j}\frac{\partial f}{\partial u_{ix_j}}, \quad \frac{\partial}{\partial x}\frac{\partial f_k}{\partial u_x} \equiv \sum_{i,j=1}^{m,s} \frac{\partial}{\partial x_j}\frac{\partial f_k}{\partial u_{ix_j}} \tag{1.85}$$

We observe that the boundary terms that arise in the integration by parts vanish if we assume that the vector function v obeys a system of boundary conditions $N^*(h)\, v = 0$ and Eq. (1.65) between the boundary-values of the variations for the state and control variables is satisfied.

From Eq. (1.85) we can conclude that the optimality condition and the relation between δJ and δh have been written, as before, in the form of Eqs. (1.68) and (1.71). The boundary-value problem for the adjoint variables v, however, assumes a different form

$$L^*(h)\,v + \frac{\partial f}{\partial u} - \frac{\partial}{\partial x}\frac{\partial f}{\partial u_x} + \sum_{i=1}^{r} \lambda_i \left(\frac{\partial f_i}{\partial u} - \frac{\partial}{\partial x}\frac{\partial f_i}{\partial u_x} \right) = 0$$
$$(N^*(h)\,v)_\Gamma = 0 \tag{1.86}$$

Even though the boundary condition in Eq. (1.86) looks like Eq. (1.70), it contains not only terms that arise from integration by parts in Eq. (1.66), but also composite terms that are generated by a similar manipulation of the variations δJ and δJ_i (during "shifting" of differentiation from δu to $\partial f_i/\partial u_x$).

Similarly, we can treat the presence of second derivatives of u in expressions for f and f_i. Without offering detailed explanations, we point out that in this case the optimality condition and the expression relating δJ and δh preserve the general form of Eqs. (1.68) and (1.71) while additional terms appear in the basic equation and in the boundary condition for the adjoint variables.

$$L^*(h)\,v + \frac{\partial f}{\partial u} - \frac{\partial}{\partial x}\frac{\partial f}{\partial u_x} + \frac{1}{2}\sum_{j,\,l=1}^{s} \frac{\partial^2}{\partial x_j \partial x_l}\frac{\partial^2 f}{\partial u_{x_j} \partial u_{x_l}}$$
$$+ \sum_{i=1}^{r} \lambda_i \left(\frac{\partial f_i}{\partial u} - \frac{\partial}{\partial x}\frac{\partial f_i}{\partial u_x} + \frac{1}{2}\sum_{j,\,l=1}^{s} \frac{\partial^2}{\partial x_j \partial x_l}\left(\frac{\partial^2 f_i}{\partial u_{x_j} \partial u_{x_l}} \right)\right) = 0$$

1.7. Extremal Conditions for Problems with Nonadditive Functionals

Apart from finding the extremum for certain integrals (with some auxiliary constraints) we may wish to consider more general problems of minimizing or maximizing nonadditive functionals that are functions of several integrals $J_1, \ldots, J_r$. Such problems arise, for example, in the optimization of natural frequencies of vibrating elastic systems or in maximization of the critical load that causes a structural element to lose its stability. Considerations of local properties of structures also lead to the study of problems with nonadditive functionals and to applications of the technique of replacing such functionals by integral functionals, described in Section 1.5. In many cases, constraints imposed on state variables and on control variables may also be represented by nonlinear functions of integral functionals.

We shall introduce below some necessary conditions for optimality in problems with nonadditive functionals, where the quality criterion that is to be minimized (maximized) and the constraints can both be represented in the form of functions depending on integral properties, such as

$$J = F(J_1, \ldots, J_r) \tag{1.87}$$

$$F_i(J_1, \ldots, J_r) = 0, \qquad i = 1, 2, \ldots, k \tag{1.88}$$

1. To begin, we shall consider a problem of the calculus of variations consisting of finding a scalar function $u(x)$, depending on the argument $x = \{x_1, \ldots, x_s\}$, that extremizes the functional of Eq. (1.87) with constraints of Eq. (1.88), where

$$J_i = \int_\Omega f_i(x, u, u_x)\, dx, \qquad i = 1, 2, \ldots, r \tag{1.89}$$

To be more specific, let us assume that the values of the function u are assigned on the boundary Γ of the region Ω. To derive Euler's equations we consider the function $u(x)$ and the function $u(x) + \delta u(x)$ [with $\delta u(x) = 0$ on Γ] and compute the variation δJ of the functional given by Eq. (1.87) corresponding to the variation δu. To accomplish this we utilize well-known formulas[50] that express the first variation of the integrals δJ_i in terms of the variation δu of the function u:

$$\delta J_i = \int_\Omega \left(\frac{\partial f_i}{\partial u} - \sum_{j=1}^{s} \frac{\partial}{\partial x_j} \frac{\partial f_i}{\partial u_{x_j}} \right) \delta u\, dx \tag{1.90}$$

Decomposing the functions F and F_i into power series in δJ_i and retaining only first order terms, we have

$$\delta J = \sum_{i=1}^{r} \frac{\partial F}{\partial J_i} \delta J_i = \int_\Omega \left[\sum_{i=1}^{r} \frac{\partial F}{\partial J_i} \left(\frac{\partial f_i}{\partial u} - \sum_{j=1}^{s} \frac{\partial}{\partial x_j} \frac{\partial f_i}{\partial u_{x_j}} \right) \right] \delta u\, dx \tag{1.91}$$

$$\sum_{i=1}^{r} \frac{\partial F_l}{\partial J_i} \delta J_i = \int_\Omega \left[\sum_{l,\, i=1}^{k,\, r} \frac{\partial F_l}{\partial J_i} \left(\frac{\partial f_i}{\partial u} - \sum_{j=1}^{s} \frac{\partial}{\partial x_j} \frac{\partial f_i}{\partial u_{x_j}} \right) \right] \delta u\, dx$$

Now, equating $\delta J + \Sigma_{l=1}^{k} \lambda_l \delta F_l$ to zero (here λ_l are Lagrange multipliers) and using Eqs. (1.90) and (1.91) and the fact that δu is arbitrary, we arrive at the following set of necessary conditions for an extremum:

$$\sum_{i=1}^{s} \frac{\partial \Phi}{\partial J_i} \left(\frac{\partial f_i}{\partial u} - \sum_{j=1}^{s} \frac{\partial}{\partial x_j} \frac{\partial f_i}{\partial u_{x_j}} \right) = 0 \tag{1.92}$$

where

$$\Phi = F + \sum_{l=1}^{k} \lambda_l F_l$$

$$\partial \Phi / \partial J_i = F_{J_i} (J_1, \ldots, J_r) + \Sigma \lambda_i F_{lJ_i} \times (J_1, \ldots J_r)$$

are computed when the functionals $J_1, \ldots, J_r$ assume the values that yield an extremum to the variational problem of Eqs. (1.87) and (1.88). Therefore Eq. (1.92) is an integro-differential equation. With the constraint of Eq. (1.88) absent, this equation was derived in Ref. 148, and for the case in which the functional to be minimized as the product of the powers of certain integrals it was derived in Ref. 201.

2. In the following[148] we consider the problem of finding a function $u(x)$ on the domain Ω that minimizes the functional of Eqs. (1.87) and (1.89) with auxiliary conditions

$$\Psi (J_1, J_2, \ldots, J_r) = C \tag{1.93}$$

$$(u)_\Gamma = 0 \tag{1.94}$$

where $C < 0$ is a given constant and Ψ is a given function. Let us suppose that F and Ψ are homogeneous functions of the variables $J_1, \ldots, J_s$, where the degree of homogeneity of the function Ψ is equal to α_i and the functions f_i are homogeneous in variables u and u_x of degrees β. Obviously, the problem of Eqs. (1.87), (1.89), (1.93), and (1.94) is a variational problem with an isoperimetric condition.

We shall indicate a device that permits us to reduce this problem to one without the isoperimetric condition. Let $u^0(x)$ be a function that solves this variational problem with the functional

$$J' = F/\Psi \tag{1.95}$$

with an auxiliary condition of Eq. (1.94) and

$$u^0(x') = 1 \tag{1.96}$$

but ignores the isoperimetric condition of Eq. (1.93). Here, x' is some arbitrary fixed point of the domain Ω. Since the functions f_i and F are homogeneous if taking into account the boundary conditions of Eq. (1.94), it follows that the function $Bu^0(x)$ (where B is a constant) is also a solution of this problem with the functional of Eq. (1.95) and with an arbitrary normalization condition replacing Eq. (1.96). Moreover, we can choose the constant B so that the condition Eq. (1.93) satisfied. Substituting Bu^0 into Eq. (1.93) we have

$$B = C^{1/\alpha\beta} [\Psi (J_1 (u^0), \ldots, J_r (u^0))]^{-1/\alpha\beta} \tag{1.97}$$

The function Bu^0 with B determined by Eq. (1.97) obviously solves the original problem of Eqs. (1.87), (1.89), (1.93), and (1.94). This remark will be useful to us in solving some eigenvalue problems.

3. We shall now consider the more general optimization problem of Eqs. (1.87), (1.88), (1.58), and (1.59) where

$$J_i = \int_\Omega f_i (x, u, u_x, h)\, dx \tag{1.98}$$

Here, $u = \{u_1(x), \ldots, u_m(x)\}$, $h = \{h_1(x), \ldots, h_n(x)\}$, $u_x = \{u_{1x_1}, \ldots, u_{mx_s}\}$, and $x = \{x_1, \ldots, x_s\}$. Making use of the results outlined in part 1 of Section 1.6, we shall obtain necessary conditions for the minimum (maximum) of the functional J with constraints of Eq. (1.88) and differential relations of Eqs. (1.58) and (1.59) that relate the state variables u and the control variables h.

Without getting involved in details, we shall write only the resulting formulas. If we consider the differential relations and the boundary conditions $(N^*(h)\, v)_\Gamma = 0$ for the adjoint function v, the formula for the variation of the functional J assumes the form

$$\delta J = \int_\Omega \Big\{ \Big[\sum_{l=1}^{r} \frac{\partial \Phi}{\partial J_l} \Big(\frac{\partial f_l}{\partial u} - \sum_{j=1}^{s} \frac{\partial}{\partial x_j} \frac{\partial f_l}{\partial u_{x_j}} \Big) + L^* (h)\, v \Big] \delta u + \Big[M^* (u, h)\, v + \sum_{l=1}^{r} \frac{\partial \Phi}{\partial J_l} \frac{\partial f_l}{\partial h} \Big] \delta h \Big\} dx \tag{1.99}$$

where Φ has the same meaning as in Eq. (1.92), $\partial f_l/\partial u = \{\partial f_l/\partial u_1, \ldots, \partial f_l/\partial u_m\}$, $\partial f_l/\partial u_{xj} = \{\partial f_l/\partial u_{1x_j}, \ldots, \partial f_l/\partial u_{mx_j}\}$, and $(\partial f_l/\partial h)\,\delta h = (\partial f_l/\partial h_1)\,\delta h_1 + \cdots + (\partial f_l/\partial h_n)\,\delta h_n$.

Setting $\delta J = 0$ in Eq. (1.99) and regarding the variations of the state and control variables as completely arbitrary, we derive the (vector) equation for the adjoint variable, an optimality criterion, and an expression that relates the variation δJ to the variation δh

$$L^*(h)\,v + \sum_{l=1}^{r} \frac{\partial \Phi}{\partial J_l}\left(\frac{\partial f_l}{\partial u} - \sum_{j=1}^{s} \frac{\partial}{\partial x_j}\,\frac{\partial f_l}{\partial u_{x_j}}\right) = 0 \tag{1.100}$$

$$M^*(u, h)\,v + \sum_{l=1}^{r} \frac{\partial \Phi}{\partial J_l}\,\frac{\partial f_l}{\partial h} = 0 \tag{1.101}$$

$$\delta J = \int_{\Omega} \left[M^*(u, h)\,v + \sum_{l=1}^{r} \frac{\partial \Phi}{\partial J_l}\,\frac{\partial f_l}{\partial h}\right] \delta h\, dx \tag{1.102}$$

Similarly, it is possible to derive necessary conditions for optimality in the case where f_l depends on higher order derivatives of the state variables. The optimality criteria and the formulas connecting δJ and δh retain the general form of Eqs. (1.101) and (1.102) and only the basic equation and the boundary terms that are used to define the adjoint variable need to be revised.

If the integrand in Eq. (1.98) contains second order derivatives of the variable, $\partial^2 u_l/\partial x_i\,\partial x_j (l = 1, 2, \ldots, m; i, j = 1, \ldots, r)$, the Eqs. (1.100) for v is given in the form

$$L^*(h)\,v + \sum_{l=1}^{r} \frac{\partial \Phi}{\partial J_l}\left(\frac{\partial f_l}{\partial u} - \sum_{j=1}^{s} \frac{\partial}{\partial x_j}\,\frac{\partial f_l}{\partial u_{x_j}} + \frac{1}{2}\sum_{j,\,d=1}^{s} \frac{\partial^2}{\partial x_j \partial x_d}\,\frac{\partial f_l}{\partial u_{x_j x_d}}\right) = 0 \tag{1.103}$$

4. To illustrate our discussion we offer the example of optimal design of a compressed elastic rod for stability. We display the basic relations in dimensionless form as

$$J_* = \max_h \min_w \left(\frac{J_1}{J_2}\right), \qquad J_3 = 1$$

$$J_1 = \int_0^1 h^2 w_{xx}^2\, dx, \qquad J_2 = \int_0^1 w^2\, dx, \qquad J_3 = \int_0^1 h\, dx$$

$$w(0) = (h^2 w_{xx})_{x=0} = 0, \qquad w(1) = (w_x)_{x=1} = 0$$

At $x = 0$ the rod has a swivel support, and at $x = 1$ it is clamped. We discuss here the case in which all cross sections of the rod have similar geometric shapes. Regarding h as the control variable, we maximize the functional $\min_w(J_1/J_2)$ which is equal to the critical value of the applied load, i.e., to the load that causes loss of stability. The constraint $J_3 = 1$ determines the volume of the rod.

For this problem the optimality criterion is given by $hw_{xx}^2 = \text{const}$.

1.8. Problems with Unknown Boundaries

In the previous section, we discussed cases in which the control variables are defined as functions on a fixed domain Ω. The functionals considered in these problems, the coefficients of the differential equations, and of the boundary conditions that determine the state $u(x)$ may all depend on the control variables. The boundary of the domain Ω is regarded as fixed and is not subjected to variations. However, many problems of optimal design can be restated as problems with unknown boundaries, i.e., the shape of the domain Ω can be regarded as the control variable. In a rigorous formulation of the problem of optimization of the shape of elastic bodies, specifically of thin-walled structural members, the control variables appear in the coefficients of the differential equations and in the boundary conditions. This feature explains the necessity of replacing the exact three-dimensional model of an elastic body by an approximate lower-dimensional model. In such approximate models, we average the control variables over one or two of the spatial coordinates. For example the reason why the thickness h enters into the basic equations of state and into the boundary conditions in the theory of bending of elastic beams and plates is the modeling procedure whereby we average the values of the state variable over the entire thickness.

We approach problems of "boundary control" by finding the geometric location of the surfaces separating materials with different mechanical properties in inhomogeneous or piecewise-homogeneous bodies. In many cases the existence of unknown surfaces or lines separating various segments of an elastic body comes about not as a direct consequence of the optimization requirements, but as a result of specific characteristics of the optimum solutions and of the techniques of finding these solutions. For example, when we design plates of minimum weight, subjected to bending, it may happen that the optimal distribution of plate thickness $h(x,y)$ assumes a zero value on some *a priori* unknown lines Γ_s in the domain Ω. Hence, we need to integrate the equations of equilibrium separately in regions whose boundaries may contain Γ_s and then "glue" together along Γ_s solutions, or their asymptotic representations. Thus we need

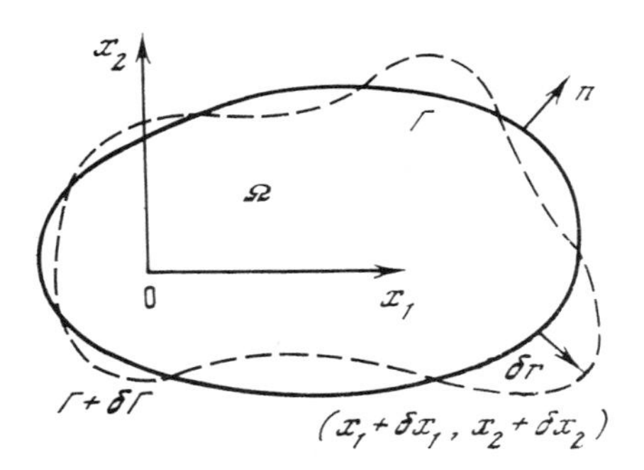

Figure 1.3

to find the location of the lines Γ_s. Here, we can make use of auxiliary relations that are generalizations of the Weierstrass–Erdmann condition.

1. In this section we introduce the necessary optimality conditions for optimization problems with unknown boundaries. Let the functional

$$J = \int_{\Omega} f(x,\ u,\ u_x)\, dx \tag{1.104}$$

be defined in the region Ω with the boundary Γ. (See Fig. 1.3).

Here $x = \{x_1, \ldots, x_s\}$, $u = \{u_1(x), \ldots, u_m(x)\}$, and $u_x = \{u_{1x_1}, \ldots, u_{mx_s}\}$. Consider first the problem of extremizing our functional with respect to u and with respect to the boundary Γ. The first variation of J in terms of the variations of Γ and u is given by the formula (derived in Refs. 50 and 75):

$$\begin{aligned}\delta J = &\int_{\Omega} \sum_{i=1}^{m} \left(\frac{\partial f}{\partial u_i} - \sum_{j=1}^{s} \frac{\partial}{\partial x_j} \frac{\partial f}{\partial u_{ix_j}} \right) \left(\delta u_i - \sum_{j=1}^{s} u_{ix_j} \delta x_j \right) dx \\ &+ \int_{\Gamma} \sum_{i=1}^{m} \left(\sum_{j=1}^{s} \frac{\partial f}{\partial u_{ix_j}} \frac{\partial x_j}{\partial n} \right) \left(\delta u_i - \sum_{j=1}^{s} u_{ix_j} \delta x_j \right) d\Gamma \\ &+ \int_{\Gamma} \left(\sum_{j=1}^{s} \frac{\partial x_j}{\partial n} \delta x_j \right) f\, d\Gamma\end{aligned}$$

where $\partial x_j/\partial n = n_j$, $n = \{n_1, \ldots, n_s\}$, and $\|n\| = 1$. For the sake of convenience, we introduce the notation $Z_i = \{\partial f/\partial u_{ix_1}, \ldots, \partial f/\partial u_{ix_s}\}$, $i = 1, 2, \ldots, m$. We make the assumption that u satisfies the necessary conditions (i.e., Euler's equations) for an extremum in the region Ω

$$\frac{\partial f}{\partial u_i} - \sum_{j=1}^{s} \frac{\partial}{\partial x_j} \frac{\partial f}{\partial u_{ix_j}} = 0, \qquad i = 1, \ldots, m \tag{1.105}$$

Then the expression for δJ becomes

$$\delta J = \int_\Gamma \sum_{i=1}^{m} (Z_i, n)\, \delta u_i \, d\Gamma + \int_\Gamma \left\{(n, \delta r) f - \sum_{i=1}^{m} (Z_i, n)(\nabla u_i, \delta r)\right\} d\Gamma \tag{1.106}$$

Brackets such as (Z_i, n), $(n, \delta r)$, and $(\nabla u_i, \delta r)$ in these formulas and in our future discussion denote a scalar product of two vectors. We shall obtain various optimality conditions by making different assumptions concerning the boundary conditions assigned to the function u.

To begin with, consider the case in which $(u)_\Gamma$ is not given and the variation $(\delta u)_\Gamma$ is an arbitrary vector function. Since $(\delta u)_\Gamma$ is arbitrary, we can derive from the equality $\delta J = 0$ a system of transversality conditions $(Z_i, n) = 0$ $(i = 1, \ldots, m)$, which can be written by using the original notation as

$$\left(\sum_{j=1}^{s} \frac{\partial f}{\partial u_{ix_j}} n_j\right)_\Gamma = 0, \qquad i = 1, \ldots, m \tag{1.107}$$

Assuming that Eq. (1.107) is satisfied, we can derive the variation of our cost functional in terms of the variation of Γ as

$$\delta J = \int_\Gamma (n, \delta r)\, f \, d\Gamma \tag{1.108}$$

from which we derive the necessary condition for optimality of Γ

$$(f)_\Gamma = 0 \tag{1.109}$$

Let us consider now a different case. We assume that the function u is given on Γ, so $(\delta u)_\Gamma = 0$. For a variation of the boundary, written in the form $\delta r = tn$, where t is a parameter and n is the vector normal to Γ, Eq. (1.106) allows us to conclude that

$$\delta J = \int_\Gamma t \left[f - \sum_{i=1}^{m} (Z_i, n)(\nabla u_i, n)\right] d\Gamma \tag{1.110}$$

Using the condition $\delta J = 0$ and the fact that t is arbitrary, we obtain the following necessary condition for optimality of the boundary Γ, with given boundary conditions assigned to the vector function u:

$$\left(f - \sum_{i=1}^{m} (Z_i, n)(\nabla u_i, n)\right)_\Gamma = 0 \tag{1.111}$$

In the specific case for which $(u)_\Gamma = \text{const}$, the vectors ∇u_i are defined at points $x \in \Gamma$ and are parallel to the normal vectors n and we have $(Z_i, n)(\nabla u_i, n) = (Z_i, \nabla u_i)$. The optimality condition of Eq. (1.111) for the boundary, on which the vector function u assumes a constant value, takes the form

$$\left(f - \sum_{i,j=1}^{m,s} \frac{\partial f}{\partial u_{ix_j}} u_{ix_j}\right)_\Gamma = 0 \tag{1.112}$$

If f is a homogeneous function of degree β in the variables u_{ix_j}, then according to a theorem of Euler the summation term in Eq. (1.112) must be equal to βf. Consequently, we again arrive at the optimality condition of Eq. (1.109).

We observe that in the general case for which the first m_1 $(m_1 \leqslant m)$ components of $u(u_1, \ldots, u_{m_1})$ are assigned on the variable boundary Γ, while the remaining components $(u_{m_1+1}, \ldots, u_m)$ are completely free and are to be determined while we solve the problem of extremizing the integral of Eq. (1.104), the condition of optimality for the boundary can be written in the following form:

$$\left(f - \sum_{i=1}^{m_1} (Z_i, n)(\nabla u_i, n)\right)_\Gamma = 0 \tag{1.113}$$

2. Let J be the functional to be optimized and the constraints assigned to the problem be some given functions of the integral functionals, i.e.,

$$\begin{gathered} J = F(J_1, \ldots, J_r), \qquad J_j = \int_\Omega f_j(x, u, u_x)\, dx \\ F_i(J_1, \ldots, J_r) = c_i, \quad i = 1, \ldots, k \end{gathered} \tag{1.114}$$

Using arguments analogous to those given in part 1 of this section and in Section 1.7, we can derive a formula for the first variation δJ in terms of the variation of the function u and of the variation of the contour Γ

$$\delta J = \int_{\Omega} \sum_{l=1,\, i=1}^{r,\, m} \frac{\partial \Phi}{\partial J_l} \left(\frac{\partial f_l}{\partial u_i} - \sum_{j=1}^{s} \frac{\partial}{\partial x_j} \frac{\partial f_l}{\partial u_{ix_j}} \right) \delta u_i \, dx$$
$$+ \int_{\Gamma} \sum_{l=1}^{r} \frac{\partial \Phi}{\partial J_l} \left[(n, \delta r) f_l + \sum_{i=1}^{m} (Z_i, n) (\delta u_i - (\nabla u_i, \delta r)) \right] d\Gamma \qquad (1.115)$$

where

$$\Phi = F - \sum_{i=1}^{k} \lambda_i F_i$$

Supposing that u is not given on Γ, we make use of Eq. (1.115) and the relation $\delta J = 0$ to derive a system of equalities that must be satisfied by the extremizing functions u_i and boundary Γ. These are the equations of Euler, the transversality conditions on Γ, and the optimality conditions for Γ

$$\sum_{l=1}^{r} \frac{\partial \Phi}{\partial J_l} \left(\frac{\partial f_l}{\partial u_i} - \sum_{j=1}^{s} \frac{\partial}{\partial x_j} \frac{\partial f_l}{\partial u_{ix_j}} \right) = 0, \qquad i = 1, \ldots, m \qquad (1.116)$$

$$\sum_{l=1,\, j=1}^{r,\, s} \frac{\partial \Phi}{\partial J_l} \frac{\partial f_l}{\partial u_{ix_j}} n_j = 0, \qquad i = 1, \ldots, m \qquad (1.117)$$

$$\left(\sum_{l=1}^{r} \frac{\partial \Phi}{\partial J_l} f_l \right)_{\Gamma} = 0 \qquad (1.118)$$

In case some components $(i = 1, \ldots, m_1)$ of the vector function u are given, the transversality conditions of Eq. (1.117) determine the remaining $m - m_1$ components, i.e., for $i = m_1 + 1, \ldots, m$, while the condition that Γ is an optimum boundary assumes the form

$$\left(\sum_{l=1}^{r} \frac{\partial \Phi}{\partial J_l} \left\{ f_l - \sum_{i=1}^{m_1} (Z_i, n) (\nabla u_i, n) \right\} \right)_{\Gamma} = 0 \qquad (1.119)$$

If the functions $u_i (i = 1, 2, \ldots, m_1)$ are constant on Γ, Eq. (1.119) is simplified to

$$\left(\sum_{l=1}^{r} \frac{\partial \Phi}{\partial J_l} \left\{ f_l - \sum_{i=1,\, j=1}^{m_1,\, s} \frac{\partial f_i}{\partial u_{x_j}} u_{x_j} \right\} \right)_{\Gamma} = 0 \qquad (1.120)$$

In the study of variation of the boundaries, much more complex problems arise in analysis of the local properties (local functionals), such as the maximum stress intensity or the maximum displacement in an elastic medium. The study of such problems generally utilizes approximate techniques outlined above in Section 1.5.

3. We now offer some examples of applications for the formulas derived in this section.

Example 1. Consider the problem of maximizing the rigidity of the strip that was considered in the second part of Section 1.4. Interpreting Eq. (1.114) for this problem, we make the identification

$$F = \Pi, \qquad F_1 = S = 1, \qquad S = \iint_{\Omega} dx\, dy$$

where Π is given by Eq. (1.25). Recalling that the displacements u_1 and u_2 are not assigned on the part of the boundary Γ_q that is subjected to a variation, and applying Eq. (1.18), which is appropriate for this case, we obtain the following necessary conditions for optimality of the unknown boundary:

$$a\,(u_{1x}^2 + u_{2y}^2) + 2bu_{1x}u_{2y} + \frac{1}{2}\,c\,(u_{1y} + u_{2x})^2 = \text{const}, \qquad (x,\, y) \in \Gamma_q \tag{1.121}$$

Example 2. As our second example, we offer the problem formulated in Section 1.4 (part 3) of maximizing the torsional rigidity of a bar by choosing the optimum shape of the cross-sectional contour Γ. Applying Eq. (1.114) to this case, we substitute $F = \Pi$ and $F_1 = S = 1$, where Π is determined by Eq. (1.33) and S is given by the same formula as in the first example. Noting that the stress function ϕ that occurs in Eq. (1.33) has a constant value on the varied contour Γ, and utilizing Eq. (1.120), we arrive at a necessary condition for the optimality of the contour Γ

$$a\varphi_x^2 + b\varphi_y^2 - 2c\varphi_x\varphi_y = \text{const}, \qquad (x,\; y) \in \Gamma \tag{1.122}$$

Example 3. Using dimensionless variables, we study the problem of optimizing the fundamental frequency of a vibrating membrane that is supported along the planar closed curve $\Gamma = \Gamma_1 + \Gamma_2$. Part Γ_1 of this curve is given, while part Γ_2 is to be determined from the extremal property of the fundamental frequency ω. We assume that the domain Ω, bounded by Γ, has a unit surface area. For this problem we have

$$J = F(J_1, J_2) \equiv \frac{J_1}{J_2}, \qquad J_1 = \iint_\Omega (u_x^2 + u_y^2)\, dx\, dy$$

$$J_2 = \iint_\Omega u^2 dx\, dy, \quad F_1 = S = 1, \quad S = \iint_\Omega dx\, dy$$

The amplitude function for the displacement of the membrane $u(x, y)$ satisfies the boundary condition $(u)_\Gamma = 0$. Making use of Eq. (1.120), we arrive at the following condition for optimality of the unknown part of the boundary:

$$u_x^2 + u_y^2 - \omega^2 u^2 = \text{const}, \qquad (x, \ y) \in \Gamma_2 \tag{1.123}$$

which was first derived in Ref. 181.

1.9. Dual Problems

We frequently consider the so-called "dual problems" in the theory of optimal design. The solutions of such problems differ only by scale factors. In many cases, a direct solution of a problem may be replaced by the formulation of a dual problem, if that problem has already been studied; the solution of a direct problem can be obtained by the technique of scaling of the solution. Thus an introduction of duality is useful in optimal design theory and permits us to reduce the number of problems that need to be solved. The idea of duality plays an important part in securing two-sided estimates on optimal solutions and in constructing numerical algorithms that "meet at a point."

The concept of duality has been utilized in solving some specific problems in earlier research articles, but a rigorous study of this concept was undertaken only recently,[227] when it was applied to problems with homogeneous functionals.

Consider such a problem with a homogeneous functional. Let $\mathfrak{A}$ be a linear vector space and $\mathfrak{K}$ be a subset which is a cone. This means that for any element $h \in \mathfrak{K}$, $\lambda h \in \mathfrak{K}$ for any $\lambda = \text{const} > 0$.

Let J_1 and J_2 be homogeneous functionals defined on $\mathfrak{A}$, whose degrees of homogeneity are α and β, respectively, i.e., $J_1(\lambda h) = \lambda J_1^\alpha(h)$, and $J_2(\lambda h) = \lambda^\beta J_2(h)$ for any $\lambda > 0$. We assume that for any $h \in \mathfrak{A}$ these functionals are positive and we consider the problem of finding an extremum

$$\begin{aligned} &\min \ J_1(h), \qquad h \in \mathscr{K}, \\ &J_2(h) \geqslant c_2 > 0 \end{aligned} \tag{1.124}$$

where c_2 is a given constant.

It is clear that this problem makes sense if α and β are of the same sign. Otherwise, no solution exists. To be more specific, assume that α and β are both positive. We shall show that if h^* solves the problem of Eq. (1.124) then $J_2(h^*) = c_2$, i.e., the minimum of J_1 is attained on the boundary of the admissible region. Assuming by way of contradiction that $J_2(h^*) > c_2$, we can choose a constant multiplier $\lambda = [c_2/J_2(h^*)]^{1/\beta} < 1$. The vector h^* is admissible since $J_2(\lambda h^*) = \lambda^\beta J_2(h^*) = c_2$. But then $J_1(\lambda h^*) \leqslant J_1(h^*)$, since $\lambda < 1$ and $\alpha > 0$. This contradicts the optimality of h^*.

Let us consider a different optimization problem with identical functionals (where c_1 is a given constant)

$$\begin{aligned} &\max\ J_2(h), \quad h \in \mathscr{K} \\ &J_1(h) \leqslant c_1 > 0 \end{aligned} \tag{1.125}$$

For this problem, the extremum is attained on the boundary $J_1(h^{**}) = c_1$, where h^{**} denotes the solution of Eq. (1.125). We shall prove the following assertion:

1. If h^* solves the problem of Eq. (1.124), then the vector $h^{**} = \gamma h^*$, where $\gamma = [c_1/J_1(h^*)]^{1/\alpha}$ solves the problem of Eq. (1.125).

To prove this assertion, we consider an arbitrary vector $h \in K$ such that $J_1(h) \leqslant c_1$ and we shall demonstrate that $J_2(h) \leqslant J_2(h^{**})$. As before, let us choose a multiplier $\chi = [c_2/J_2(h)]^{1/\beta}$. The vector χh is an admissible vector because $\chi h \in K$ and $J_2(\chi h) = c_1$. Because h^* is optimal, the following inequality is true: $J_1(\chi h) \geqslant J_1(h^*)$. That is, $[c_2/J_2(h)]^{\alpha/\beta} J_1(h)] \geqslant J_1(h^*)$. We have already shown that $J_2(h^*) = c_2$. Hence, using the above inequality, we have

$$\begin{aligned} J_2(h) &\leqslant J_2(h^*) \left[\frac{J_1(h)}{J_1(h^*)}\right]^{\beta/\alpha} \leqslant J_2(h^*) \left[\frac{c_1}{J_1(h^*)}\right]^{\beta/\alpha} \\ &= J_2\left(\left[\frac{c_1}{J_1(h^*)}\right]^{1/\alpha} h^*\right) = J_2(\gamma h^*) = J_2(h^{**}) \end{aligned} \tag{1.126}$$

Note that we did make use of the inequality $J_1(h) \leqslant c_1$. Thus, assertion 1 is proved. The converse is also true.

2. If h^{**} solves the problem of Eq. (1.125), then the vector $h' = \mu h^{**}$, with $\mu = [c_2/J_2(h^{**})]^{1/\beta}$ solves the problem of Eq. (1.124). The proof is a complete repetition of the one given above. Now let us show that $\gamma\mu = 1$. Indeed, making use of the equalities $c_2 = J_2(h^*)$ and $c_1 = J_1(h^{**})$, we have

$$\begin{aligned} \gamma\mu &= \left[\frac{c_1}{J(h^*)}\right]^{1/\alpha} \left[\frac{c_2}{J_2(h^{**})}\right]^{1/\beta} = \left[\frac{J_1(h^{**})}{J_1(h^*)}\right]^{1/\alpha} \left[\frac{J_2(h^*)}{J_2(h^{**})}\right]^{1/\beta} \\ &= \mu^{-1} \left[\frac{J_1(h')}{J_1(h^*)}\right]^{1/\alpha} \mu \left[\frac{J_2(h^*)}{J_2(h')}\right] = 1 \end{aligned} \tag{1.127}$$

In deriving the last estimate in Eq. (1.127), we made use of the fact that h^* and h' are optimum vectors and that consequently $J_1(h^*) = J_1(h')$ and $J_2(h^*) = J_2(h')$.

The problems of Eqs. (1.124) and (1.125) shall be called dual with respect to each other. For any dual problems, the following assertion is true:

3. If a solution exists and is unique to either the problem of Eqs. (1.124) or (1.125), then the solution of the dual problem also exists and is also unique. Moreover, these solutions obey the following relations:

$$h^{**} = \gamma h^*, \quad \gamma = [c_1/J_1(h^*)]^{1/\alpha} = [c_2/J_2(h^{**})]^{-1/\beta} \tag{1.128}$$

The proof of this assertion follows immediately from assertions 1 and 2 and Eq. (1.127). Also, the assumption of uniqueness of solutions implies that in the chain of inequalities of Eq. (1.126) strict inequalities replace the $\leqslant$ symbol.

Remark 1. Because the functionals in the problems of Eqs. (1.124) and (1.125) were homogeneous, the solutions h^* and h^{**} were representable in the form $h^* = c_2^{1/\beta} h_1$ and $h^{**} = c_1^{1/\alpha} h_2$, where h_1 and h_2 were solutions of identical problems with $c_2 = 1$ and $c_1 = 1$.

Remark 2. If the basic functionals have degrees of homogeneity of different sign, than it makes sense to consider the following problems:

$$\min J_1(h), \qquad \min J_2(h)$$
$$J_2(h) \leqslant c_2 > 0, \qquad J_1(h) \leqslant c_1 > 0$$
$$h \in \mathcal{K} \qquad h \in \mathcal{K}$$

These problems are dual with respect to each other. This is not hard to check after substituting $J_2^1(h) = 1/J_2(h)$. Then in the functionals J_1 and J_2^1 the problem is easily reduced to the one given by Eqs. (1.124) and (1.125).

We shall introduce some examples of dual problems that arise in the optimization of elastic beams. Using dimensionless variables and assuming that the beam has a unit length positioned along the x axis, we postulate a free support at the end points $x = 0$ and $x = 1$. Let $S(x)$ be the distribution of cross-sectional area along the length of the beam, while V, P, ω^2, and w denote, respectively, the volume of the beam, the critical compressive load which causes the loss of stability (equal forces are applied at both ends), the square of the fundamental frequency of free transverse vibrations, and the magnitude of the deflection at the point $x = 1/2$ when a transverse unit point load is applied to that point. The following formulas are valid for the quantities introduced here:

$$V = \int_0^1 S dx, \qquad P = \min_u \frac{\int_0^1 u_x^2 dx}{\int_0^1 S^{-2} u^2 dx}, \qquad \omega^2 = \min_u \frac{\int_0^1 S^2 u_{xx}^2 dx}{\int_0^1 S u^2 dx} \tag{1.129}$$

$$w = \int_0^1 S^{-2} \psi(x) \, dx, \qquad \psi(x) = \begin{cases} 3x^2/8, & 0 \leqslant x \leqslant 1/2 \\ 3(1 - x^2)/8, & 1/2 \leqslant x \leqslant 1 \end{cases}$$

Here we have assumed that the basic shape of the cross-sectional area does not change, but the area varies along the length of the rod (see Section 1.3). Let V_0, P_0, ω_0, and w_0 be given positive constants. We regard the function $S(x)$ as the unknown control variable. In determining $S(x)$, we consider the condition $h \geqslant 0$, as well as other constraints.

Example 1. We shall analyze the problem of minimizaton of the volume of a beam that is subjected to a transverse unit point load at the point $x = 1/2$. We impose a constraint on the magnitude of the deflection at the point of application of the load

$$V \to \min_S \quad w^{-1} \geqslant w_0^{-1} \tag{1.130}$$

The functionals V and w^{-1} defined in Eq. (1.129) are nomogeneous in S with degrees of homogeneity $\alpha = 1$ and $\beta = 2$, respectively. The following problem is the dual of the one given in Eq. (1.130): Minimize w (or maximize its inverse w^{-1}), with the following constraint assigned to the volume of the beam:

$$w^{-1} \to \max_S, \qquad V \leqslant V_0 \tag{1.131}$$

If the volume functional attains its minimum, with suitable constraints, with the control variable S^*, then the function S^{**} that mimimizes the magnitude of the deflection w while the volume of the beam does not exceed a certain given value, is given in agreement with Eq. (1.128) by the formula

$$S^{**} = V_0 S^* \Big/ \int_0^1 S^* \, dx \tag{1.132}$$

Vice versa, if the solution S^{**} of the dual problem of Eq. (1.131) is known, then the solution S^* of the original problem of Eq. (1.130) can be expressed in terms of S^{**} according to the formula

$$S^* = \frac{S^{**}}{\sqrt{w_0}} \left[\int_0^1 \psi \, (S^{**})^{-2} \, dx \right]^{1/2} \tag{1.133}$$

Example 2. Let us select the distribution of cross-sectional area that minimizes the volume of a rod, with a constraint on the critical load that causes the loss of stability

$$V \to \min_S, \qquad P \geqslant P_0 \tag{1.134}$$

The degrees of homogeneity of V and P are $\alpha = 1$ and $\beta = 2$. The problem that is dual to Eq. (1.134) is that of maximizing the value of the critical load, with a constraint assigned to the maximum volume of the rod

$$P \to \max_S, \qquad V \leqslant V_0 \tag{1.135}$$

The solutions S^* of the direct problem of Eq. (1.134) and S^{**} of the dual of Eq. (1.135) are related to each other by means of the following equalities:

$$\begin{aligned} S^{**} &= \gamma S^*, \qquad \gamma = V_0 \Big/ \int_0^1 S^* \, dx \\ S^* &= \mu S^{**}, \qquad \mu = \left[P_0 \int_0^1 u^2 (S^{**})^{-2} \, dx \Big/ \int_0^1 u_x^2 \, dx \right]^{1/2} \end{aligned} \tag{1.136}$$

which are a direct consequence of Eq. (1.128).

Example 3. Consider the problem of minimizing the volume of a beam, with a constraint assigned to the fundamental frequency of transverse vibration, i.e.,

$$V \to \min_S, \qquad \omega^2 \geqslant \omega_0^2 \tag{1.137}$$

The dual problem is that of maximizing the fundamental natural frequency, with a condition that the volume of the beam does not exceed a given value, i.e.,

$$\omega_0^2 \to \max_S, \qquad V \leqslant V_0 \tag{1.138}$$

The degrees of homogeneity of the functionals V and ω^2 with respect to S are equal, respectively, to $\alpha = 1$ and $\beta = 1$. Substituting these values of α and β into

Eq. (1.128) we derive the following formulas relating the solution S^* of the direct problem of Eq. (1.137) to the solution S^{**} of the dual problem of Eq. (1.138):

$$
\begin{aligned}
S^{**} &= \gamma S^*, \qquad \gamma = V_0 \Big/ \int_0^1 S^*\,dx \\
S^* &= \mu S^{**}, \qquad \mu = \omega_0 \int_0^1 u^2 S^{**}\,dx \Big/ \int_0^1 u_{xx}^2 (S^{**})^2\,dx
\end{aligned}
\tag{1.139}
$$

1.10. Application of Numerical Techniques in Solving Problems of Optimal Design

The problems that are studied in the theory of optimal design generally lead to extremely complex mathematical questions. In many cases, the search for an optimum shape or an optimum composition of an elastic body reduces to an equivalent formulation of solving a variational problem with unknown boundaries, or to the solution of a game-theoretic statement of the optimization problem. No standard techniques exist for solving such problems. Well-known difficulties are related to the fact that such problems of optimization for elastic bodies are closely related with some nonlinear problems of mechanics. The nonlinearity of these problems is caused by nonlinearities in criteria of optimality. For this reason, successful developments in the theory of optimal design and the effectiveness of techniques used in specific cases are closely related to advances in numerical techniques and developments of algorithms that are adapted to modern computer technology.

Research devoted to the development of numerical algorithms for optimal design, specifically tailored to a certain class of problems and exploiting properties related to that class of problems is being intensively carried out at the present time. Various numerical schemes are being worked out and tested for specific classes of problems. Such schemes are based on finite difference, variational-finite difference, purely variational techniques, and on nonlinear programming methods.

Numerous variants of the finite-element approach are also utilized. However, at the present time, the rigorous justification and proof of convergence for many algorithms are not yet developed, and no rigorous techniques exist for comparing the general effectiveness of proposed algorithms (except in certain narrow applications and test cases). For this reason, we shall refrain from judging or comparing various available techniques. We shall limit our exposition to a description of algorithms of successive optimization that were developed and applied during a period of many years at the Institute for the Problems of Mechanics of the Academy of Sciences of USSR,[25-28, 31] and use them to solve some problems posed in this monograph.

Let us describe the simplest iterative algorithm for successive optimization and point out certain variants, or modifications of this algorithm. To be specific, we consider the optimization problem of Eqs. (1.158)–(1.162) with two independent variables x and y and assume that we are permitted to use calculus of variations techniques to determine the state variable u and the conjugate variable v. The algorithm we propose consists in successive approximations of the optimal solution. It uses the assumption of "smallness" of the variation of the control variable h and is based on iterative solving of the "direct" problem, which consists of finding the functions u and v for a fixed value of h.

Essential parts of this algorithm are the following two blocks (shown in Fig. 1.4): block $B_{u,v}$ for solving "the direct" problem of determining the functions u^k and v^k for a fixed value of the function h^k and block B_h for determining a new approximation h^{k+1} for the control variable, using the current values of

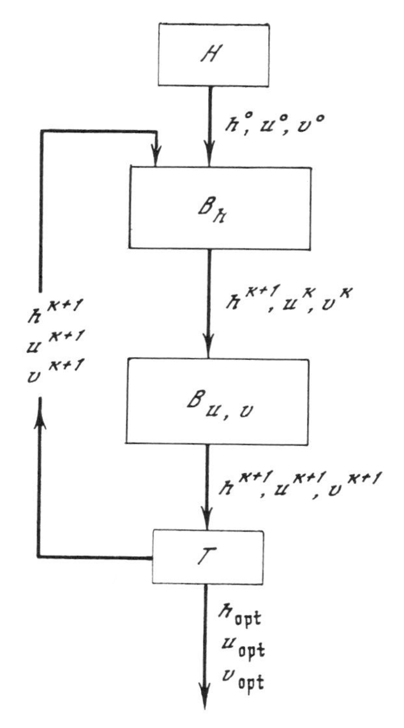

Figure 1.4

h^k, u^k and v^k (where the integer k counts the iterations). The symbols H and T denote the operation of introducing the initial data (initial approximation) and of checking the accuracy of the derived approximation, respectively. The algorithm consists of specific iterative steps.

In each step, say step $(k + 1)$, the following operations are performed: Using the results of computations that were completed in the preceding iterative step, we consider the functions $h^k(x, y)$, $u^k(x, y)$, and $v^k(x, y)$ and the value J^k of the cost functional as given. In step $(k + 1)$, we first determine a new value of the control function $h^{k+1}(x, y)$ (computing the block B_h). Here we utilize the gradient projection technique. Subsequently using the method of local variations,[148] we solve the boundary-value problem of Eqs. (1.58), (1.59), (1.69), and (1.70) (i.e., we compute the block $B_{u,v}$) for the given distribution $h^{k+1}(x, y)$, and we compute the functions u^{k+1} and v^{k+1} and evaluate the functional J^{k+1}.

Now we evaluate the error (or a violation of constraints) for the new values of $h^{k+1}(x, y)$, $u^{k+1}(x, y)$, $v^{k+1}(x, y)$, and J^{k+1} in complying with the necessary conditions for optimality. If the error is sufficiently small, the process is terminated. If the error is too large, we proceed to iterative step $(k + 2)$ and repeat an analogous sequence of operations.

In eigenvalue optimization problems (such as problems of optimizing the critical load or the critical frequency), we use the successive approximation technique inside the iteration block $B_{u,v}$, instead of the technique of local variations to determine the state variables.

1. We determine the new approximation h^{k+1} to the optimal control variable inside block B_h, utilizing the gradient projection technique. To compute h^{k+1} using this technique, we make use of the following formulas:

$$h^{k+1} = h^k - \tau\Lambda^k$$
$$\Lambda^k = M^*(u^k,\ h^k)\,v^k + \left(\frac{\partial f}{\partial h}\right)^k + \sum_{i=1}^{r} \lambda_i^k \left(\frac{\partial f_i}{\partial h}\right)^k \tag{1.140}$$

where $\tau > 0$ is the step-size in the direction of the gradient and λ_i are Lagrange multipliers that may be determined with the help of Eq. (1.62). If we vary h in agreement with Eq. (1.140), then the value of J is decreased. Indeed, Eqs. (1.68) and (1.140) (with $\delta h = \tau\Lambda^k$) imply the inequality

$$\delta J = -\tau \iint_{\Omega} (\Lambda^k)^2\, dx dy \leqslant 0 \tag{1.141}$$

Application of Eq. (1.140) is closely connected with the computation of the Lagrange multipliers λ_i at each iterative step. This may frequently lead to considerable complications. For this reason, instead of the Eq. (1.140), we may use the following trick, which permits us to retain the isoperimetric inequalities of the type of Eq. (1.62), while varying h: Let the number of integral constraints in the problem of Eqs. (1.58)-(1.62) be equal to two. The formulas for the first variation of the functionals $J(h)$, $J_1(h)$ and $J_2(h)$ are given in the from

$$\delta J = (\psi, \delta h), \ \delta J_i = (\psi_i, \delta h), \ i = 1, 2 \tag{1.142}$$

where ψ, ψ_i are known functions and the brackets denote the scalar product of the corresponding functions. Let us set

$$\delta h = \{ -\psi + \mu_1\psi_1 + \mu_2\psi_2 \} \tau \tag{1.143}$$

and determine the constants μ_1 and μ_2 by making use of the conditions $(\psi_i, \delta h) = 0$, which are a consequence of the constraints $J_1 = c_1$ and $J_2 = c_2$. We have

$$\begin{aligned} \mu_1 &= \frac{(\psi, \psi_1)(\psi_2, \psi_2) - (\psi, \psi_2)(\psi_1, \psi_2)}{(\psi_1, \psi_1)(\psi_2, \psi_2) - (\psi_1, \psi_2)^2} \\ \mu_2 &= \frac{(\psi, \psi_2)(\psi_1 \, \psi_1) - (\psi, \psi_1)(\psi_1, \psi_2)}{(\psi_1, \psi_1)(\psi_2, \psi_2) - (\psi_1, \psi_2)^2} \end{aligned} \tag{1.144}$$

If only the isoperimetric condition $J_1 = c_1$ is assigned to the problem, then the expression for δh assumes the form

$$\delta h = \tau \left\{ -\psi + \frac{(\psi, \psi_1)}{(\psi_1, \psi_1)} \psi_1 \right\} \tag{1.145}$$

Using the Cauchy-Schwartz-Bunyakovskii inequality and some rather elementary arguments, it is easy to show that for the variation of h considered here, the following conditions must be satisfied:

$$\delta J \leqslant 0, \ \delta J_i = 0, \ i = 1,2 \tag{1.146}$$

Consequently, for such choices of δh [given by Eqs. (1.143)-(1.145)] and for sufficiently small values of τ, the value of J will decrease and the values of J_i remain unchanged. This device may be utilized in the presence of r constraints, but in this case to determine the r-constants μ_i we need to solve, in each iterative step, r-linear algebraic equations obtained from the conditions $(\psi_i, \delta h) = 0$, $i = 1, \ldots, r$.

2. In the transition from one step in the algorithm to the next, we can affect small improvements due to the variation δh of the control function. If the variation $\delta h = h^{k+1} - h^k$ is sufficiently small, then in the computation of new approximate functions u^{k+1} and v^{k+1}, we have a "good" initial approximations, u^k and v^k. This circumstance justifies the use of these quantities in the iterative technique.

An example of a such a technique which could be successfully applied to the solution of a "direct" problem is the technique of local variations proposed by F. L. Chernous'ko in Ref. 145, and which was further improved in Ref. 148. We now explain this technique.

Let us suppose that we need to determine a function $u(x, y)$ of two independent variables in the domain Ω with the boundary Γ. This function satisfies the constraints

$$u^- (x, y) \leqslant u (x, y) \leqslant u^+ (x, y) \tag{1.147}$$

and minimizes the functional

$$J = \iint_\Omega f(x, y, u, u_x, u_y)\, dx dy \tag{1.148}$$

Here u^+ and u^- are functions given in $\Omega + \Gamma$ with $u^+ \geqslant u^-$ everywhere in $\Omega + \Gamma$ and f is a given function. On the contour Γ we can assign to u a boundary condition $u = \chi(x, y)$, by requiring that the given functions u^+ and u^- satisfy $u + (x, y) = u - (x, y) = \chi(x, y)$ for all $(x, y) \in \Gamma$.

We shall decompose the region Ω by drawing a grid into cells Ω_{ij}, having surface areas S_{ij}. We denote by u_{ij} the values of the function u at the grid corner points. The integral of Eq. (1.148) is replaced by an approximate sum of the integrals over the cells Ω_{ij}

$$\begin{aligned} J &\approx I = \sum I_{ij} \\ I_{ij} &= S_{ij} f (x_i^*, y_j^*, u_{ij}^*, (u_x)_{ij}^*, (u_y)_{ij}^*) \end{aligned} \tag{1.149}$$

We use as the independent variables in the domain of the function f in Eq. (1.149) the average values in each cell of the independent variables, of the unknown function, and of certain finite difference expressions for its derivatives.

The constraints in Eq. (1.147) imply inequalities that should be satisfied by the grid values of the functions,

$$u^-(x_i, y_j) \leqslant u_{ij} \leqslant u^+(x_i, y_j) \tag{1.150}$$

It should be noted that if the functional J is convex, it is necessary for the approximation scheme to be a strictly convex function of the grid values u_{ij} to insure uniqueness of the solution and to speed up convergence of the computational algorithm. Construction of strictly convex schemes, as shown in,[29] can be attained by increasing the number of values of the integrand f inside each cell. We can find in[29] some schemes that preserve strict convexity, i.e., schemes that replace a strictly convex function by a strictly convex approximation. The following is the simplest variant of the algorithm using the technique of local variations applied to minimization of I given in Eq. (1.149), with the constraints of Eq. (1.150). Fixing the size of the step δ varying the grid function u_{ij}, fixing the grid itself, and using a coordinatewise relaxation technique, we find the values of u_{ij} that satisfy the constraints of Eq. (1.150) and realize the minimum for the functional given in Eq. (1.149) in a finite number of steps. Following the convergence of these iterations, we decrease the size of the step δ and again realize the minimizing process for our sums. This decrease in δ is continued until such time when convergence is attained for some sufficiently small value $\delta = \delta^*$. Then we decrease the size of the cells, choose a new value for δ and proceed once again to decrease the value of δ, minimizing the value of I for each value of δ. In this manner the method of local variations encompasses iteration techniques, namely the process of decreasing the cell size, the process of decreasing the size of the step δ, and an iteration process for a fixed size of the grid and magnitude of δ.

3. Because in the use of such algorithms of successive optimization we need to turn repeatedly to a solution of the direct problems, application of these techniques makes sense only if we are assured of sufficiently fast convergence. One of the possible devices for speeding up the convergence of the method of local variations utilizes a variant of the variable step method (i.e., varying δ) (see[8,148]). Varying the solutions using variable step sizes is particularly effective when the unknown quantities have different orders of magnitude in different sectors of the domain of definition.

Let us omit the assumption concerning constant size of the step δ and let us choose for each point $x = x_i$ its own step $\delta = \delta_i$, derived from the condition that the increase $\Delta I = \Delta I(\delta)$ is minimized, where $\Delta I(\delta)$ denotes the sum of the terms defined by Eq. (1.149), which depends on the values u_i of the function u at corners x_i of the grid. For the sake of simplicity and briefness of notation concerning the basic relations, we shall consider the one-dimensional problem of minimizing an integral defined over a line segment, instead of over a two-dimensional domain Ω.

Let us describe this scheme in detail. First, let us suppose that either the constraints $u^-(x_i) \leqslant u_i \leqslant u_i^+(x_i)$, are entirely absent, or else that they do not influence the solution of the problem. We compute the change ΔI in the value of the functional caused by replacing the value u_i at the point x_i by $u_i + \delta_i$

$$\Delta I = I_{i-1}(u_{i-1}, u_i + \delta_i) + I_i(u_i + \delta_i, u_{i+1}) - I_{i-1}(u_{i-1}, u_i) - I_i(u_i, u_{i+1}) \tag{1.151}$$

Now we substitute into Eq. (1.151) the expression $\Delta I_i = \Delta x f(x_i + \Delta x)/2$, $(u_i + u_{i+1})/2$, $(u_{i+1} - u_i)/\Delta x$, and decompose the formula thus obtained into Taylor series in the variable δ_i, up to the order of accuracy $0(\delta_i^3)$. Thus we have

$$\begin{aligned}\Delta I = \delta_i &\left[\frac{1}{2}((f_u)_{i-1}^* + (f_u)_i^*) - \frac{1}{\Delta x}((f_{u_x})_i^* - (f_{u_x})_{i-1}^*)\right] \\ &+ \delta_i^2\left[\frac{1}{8}((f_{uu})_{i-1}^* + (f_{uu})_i^*) + \frac{1}{2\Delta x}((f_{uu_x})_{i-1}^* - (f_{uu_x})_i^*)\right. \\ &\left.+ \frac{1}{2(\Delta x)^2}((f_{u_xu_x})_{i-1}^* + (f_{u_xu_x})_i^*)\right]\end{aligned} \tag{1.152}$$

Here the notation $(\ldots)_i^*$ implies that the quantity inside the brackets is computed for the values of the arguments equal to $x = (x_i + \Delta x)/2$, $u = (u_i + u_{i+1})/2$, and $u_x = (u_{i+1} - u_i)/\Delta x$. For fixed values of the grid functions u_i, the expression for ΔI given by Eq. (1.152) becomes a quadratic formula in the variable δ_i. Choosing the step size from the condition of minimizing ΔI with respect to δ_i: $d(\Delta I)/d\delta_i = 0$ and $d^2(\Delta I)/d\delta_i^2 > 0$, we have

$$\begin{aligned}\delta_i &= -\frac{1}{2\sigma_i}\left[(f_u)_{i-1}^* + (f_u)_i^* - \frac{2}{\Delta x}((f_{u_x})_i^* - (f_{u_x})_{i-1}^*)\right] \\ \sigma_i &= \frac{1}{4}((f_{uu})_{i-1}^* + (f_{uu})_i^*) + \frac{1}{\Delta x}((f_{uu_x})_{i-1}^* - (f_{uu_x})_i^*) \\ &\quad + \frac{1}{(\Delta x)^2}((f_{u_xu_x})_{i-1}^* + (f_{u_xu_x})_i^*) > 0\end{aligned} \tag{1.153}$$

If $\sigma_i < 0$ in Eq. (1.153), then for δ_i given by Eq. (1.153), ΔI assumes a maximum; while if $\sigma_i = 0$, extremum is not attained by values given in Eq. (1.153). We observe that Eq. (1.153) prescribing δ_i contains in the square brackets a finite difference approximation to the Euler equation $f_u - (f_{u_x})_x = 0$. For this reason—the equality $\delta_i = 0$ may occur at some point $x = x_i$ only if for the given approximation u_i on the interval (x_{i-1}, x_{i+1})—the finite difference analogue of Euler's equation holds. Consequently, if u_i solves this system of finite difference equations, then by invoking Eq. (1.153), we obtain δ_i for all $x = x_i$. The algorithm of local variations, utilizing the above step size δ_i, is as follows: Let the sth approximation be available, i.e., the numbers u_i^s have been determined. Let the $(s + 1)$ approximation be defined in the following manner

$$u_i^{s+1} = \begin{cases} u_i^s + \delta_i, & \sigma_i > 0 \\ u_i^s, & \sigma_i \leqslant 0 \end{cases} \tag{1.154}$$

The quantities δ_i and σ_i in Eq. (1.154) are computed by making use of Eq. (1.153) with $u_{i-1} = u_{i-1}^{s+1}$, $u_i = u_i^s$ and $u_{i+1} = u_{i+1}^s$. Iteration using a fixed grid, i.e., fixed Δx, is continued as long as the condition max $|\delta_i| \leqslant \epsilon$ is violated. When this condition is satisfied, the quantity Δx is reduced and the iteration continues in an analogous manner, on the smaller size grid. The computation is terminated if the quantity Δx is small and satisfies the inequality $\epsilon \ll \Delta x$. It is easy to see, by considering Eq. (1.153), that the finite difference version of Euler's equations is satisfied with an error of $\epsilon/(\Delta x)^2$.

The algorithm that we have described above may be easily generalized for the case in which the quantities u_i satisfy the inequalities $u_i^-(x_i) \leqslant u_i \leqslant u_i^+(x_i)$. In solving this problem, let us assume that at some point $x = x_i$ we have obtained either $u_i^s + \delta_i > u_i^+$ or $u_i^s - \delta_i < u_i$. Then we can take for u_i as the $(s + 1)$ approximation, $u_i^{s+1} = u_i^+$ or $u_i^{s+1} = u_i^-$, respectively. That is, we realize the variation at that point with the greatest possible step size; $\delta_i = u_i^+ - u_i^s$, $\delta_i = u_i^- - u_i^s$. The correct approach in this operation can be seen as a consequence of the monotone decrease in ΔI with the change of the step size from zero to δ_i in accordance with Eq. (1.153). Thus the successive approximations formula for the problem with active constraints takes the form

$$u_i^{s+1} = \begin{cases} u_i^s + \delta_i, & u_i^- \leqslant u_i^s + h_i \leqslant u_i^+, \quad \sigma_i > 0 \\ u_i^s, & \sigma_i \leqslant 0, \\ u_i^+, & u_i^s + \delta_i > u_i^+, \quad \sigma_i > 0 \\ u_i^-, & u_i^s + \delta_i < u_i^-, \quad \sigma_i > 0 \end{cases} \tag{1.155}$$

Specific examples of applications of the technique of local variations with variable step sizes may be found in.[25-28, 64, 148]

4. A wide variety of problems in optimal design is connected with optimi-

zation of eigenvalues (critical loads and natural frequencies) of linear self-adjoint boundary-value problems for the differential equation

$$L(h)\, u = \lambda M(h)\, u \tag{1.156}$$

with boundary conditions

$$N(h)\, u = 0 \tag{1.157}$$

The symbols $L(h)$, $M(h)$ and $N(h)$ in Eqs. (1.156) and (1.157) denote linear differential operators whose coefficients depend on h, while λ is an eigenvalue. In solving eigenvalue optimization problems for Eqs. (1.156) and (1.157), while utilizing successive optimization algorithms, we demand that the quantities u and λ satisfying Eqs. (1.156) and (1.157) are computed in each step of the algorithm. For this reason, we use in these algorithms the technique of successive approximations (as given in Ref. 67). This technique is sufficiently simple and is convenient for solving problems of the type given in Eqs. (1.156) and (1.157). The technique consists of successive iterations that yield the quantities $u^0, u^1, \ldots, u^s, u^{s+1}, \ldots$ and $\lambda^0, \lambda^1, \ldots, \lambda^s, \lambda^{s+1}, \ldots$. Let us suppose that the function u^s has been found at the conclusion of the sth iteration. To find the function u^{s+1} and the eigenvalue λ^{s+1}, it is necessary to solve the boundary-value problem

$$L(h)\, u^{s+1} = M(h)\, u^s, \quad N(h)\, u^{s+1} = 0 \tag{1.158}$$

and to calculate λ^{s+1} using the formula

$$\lambda^{s+1} = \frac{(u^{s+1},\ L(h)\, u^{s+1})}{(u^{s+1},\ M(h)\, u^{s+1})} \tag{1.159}$$

where the round brackets denote an inner product.

References 67 and 68 contain certain generalizations and modifications of the schemes of the type reflected in Eqs. (1.158) and (1.159) and of their applications to solutions of eigenvalue problems.

5. We proceed to make some observations. Applying the algorithm of successive optimization to minimize (maximize) a local functional, it is necessary to utilize the technique described in Section 1.5 to reduce the problem to one with an integral functional. The algorithm of successive approximations permits us to make an improvement in the quality criterion. This process of solution can be terminated when we complete the computation of an arbitrary (kth) step. Thus

we obtain a new distribution h^k for the values of the control variable that is better in the sense of optimizing the cost functional than the originally assumed (initial) distribution h^0. Moreover, we now have the distribution of values for the function u^k, derived from the values of h^k, that describes the true state of the structure. In this manner, at each step of the algorithm, we secure some useful information that may be utilized in revealing trends emerging in the gradual process of optimizing the shape and the interior composition of the structure. For example, in problems of optimal design that result in singular solutions such application of algorithms of successive approximations allows us to note the formation of singularities before the total run of the algorithm is completed. It thus permits us to introduce additional constraints that allow only regular solutions for that problem to exist. Thus, utilizing algorithms of successive optimization, we arrive at a convenient application of a search procedure. This permits an active interaction with the computer in the midst of a solution run.

2

One-Dimensional Optimization Problems

As we have previously stated, questions of optimal design lead to very complex nonlinear problems. Even to formulate problems of optimal design presents a difficult task. Therefore, formulation of the simplest problems that allow for a purely analytic study and derivation of an exact solution assumes a considerable importance. Such studies may be carried out in certain one-dimensional problems concerning structural elements (beams, columns, curved rods, and systems of beams), and some realistic limitations may be stated, permitting us to deduce qualitative properties of optimal shapes and internal composition of structures and to compare the effectiveness of various techniques of optimization. Of considerable importance is the fact that we may conduct tests that are essential in validation of numerical algorithms and in approximate techniques that are intended for use in two-dimensional problems.

In this chapter, we offer an exposition of certain one-dimensional problems of optimization. This exposition is based on research results given in Refs. 11, 12, 13, 15, 24, 30, 147, 223. In Section 2.1 we offer solutions of some problems of optimal design of beams in transverse bending. Design of rods that may be subjected to the greatest axial load without loss of stability is discussed in Section 2.2. In Section 2.3, we study the optimum shapes of branched systems of rods. In Section 2.4, we construct solutions for the optimization problems of optimum distribution of thickness and the shape of the center line for curved rods and beams. In Section 2.5, we consider some problems of beam design, taking into account heating and an initial stress distribution.

2.1. Optimization Problems for Beams Subjected to Bending

We have already made the comment in the Introduction that research into optimization of structures originated in the classical work of Galileo on the design of the shape of a beam. Subsequently, many problems concerning the optimization of bent beams have been solved. Nevertheless, a considerable portion of modern research in optimal design uses the classical model of beam theory. The equations describing the bending of beams are among the simplest equations in the strength of materials theory and are suitable for considering alternate formulations of such problems and for comparing algorithms and techniques. In this monograph we consider, within the framework of classical beam theory, certain typical features of optimization processes that concern the optimization of stiffness or of strength.

1. Let a beam be freely supported at the points $x = \pm l$ on the x axis and be subjected to bending by a transverse point load applied at the point $x = 0$, i.e., $q = P\delta(x)$. The length $2l$ of the beam and its volume V are considered as given, thus constituting a constraint on the distribution of cross-sectional area S.

We shall consider the problem of minimizing the magnitude of the deflection w at the point of application of the point load ($x = 0$) by optimizing the shape of the beam, i.e., the problem of finding the control variable $h(x)$ that determines the cross-sectional area or that determines the magnitude of the functions S and $D = EI$ (where $S = B_\alpha h$, $D = EI = A_\alpha h^\alpha$, as in Section 1.3). Here E stands for Young's modulus and I for the moment of inertia of the cross-sectional area of the beam about the neutral axis, which is perpendicular to the plane of bending and passes through the neutral plane of bending stress. The constants A_α and B_α depend on the geometry of the cross section of the beam and α is a parameter that may take values 1, 2, or 3. We introduce the basic relations of our problem

$$(Dw_{xx})_{xx} = q, \qquad (w)_{x=\pm l} = (Dw_{xx})_{x=\pm l} = 0$$
$$\int_{-l}^{l} S\, dx = V, \qquad J \equiv w(0) \to \min_h \tag{2.1}$$

We observe that the problem that is dual to Eq. (1.1) is that of minimizing the volume of the beam for a given deflection $w(0)$. The condition of optimality (see Section 1.6) both for the direct and for the dual problem have the general form

$$h^{\alpha-1} w_{\lambda x}^2 = \lambda^2 \tag{2.2}$$

where λ^2 is an unknown constant. To find λ we use the equality between the work performed by the load and the strain energy of the deflected beam

$$P = \frac{A_\alpha}{w(0)} \int_{-l}^{l} h^\alpha w_{xx}^2 \, dx = \frac{A_\alpha \lambda^2 V}{B_\alpha w(0)} \tag{2.3}$$

We shall consider the case in which $\alpha = 1$, which corresponds to a trilayer beam having a variable thickness of the reinforcing layers equal to $h/2$. In this case, S denotes the total area of the reinforcing layers; i.e., $S = bh$, where b is the width of the rectangular beam. Problems of optimal design of trilayer beams are studied in Ref. 223, where the authors pay attention to an interesting feature of this problem, namely that the optimality condition for such structures (beams and plates) does not directly involve the control variable. If in a given problem a sufficient number of boundary constraints are assigned to the state function w which do not depend on the control variable h, then the optimality condition and the boundary conditions may be used to find the state function w. To determine the control variable h (the thickness of the reinforcing layers), we need to use the equations of equilibrium and the remaining boundary conditions, which depend on both w and h. The scheme suggested in[(223)] for solving problems of optimizing trilayer structures is now applied to study of optimal design of a trilayer simply supported beam.

Using the optimality condition $w_{xx}^2 = \lambda^2$ and the boundary conditions $w(-l) = w(l) = 0$, we obtain $w = \frac{1}{2}(\lambda^2 - x^2)$. Applying Eq. (2.3), we have $\lambda^2 = Pbw(0)/(A_1 V) = [Pbl^2/(A_1 V)]^2$. Invoking the optimality condition, the equilibrium equation of Eq. (2.1), and the boundary conditions $hw_{xx} = 0$ at $x = \pm l$, we arrive at the boundary-value problem $h_{xx} = 0$, which in conjunction with the isoperimetric condition of Eq. (2.1) determines the function $h(x)$. After some elementary substitutions, we derive

$$\begin{gathered} h = \frac{V(l+x)}{bl^2} \quad (-l \leqslant x \leqslant 0), \qquad h = \frac{V(l-x)}{bl^2} \quad (0 \leqslant x \leqslant l) \\ w = \frac{Pbl^2}{4A_1 V}(l^2 - x^2), \qquad J_* = w(0) = \frac{Pbl^4}{4A_1 V} \end{gathered} \tag{2.4}$$

The distribution of the thickness and displacements for the optimum beam [with $l = 1$, $V/b = 0.1$ and $Pb/(4A_1 V) = 0.3$] is illustrated in Fig. 2.1.

It is clear from Eq. (2.4), and can be seen in Fig. 2.1, that the thickness of the optimum beam attains its maximum at the point of application of the load, decreases towards the ends, and is equal to zero at the simple supports. This differs from the cases of beams of constant thickness and in general from beams for which $h(x) \neq 0$, where the condition $hw_{xx} = 0$ is satisfied by making the curvature of the beam axis equal to zero ($w_{xx} = 0$ at $x = \pm l$). For optimum

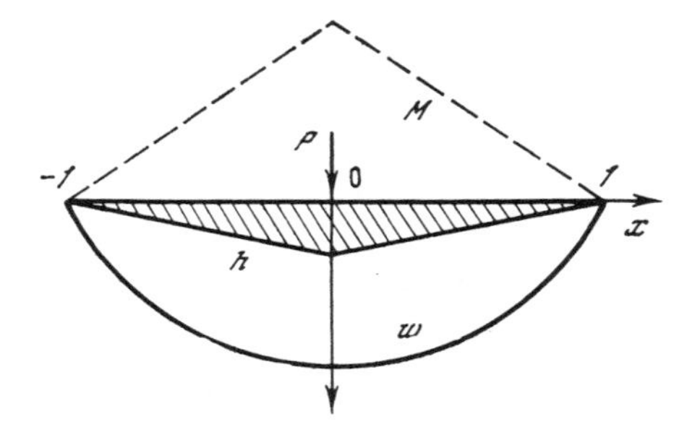

Figure 2.1

beams, this condition is satisfied by equating to zero the thickness of the beam at the end points. The curvature of the optimum beam remains constant along the entire length $-l \leqslant x \leqslant l$. The bending moment M (indicated by the dotted line in Fig. 2.1) and the strain energy density for the element of the beam having coordinate x vary as functions of x in a manner similar to $h(x)$, i.e., they attain their maximum for $x = 0$ and are equal to zero at the points $x = \pm l$.

We shall estimate the effectiveness of the optimal solution by comparing it with the beam of constant thickness. For the beam of constant thickness, having the same volume $J = w(0) = Pl^4/(3A_1 V)$, the gain in the value of the functional for the optimum beam is given by $(J - J_*)/J = 1/4$, i.e., 25%.

With the same assumptions made above, we shall consider an arbitrary solid beam having variable cross sectional height $h(x)$. Assuming that $\alpha = 3$ (see Section 1.3) and making use of the optimality condition of Eq. (1.2), we transform the deflection equation and the boundary conditions to the form $(w_{xx}^{-2})_{xx} = 0$, for $-l < x < 0$, $0 < x < l$, and $w(-l) = w(l) = w_{xx}^{-2}(-l) = w_{xx}^{-2}(l) = 0$. Adjoining the condition $w_x(0) = 0$, we obtain a solvable boundary-value problem for determining the deflection function. The boundary condition $w_x(0) = 0$ is a consequence of symmetry of the problem, relative to the point $x = 0$, and our assumption that $h(0) \neq 0$. Because of the above mentioned symmetry, we only need to determine all unknown quantities on the interval $0 < x < l$. Integrating the equation $(w_{xx}^{-2})_{xx} = 0$, and computing three constants of integration by utilizing the conditions $w_x(0) = 0$ and $w(l) = w_{xx}^{-2}(l) = 0$, we derive an expression for w which contains only a single unknown constant.

To find the distribution of thickness and the values of λ and the unknown constant, we apply the optimality conditions; Eq. (2.3) and the isoperimetric condition of Eq. (2.1). We have

$$h = \frac{3V}{4bl}\left(1 - \frac{x}{l}\right)^{1/2}, \qquad J_* = w(0) = \frac{64Pl^6b^3}{81A_3V^3}$$
$$w = \frac{64Pb^3l^6}{81A_3V^3}\left(1 - \frac{x}{l}\right)\left(3 - 2\sqrt{1 - \frac{x}{l}}\right), \qquad 0 \leqslant x \leqslant l \tag{2.5}$$

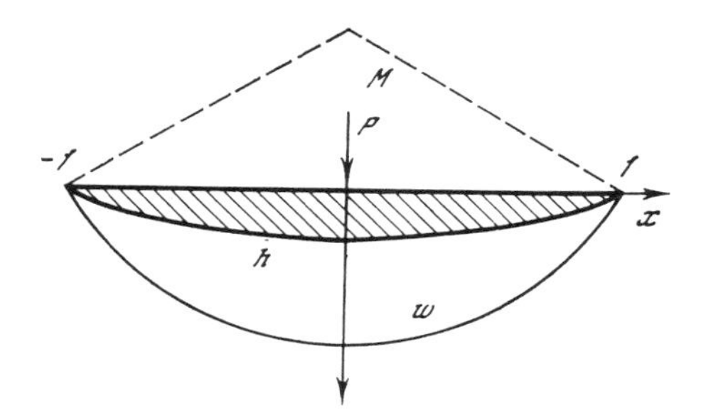

Figure 2.2

The functions $h(x)$ and $w(x)$ are graphed in Fig. 2.2 [$l = 1.3V/(4b) = 0.1$, and $64Pb^3/(81A_3V_3) = 0.3$].

The optimum distribution of thickness for both the uniform beam and the trilayer beam attains a maximum at the center of the beam ($x = 0$) and is equal to zero at both ends. In the case $\alpha = 1$, the thickness h decreases in proportion to the distance from the end of the beam, while for $\alpha = 3$, the thickness h behaves like $\sqrt{t}$, where $t = l - x$. Hence, we have a qualitative difference in the behavior of the deflection function for $\alpha = 1$ and $\alpha = 3$. The curvature of the center line for the uniform beam varies along the length of the beam and becomes unbounded in the neighborhood of the free supports.

Comparing the magnitude of the deflection $w(0)$ from Eq. (2.5) with the value $w(0) = 4Pb^3l^6/(3A_3V^3)$ for the beam with constant height and the same volume we find that the gain attained by optimizing the shape amounts to 40.7%.

2. For the solutions discussed in the preceding paragraph, the curvature w_{xx} of the center line of the bent beam was a continuous function of x on $-l < x < l$. However, optimum beams do not necessarily have this property for other boundary conditions. For example, if we wish to find the optimum shape of a trilayer beam that is built-in at both ends, there is no optimal solution for which w_{xx} is continuous for $-l < x < l$. Indeed, the optimality condition $w_{xx}^2 = \lambda^2$, combined with the boundary conditions

$$w(-l) = w_x(-l) = w(l) = w_x(l) = 0 \tag{2.6}$$

that correspond to built-in ends, results in a boundary-value problem that does not have any solution if w_{xx} is assumed to be continuous on the interval $-l < x < l$, and $\lambda^2 \neq 0$. Hence, in the case of built-in ends, we must look for solutions that allow discontinuities in the function w_{xx}.

We shall analyze the properties of the solution of this problem that exhibit such discontinuities for arbitrary values of α. Let us write the necessary conditions for the existence of a extremum (i.e., the Weierstrass–Erdmann condition) at the point x_* at which the curvature goes through a discontinuity: $[h^\alpha w_{xx}]_-^+ = 0$, $[(h^\alpha w_{xx})_x]_-^+ = 0$, and $[h^\alpha w_{xx}^2]_-^+ = 0$. Here, the square brackets with the (+) and (−) super and subscripts denote the limits of the quantities enclosed in the square brackets at $x = x_* + 0$ and $x = x_* - 0$, respectively. Consequently, we conclude from these limit relations that $(h^\alpha w_{xx})[w_{xx}]_-^+ = 0$. Therefore, if w_{xx} has a discontinuity, $h^\alpha w_{xx}$ must be equal to zero at the point of discontinuity. This condition implies that at such singular points the bending moment $M = h^\alpha w_{xx}$ must be equal to zero. The condition $[(h^\alpha w_{xx})_x]_-^+ = 0$ implies that at any such singular point the shear force $Q = (h^\alpha w_{xx})_x$ is continuous. Auxiliary conditions that may be used to determine the existence of discontinuous solutions, which satisfy optimality conditions of Eq. (2.2), assume the following form:

$$\begin{aligned} &w^+ = w^-, \quad w_x^+ = w_x^- \\ &h^+ = h^- = 0, \quad (h^{(\alpha+1)/2})_x^- = -(h^{(\alpha+1)/2})_x^+ \end{aligned} \tag{2.7}$$

Let us first consider a trilayer beam ($\alpha = 1$) whose ends are rigidly clamped at $x = \pm l$. As before, let us assume that a concentrated load is applied at the center of the beam. From symmetry of the problem we conclude that the singular points are symmetrically distributed about the origin, i.e., $x_{*1} = -x_{*2}$, where $x_{*1} \in [-l, 0]$. The distribution of displacements and the location of the singular point x_{*1} can be found by utilizing Eqs. (2.2), (2.3), and the conditions listed on the first line of Eq. (2.7). To find the distribution of thickness we use the condition $h_{xx} = 0$ $(-l < x < 0)$, the isoperimetric condition of Eq. (2.1), and the equalities given on the second line of Eq. (2.7). We have

$$h = \begin{cases} -\dfrac{V}{bl^2}\left(x + \dfrac{l}{2}\right), & -l \leqslant x \leqslant -\dfrac{l}{2} \\ \dfrac{V}{bl^2}\left(x + \dfrac{l}{2}\right), & -\dfrac{l}{2} \leqslant x \leqslant 0 \end{cases}$$

$$w = \begin{cases} \dfrac{Pbl^4}{8A_1V}\left(1 + \dfrac{x}{l}\right)^2, & -l \leqslant x \leqslant -\dfrac{l}{2} \\ \dfrac{Pbl^4}{16A_1V}\left(1 - \dfrac{2x^2}{l^2}\right), & -\dfrac{l}{2} \leqslant x \leqslant 0 \end{cases} \tag{2.8}$$

$$J_* = w(0) = Pbl^4/16A_1V$$

The distribution of thickness given by Eq. (2.8) and the distribution of displacements of a trilayer beam are given in Fig. 2.3 [$l = 1$, $V/g = 0.1$, and $w(0) = 0.3$].

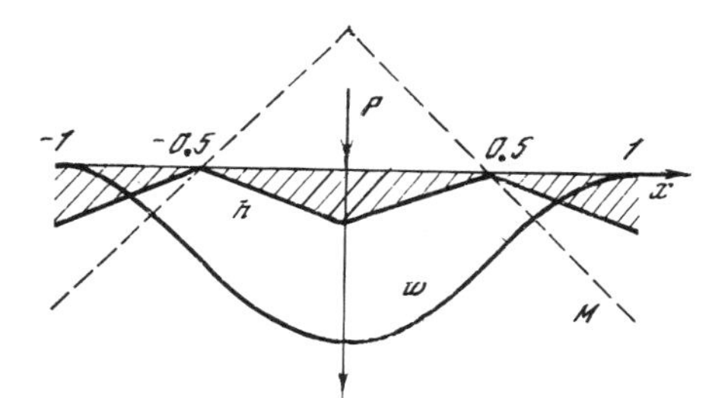

Figure 2.3

For uniform beams with built-in ends, the use of the singular point relations given above permits us to derive the solution of the optimal design problem. For beams of variable thickness having a rectangular cross section ($\alpha = 3$), the optimum distribution of thickness and deflection are described by the formulas

$$h = \begin{cases} \frac{3V}{4bl}\left(-1-\frac{2x}{l}\right)^{1/2}, & -l \leqslant x \leqslant -\frac{l}{2}, \\ \frac{3V}{4bl}\left(1+\frac{2x}{l}\right)^{1/2}, & -\frac{l}{2} \leqslant x \leqslant 0, \end{cases} \quad J_* = w(0) = \frac{4Pb^3l^6}{81A_3V^3}$$

$$w = 3w(0)\left[\left(1+\frac{x}{l}\right)+\frac{1}{3}\left(\left(-1-\frac{2x}{l}\right)^{3/2}-1\right)\right], \quad -l \leqslant x \leqslant -\frac{l}{2} \tag{2.9}$$

$$w = 3w(0)\left[\frac{x}{l}+\frac{1}{3}\left(1-\left(1+\frac{2x}{l}\right)^{3/2}\right)\right]-w(0), \quad -\frac{l}{2} \leqslant x \leqslant 0$$

Graphs of the functions $h(x)$, $w(x)$, and the bending moment distribution $M(x)$ are shown in Fig. 2.4 [$l = 1$, $3V/(4b) = 0.1$, and $w(0) = 0.3$].

3. Let us now assume that the beam is simply supported at the end points $x = \pm l$ (Fig. 2.5) and is fastened to an elastic foundation ($z \geqslant 0$). Let us assume that the distributed reaction force at the point x is proportional to the deflec-

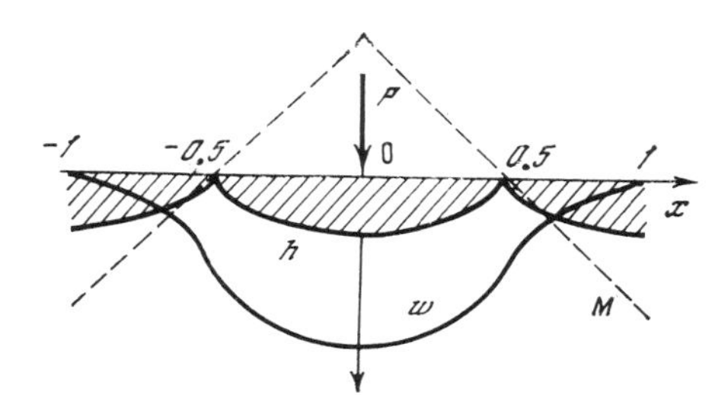

Figure 2.4

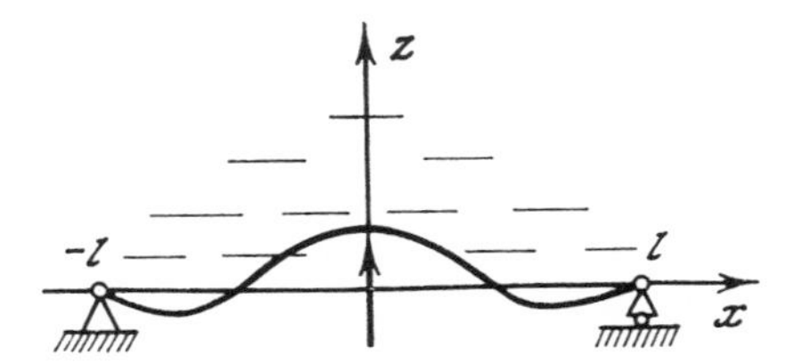

Figure 2.5

tion $w(x)$. A concentrated load is applied to the beam at the point $x = 0$. In this case, the deflection equation and the equation giving the force P as a function of the strain energy of the beam assume the form

$$(Dw_{xx})_{xx} + cw = q, \qquad P = \frac{1}{\delta} \int_{-l}^{l} (Dw_{xx}^2 + cw^2)\, dx$$

where c is the rigidity coefficient for the foundation.

We pose the problem of maximizing the force P that produces a given deflection $w(0) = \delta$, by choosing the optimum shape of the beam. It is clear that this problem is equivalent to the problem of finding the distribution of beam thickness that minimizes the deflection $w(0)$ for a given load P. We attempt to solve this problem on the interval $0 < x < l$, making use of the boundary conditions $w(0) = \delta$, $w(l) = 0$, and $(h^\alpha w_{xx})_{x=l} = 0$. We shall first consider the case of small rigidity coefficient c and apply a perturbation technique to determine the unknown quantities. Therefore, we expand the unknown quantities w, h, P, and λ in power series of a small parameter ϵ

$$w = w_0 + \varepsilon w_1 + \ldots, \; h = h_0 + \varepsilon h_1 + \ldots$$
$$P = P_0 + \varepsilon P_1 + \ldots, \; \lambda = \lambda_0 + \varepsilon \lambda_1 + \ldots$$

Substituting these series for w, h, P, and λ into Eq. (2.9), (with $\alpha = 3$), we equate coefficients of equal powers of ϵ. To find the unknown quantities in the zeroth-order approximation, which corresponds to the absence of reaction from the foundation (i.e., $c = 0$), we have to solve a system of equations. The solution (see the first paragraph of this chapter) has the form

$$w_0 = \delta(1 - x/l)(3 - 2\sqrt{1 - x/l}), \qquad \lambda_0 = -\frac{9}{8}\frac{\delta V}{bl^3}$$
$$h_0 = \frac{3V}{4bl}\left(1 - \frac{x}{l}\right)^{1/2}, \qquad P_0 = \frac{81\delta A_3 V^3}{64 b^3 l^6}$$

Taking into account the properties of the zeroth-order approximation, we can formulate the boundary-value problem for the unknown quantities in the first-order approximation

$$A_3\left[h_0^2(h_0 w_{1xx} + 3h_1 w_{0xx})\right]_{xx} + w_0 = 0, \quad 0 < x < 1$$

$$h_1 w_{0xx} + h_0 w_{1xx} = \lambda_1, \qquad 0 \leqslant x \leqslant l, \qquad w_1(0) = 0,$$

$$w_1(l) = 0, \qquad w_{1x}(0) = 0, \qquad (w_{1xx} h_0^3 + 3h_0^2 h_1 w_{0xx})_{x=l} = 0$$

$$\int_0^l h_1\, dx = 0, \qquad P_1 = \frac{2}{\delta}\left(\frac{\lambda_0 \lambda_1 V A_3}{b}\right) + \int_0^l w_0^2\, dx\Big)$$

Let us integrate both terms of the equation given on the first line of these relations and determine the four constants of integration and the constant λ_1 by using the isoperimetric condition and four boundary conditions. Thus, we obtain corrections to the distribution of thickness and to the magnitude of the force that are given in the first-order approximation

$$h_1 = \frac{8b^2 l^6}{27V^2 A_3}\left(1 - \frac{x}{l}\right)^{1/2}\left[\left(1 - \frac{x}{l}\right)^2\left(1 - \frac{16}{35}\left(1 - \frac{x}{l}\right)^{1/2}\right) - \frac{9}{35}\right]$$

$$P_1 = \tfrac{8}{7} l\delta, \quad \lambda_1 = 0$$

Figure 2.6 illustrates the optimum distribution of thickness for the beam, using dimensionless coordinates $x' = x/l$ and $h' = 2hbl/V$, where the curves 1 and 2 correspond to the values of the parameter $c = 0$ and $c = 0.2$, respectively.

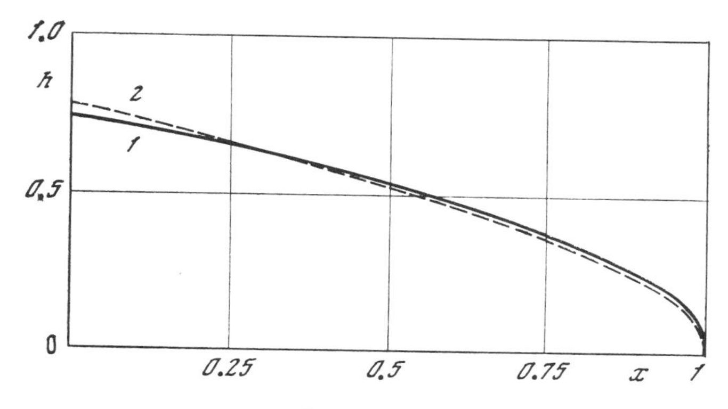

Figure 2.6

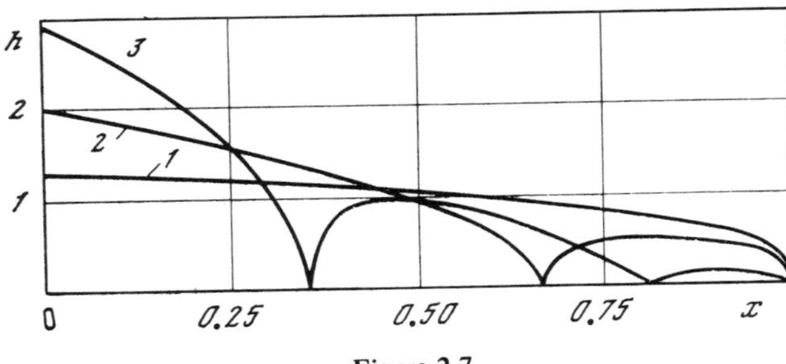

Figure 2.7

It should be clear from these graphs, and from the formulas, that the thickness at the center of the optimum beam increases with an increase in the value of the parameter c, while the thickness near the ends of the beam decreases (i.e., the ends begin to resemble sharp points).

The solution of this problem for large values of c can be obtained numerically using the technique described in Section 1.10, i.e., the algorithm of successive optimization. Results of numerical computations for the values $c = 0$, $c = 16$, and $c = 512$ are shown (for one-half of the beam length) in Fig. 2.7, curves 1 to 3, respectively. As can be seen from the graph corresponding to the case $c = 16$, the optimum thickness is equal to zero at some interior point of the interval $[0, 1]$. For $c = 512$, two points exist on the interval $[0, 1]$ at which the thickness h is equal to zero (the optimum beam is partitioned into five beams that are connected by pin joints). The relation between P_* and c is illustrated in Fig. 2.8.

Results given in this section have been derived in Ref. 30, where the author considered not only the design of a homogeneous beam, but also the optimum

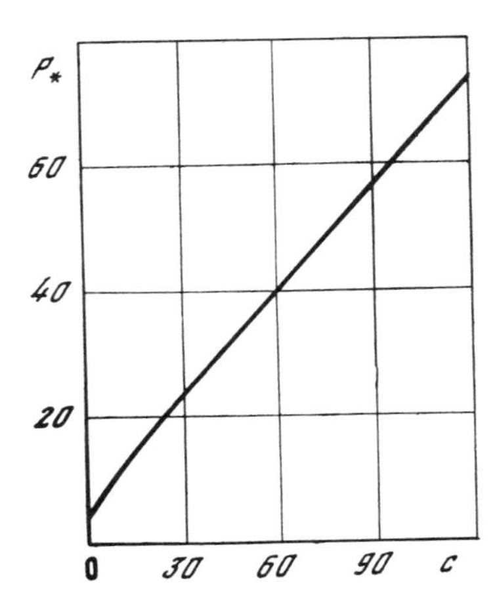

Figure 2.8

distribution of thickness of the load-carrying layers of a trilayer beam that is fastened to the elastic foundation.

4. In the cases discussed above, we have considered properties of beams such as the total volume (or weight), or the magnitude of the deflection at the point of application of the force. Considering only these two functionals and using the distribution of thickness as our control variable, we were able to formulate the problem of minimizing the volume, for a given magnitude of displacement, or of minimizing the displacement, for a given volume of the beam. It is important to note that these problems are dual to each other and it is not necessary to solve both of them independently. For this reason, we have studied only the problem of optimizing the rigidity of beams.

Similarly, if we take as the pay-off functional either the volume or the magnitude of the stress ψ, we can formulate the problem of minimizing the maximum value of the stress ψ, for a given volume of the beam, or of minimizing its volume, for a given value of the maximum stress intensity. These problems also turn out to be dual. Therefore, in our study we shall follow Ref. 12 and consider only the problem of minimizing the volume of the beam, for a given strength criterion.

We shall assume that the beam axis follows the coordinate x axis and that the beam has a rectangular cross section with variable height $h = h(x)$ and width $b = b(x)$. The length of the beam is l. Here we consider only the statically determinate cases (such as cantilevered or simply supported beams subjected to transverse loads) for which the distribution of bending moment $M = M(x)$ acting on the cross-sectional area of the beam is independent of the elastic properties or the shape of the cross section. The function $M(x)$ can be regarded as given for $0 \leqslant x \leqslant l$. The functions $b = b(x)$ and $h = h(x)$ that determine the shape of the beam are in turn regarded as unknown and are determined by solving the following variational problem:

$$V = \int_0^l bh\,dx \to \min, \quad \psi = \sigma_x^2 + \beta\tau_{xy}^2 \leqslant k^2 \quad (\beta = 3,4)$$

$$\sigma_x = \frac{My}{I}, \quad \tau_{xy} = \frac{1}{b}\left(\frac{Md}{I}\right)_x, \quad I = \frac{bh^3}{12}, \quad d = \frac{b}{2}\left(\frac{h^2}{4} - y^2\right) \tag{2.10}$$

Here, I stands for the moment of inertia of the cross-sectional area about the axis perpendicular to the plane of bending and passing through the neutral axis of the beam, d is the first (static) moment, and k is a given constant. With the exception of the components of the stress tensor, the subscript x denotes partial differentiation of a given quantity with respect to the indicated variable.

The coordinate y measures the distance from the center of gravity of the cross section and varies between the limits $-h/2 \leqslant y \leqslant h/2$. The constraint of Eq. (2.10) is a bound on the maximum value of the shear stress when $\beta = 4$. In the case $\beta = 3$, this inequality is equivalent to assigning a bound to the maximum value of the potential energy due to the deviating components of the stress tensor (i.e., the energy due to the change of shape alone). Making use of the equations given in Eq. (2.10) that relate σ_x, τ_{xy}, I, and d, we obtain a formula for ψ that expresses its explicit dependence on variables b, h, and M.

The function ψ depends not only on b, h, and M

$$\psi = \frac{144}{b^2h^6}\left\{M^2y^2 + \frac{\beta}{4}\left[\left(\frac{h^2}{4} - y^2\right)M_x + \frac{M}{h}\left(3y^2 - \frac{h^2}{4}\right)h_x\right]^2\right\}$$

but also on the coordinate y, which varies with the height of the cross-sectional area of the beam ($|y| < h/2$). We need not consider dependence on this variable in our future discussion. Clearly, ψ depends on y, but the values of ψ are only bounded from above. Introducing an auxiliary function $\Psi = \max_y \psi$, where $-h/2 \leqslant y \leqslant h/2$, we change the strength constraint condition to an equivalent form $\Psi \leqslant k^2$. Now we write a formula for Ψ. In computing Ψ, we observe that, as a function of the variable y^2, the function ψ is given by a quadratic formula and the coefficient of y^4 is nonnegative. Hence, the maximum of ψ, regarded as a function of y, is attained on the interval $-h/2 \leqslant y \leqslant h/2$, either when $y = 0$ or when $y = \pm h/2$. In computing this maximum, we obtain the required formula

$$\Psi = \max\left[\frac{36\,M^2}{b^2h^4}\left(1 - \frac{\beta}{4}\,h_x^2\right),\quad \frac{9\beta}{4b^2h^2}\left(M_x - \frac{M}{h}\,h_x\right)^2\right] \tag{2.11}$$

The operation max in Eq. (2.11) consists in choosing the larger of the two quantities given inside the square brackets.

5. Let us determine the shape of the beam for a given function $h = h(x)$, while the optimum distribution of the thickness of the cross section is to be determined. We substitute Eq (2.11) into the strength constraint $\Psi \leqslant k^2$ and discuss separately the cases in which the maximum in Eq. (2.11) is attained by the first or the second of the quantities inside the square brackets. Consequently, we derive inequalities that must be obeyed by the function $b(x)$, if the strength criterion is to be satisfied

$$b \geqslant \frac{6M}{kh^2}\left(1 + \frac{\beta}{4}\,h_x^2\right)^{1/2}, \qquad b \geqslant \frac{3\sqrt{\beta}}{2k}\left|\left(\frac{M}{h}\right)_x\right| \tag{2.12}$$

We have thus reduced our problem to that of finding a function $b = b(x)$ that satisfies the inequalities of Eq. (2.12) and minimizes the integral of Eq. (2.10).

It is clear that the minimizing function must be of the form

$$b = \max\left[\frac{6\,|M|}{kh^2}\left(1+\frac{\beta}{4}h_x^2\right)^{1/2}, \quad \frac{3\sqrt{\beta}}{2k}\left|\left(\frac{M}{h}\right)_x\right|\right] \tag{2.13}$$

In particular, for beams having a constant height (h = const), Eq. (2.13) indicates that $b = \max\,[6M/(kh^2),\ 3\sqrt{\beta}\,|M_x|/(2kh)]$. In the specific case $\beta = 4$, $k = 1$, $h(x) = 1$, $M = x(1 - x)$, and $l = 1$, the function $b(x)$ is illustrated in Fig. 2.9.

6. The next problem is to find $h = h(x)$ with constant width, b = const. We consider a cantilevered beam, clamped at the left end, i.e., at the point $x = 0$. A point load is applied at the right end, i.e., at the point $x = l$. In this case, $M = -P(l - x)$. We examine the inequalities that were imposed on the function $h(x)$ because of strength considerations. If the maximum in Eq. (2.11) is realized by the first term inside the square brackets, then the condition $\Psi \leqslant k^2$ can be used to derive the following two inequalities:

$$h \geqslant \chi(x) \equiv [6P(l - x)/kb^2]^{1/2} \tag{2.14}$$

$$|h_x| \leqslant \left[\frac{4}{\beta}\left(\frac{k^2b^2h^4}{36\,P^2(l-x)^2} - 1\right)\right]^{1/2} \tag{2.15}$$

We observe that the expression enclosed inside the round brackets in Eq. (2.15) is always positive if the inequality of Eq. (2.14) is satisfied.

We shall prove that the admissible solutions $h(x)$ (i.e., satisfying the condition $\Psi \leqslant k^2$) must satisfy the inequality

$$h \geqslant 3P\sqrt{\beta}/2kb \tag{2.16}$$

Let us assume to the contrary that Eq. (1.16) is violated and that the solution $h(x)$ passes through a point x_0, $h(x_0)$, such that $h(x_0) < 3P\sqrt{\beta}/(2bk)$.

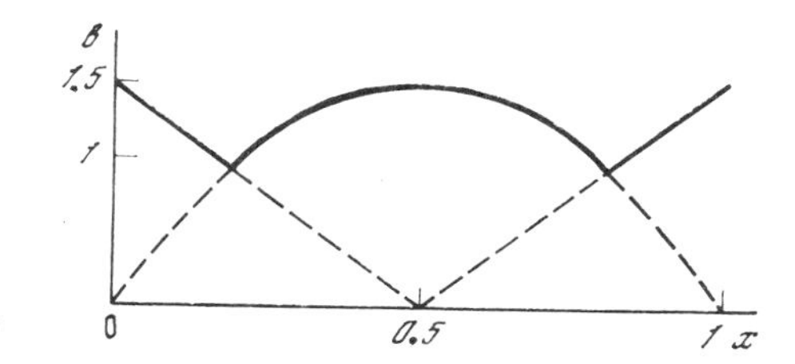

Figure 2.9

Using this inequality and the inequality $[h^{-1}(l-x)]_x \leqslant 2kb/(3\sqrt{\beta}P)$, which is a consequence of our strength constraints and of our assumption that the maximum of Eq. (2.11) is attained by the second term inside the square brackets. With the help of these two inequalities, after some integration and elementary manipulations we obtain the two-sided estimates

$$\frac{(l-x)\,h\,(x_0)}{l-x_0\,(1+t)+tx} \leqslant h \leqslant \frac{(l-x)\,h\,(x_0)}{l-x_0\,(1-t)-tx}, \qquad t \equiv \frac{2kbh\,(x_0)}{3\sqrt{\beta}\,P} \qquad (2.17)$$

The right-hand side inequality in Eq. (2.17) implies that an admissible function $h=h(x)\to 0$ as $x\to l$, since in a small neighborhood of $x=l$ the principal term of series representing the right-hand side majorant in Eq. (2.17) assumes the form $c_1(l-x)$ (with $c_1>0$). Observing that the function χ in Eq. (2.14) behaves like $c_2(l-x)^{1/2}$ as $x\to l$ (with $c_2>0$), we conclude that the graphs of functions (i.e., of the majorant functions and the function χ) intersect each other at some point on the interval $[0,l]$. Consequently, the function $h(x)$ that satisfies Eq. (2.17) must violate the inequality of Eq. (2.14). This contradiction proves the validity of Eq. (2.16). Let us denote by $H(x)$ (with $0\leqslant x\leqslant l$) the solution of the problem

$$H_x = -\left[\frac{4}{\beta}\left(\frac{k^2b^2H^4}{36P^2\,(l-x)^2}-1\right)\right]^{1/2}, \qquad H\,(0)=\left[\frac{6Pl}{kb}\right]^{1/2} \qquad (2.18)$$

Identifying $H=\chi$ at $x=0$ and letting the derivatives of these functions be equal to zero at the initial point $H_x=0$, we obtain $\chi_x = [3P/(2kbl)]^{1/2}$. Hence, for any trajectory of a solution of Eq. (2.18) originating at the initial point, there exists an interval $0<x<\epsilon(\epsilon>0)$ on which $H(x)>\chi(x)$.

We shall demonstrate that $H(x)>\chi(x)$ for all x, $0<x\leqslant l$. We examine the conditions that would have to exist if the curves $H(x)$ and $\chi(x)$ intersect each other. At the point of intersection, we clearly have the relations $H(x)=\chi(x)$ and $H_x\leqslant\chi_x$. However, Eq. (2.18) indicates that if $H=\chi$ then $H_x=0$, while $\chi_x<0$ on the entire interval $[0,l]$ and in particular at the hypothetical point of intersection of these curves. Consequently, the second of the above stated conditions (i.e, $H_x\leqslant\chi_x$) cannot be satisfied, and the curves $H(x)$ and $\chi(x)$ cannot intersect each other for any x such that $0<x\leqslant l$. Therefore, $H(x)\geqslant\chi(x)$ on the entire interval $0\leqslant x\leqslant l$.

Now, let us prove that $H(x)\to 0$ as $x\to l$. We shall prove it by contradiction. Let us assume that there exists a positive constant $\epsilon>0$ such that

$H(x) > \epsilon$ for $0 \leqslant x \leqslant l$. We observed that as $x \to l$, Eq. (2.18) has an asymptotic representation $H_x = kbH^2/[3P\sqrt{\beta}(l - x)]$. A solution of this equation $H = -3P\sqrt{\beta}/[kb \ln c(l - x)]$ (where $c > 0$ is a constant) tends to zero as $x \to l$. For sufficiently small values of $l - x$ we have $H(x) < \epsilon$. This contradiction completes our proof.

It follows directly from the above argument that the function

$$h = \max\left[\frac{3P\sqrt{\beta}}{2bk}, \quad H(x)\right] \tag{2.19}$$

satisfies the strength requirements and assigns a minimum to the integral of Eq. (2.10). Consequently it is the solution of our optimization problem. The optimal solution of Eq. (2.19) can also be represented in the form $h = H(x)$ for $0 \leqslant x \leqslant x_*$ and $h = 3P\sqrt{\beta}/(2bk)$ for $x_* \leqslant x \leqslant l$. The magnitude of x_* can be computed by using the assumption that h is a continuous function. We observe that the function $H(x)$ considered on the interval $0 \leqslant x \leqslant l$ attains its maximum at $x = 0$. Therefore, if $H(0) \leqslant 3P\sqrt{\beta}/(2b)$, which is true if $P \geqslant 8kbl/(3\beta)$, then the optimum beam is the beam of constant height $h = 3P\sqrt{\beta}/(2kb)$.

We determine the function $H(x)$ by numerically integrating the function in Eq. (2.13). For numerical convenience, we change to dimensionless variables $H' = kbH/(3P\sqrt{\beta})$ and $t = 2(l - x)\, kb/(3P\beta)$. In terms of these new variables (we shall omit the primes) our equation and the initial condition of Eq. (2.13) assume the form

$$H_t = \left[\frac{H^4}{t^2} - 1\right]^{1/2}, \quad H(t_1) = \sqrt{t_1} \tag{2.20}$$

Our problem can be regarded as a one-parameter problem, with the parameter t_1. We shall integrate Eq. (2.20) on the interval $t_0 \leqslant t \leqslant t_1$ from right to left starting at the point $t = t_1$ and taking as our initial condition $H(t_1) = \sqrt{t_1}$, and determining the value of t_0 from the condition $H(t_0) = 1/2$, which is a consequence of Eq. (2.16).

Numerical integration of Eq. (2.20) was performed by a computer for various values of the parameter t_1. The computed relations $H = H(t)$ for $t_1 = 0.3$, 0.4, 0.5, 1, and 2 are illustrated, respectively, by curves 1 to 5 in Fig. 2.10. The dotted curve in Fig. 2.10 is the graph of the function $H = \sqrt{t}$.

The solution of an analogous problem for a beam of circular cross section is given in Ref. 30.

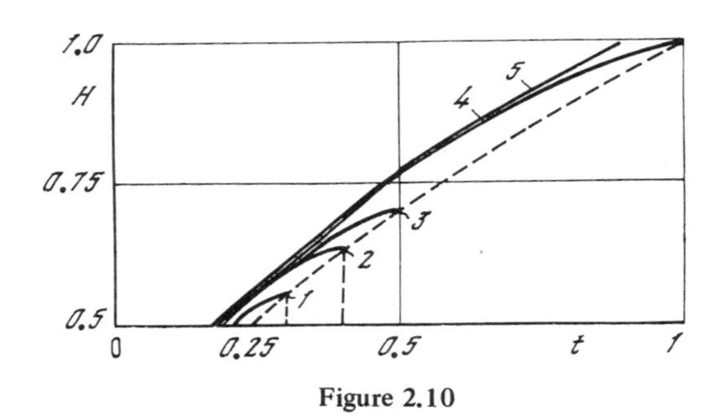

Figure 2.10

2.2. Optimization of Stability for Elastic Beams

Problems of optimizing the stability of elastic beams are closely related to many classical problems of optimal design. In studies of these problems given in Refs. 107, 142, 171, 184, 185, 186, 188, 233, and 235, it was generally shown that when we optimize the design we establish perspectives for further basic research along the same line. The techniques developed for the solution of such problems are themselves of considerable interest. We should remark that both the research results accomplished and the techniques developed here relate to optimization of stability of conservative elastic systems that are described by eigenvalue formulations of self-adjoint boundary-value problems. Problems of optimal design for nonconservative systems, in particular of structural designs with follower-type loads, are now in the beginning stages of investigation.

In the discussion given below, following the work in Ref. 11, we shall consider the problems of stability for a compressed rod. We shall, in the course of this investigation, generalize the well-known solution given in Ref. 171 for the specific case of an elastic support.

We assume that a cantilevered column is elastically held at one end and a compressive force P is applied to the free end (as shown in Fig. 2.11). For sufficiently large values of P, while the column loses its stability and buckles, the force vector retains its original direction, but its line of action is displaced in a parallel manner, and the point of application remains to be the free end of the column. We adopt rectangular coordinates x and z, where the end of the column is the origin o of this coordinate system and the axis x coincides with the direction of the force P (see Fig. 2.11). The axis z lies in the buckling plane for the beam. Its direction is indicated in Fig. 2.11. We restrict our discussion to small deformations of the beam and we study its equilibrium within the framework of the linear theory of elasticity. The magnitude of the deflection

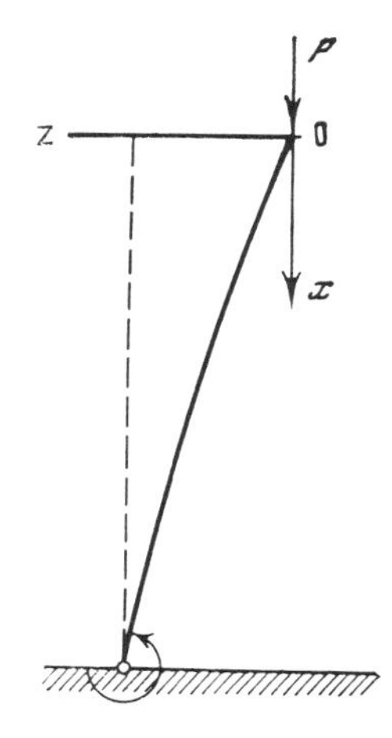

Figure 2.11

of the bent axis of the beam, denoted $w(x)$, is measured from the line of action of the force, while l denotes the length of the beam. By assuming that the cross sections of the beam are of similar shape, we shall derive the basic relationships arising in the problem of maximizing the load causing the loss of stability. We shall find the optimum (in the above sense) distribution of the cross-sectional area $S = S(x)$ as a function of the distance x measured along the axis of the beam

$$\begin{aligned} &EIw_{xx} + Pw = 0, \quad EI = C_2 S^2 \\ &w\,(0) = 0, \quad (cw_x - Pw)_{x=l} = 0 \\ &\int_0^l S\,dx = V, \qquad P \to \max_S \end{aligned} \tag{2.21}$$

where E, c, V, and $I = I(x)$ are, respectively, Young's modulus, the coefficient of rigidity of the foundation, the given volume of the beam, and the moment of inertia of the cross-sectional area, relative to the axis perpendicular to the bending plane xz and intersecting the neutral axis of the beam at the point with the coordinate x. The load P denotes the eivenvalue of the boundary-value problem of Eq. (2.21).

A necessary condition for optimality of compressed columns was derived earlier by following different techniques in Refs. 93, 107, 142, 171, 184, 239. It assumes the form

$$w^2 = S^3 \tag{2.22}$$

In these references, this condition was derived in considering other types of boundary conditions. However, it is not hard to show that it remains valid in the case considered above. The necessary condition for optimality of Eq. (2.22) relates the functions w and S and, together with basic relations of Eq. (2.21), leads to a solvable eigenvalue boundary-value problem. Using Eq. (2.22) to eliminate the function S from the bending equations and from the isoperimetric condition, and introducing dimensionless variables $x' = x/l$, $w' = wC_2/(Pl^2)$, $S' = lS/V$, and $P' = Pl/c$, we obtain the following relations:

$$w_{xx} + w^{-1/3} = 0,\ w(0) = 0,\ (w_x - Pw)_{x=1} = 0$$
$$\int_0^1 w^{2/3}\, dx = \frac{\gamma}{\sqrt{P}}, \qquad \gamma = \frac{V}{l}\left(\frac{C_2}{cl}\right)^{1/2} \tag{2.23}$$

The general solution of the differential equation of Eq. (2.23) may be represented in parametrized form (see Ref. 233) as

$$x = {}^1/_2 \sqrt{3}\, \mu_1^{2/3} (\theta - {}^1/_2 \sin 2\theta) + \mu_2, \qquad w = \mu_1 \sin^3 \theta$$
$$\theta_0 \leqslant \theta \leqslant \theta_1 \tag{2.24}$$

where μ_1, μ_2, θ_0, and θ_1 are constants that are to be determined. To find the values of these constants, we make use of the boundary conditions at $x = 0$ and $x = 1$ and the conditions $x(\theta_0) = 0$ and $x(\theta_1) = 1$. After substituting these values into the solution of Eq. (2.24), we have

$$\sin \theta_0 = 0,\ \ \mu_2 = -{}^1/_2 \sqrt{3}\, \mu_1^{2/3} \theta_0 \tag{2.25}$$

$${}^1/_2 \sqrt{3}\, \mu_1^{2/3} (\theta_1 - \theta_0 - {}^1/_2 \sin 2\theta_1) = 1 \tag{2.26}$$

$$\mu_1^{2/3} = \sqrt{3} \cos \theta_1 / P \sin^3 \theta_1 \tag{2.27}$$

Equations (2.25)-(2.27) are satisfied if we substitute $\theta_0 = 0$, $\mu_2 = 0$, $\theta_1 = \theta_*$ and $\mu_1^{2/3} = \sqrt{3} \cos \theta_* / (P \sin^3 \theta_*)$, where θ_* denotes a root of the equation

$$\varphi(\theta) \equiv 3 \cos \theta (\theta - {}^1/_2 \sin 2\theta)/2 \sin^3 \theta = P \tag{2.28}$$

that satisfies the inequality $0 \leqslant \theta_* \leqslant \pi/2$. Equation (2.28) may be obtained by substituting into Eq. (2.26) the expression for $\mu_1^{2/3}$, which satisfies Eq. (2.27), and setting $\theta_0 = 0$.

If we add to $\theta_0 = 0$ and to $\theta_1 = \theta_*$ a quantity that is a multiple of π [while the constants μ_1 and μ_2 are still computed by the use of Eqs. (2.25) and (2.27)], this can result only in the change of sign in Eq. (2.24) for w. Since w is determined by Eq. (2.23) only up to its sign, this translation along the θ axis does not result in a different solution. It is theoretically possible to find the solution of Eqs. (2.25)–(2.27) with $\theta_0 = 0$ and $\pi i < \theta_1 < \pi \times (i + 0.5)$, where $i = 1, 2, \ldots$. At the same time, it is easy to see by looking at Eq. (2.24) that the function w has additional loops. Since we are computing the eigenvalue P and the corresponding eigenfunction has a minimum number of loops, it is not necessary to consider such (more complex) solutions. The solution of the problem posed in Eqs. (2.25)–(2.27), with $\pi(i - 0.5) < \theta_1 < \pi i$, $i = 1, 2, \ldots$, does not exist, since in this case the right-hand side of Eq. (2.27) is negative, while the left-hand side is positive.

Let us study some properties of the function $\phi(\theta)$ that will be needed in our future discussion. For small values of θ, we represent $\phi(\theta)$ by a Taylor series, with accuracy up to terms of the order θ^3. Then we have $\phi(\theta) \approx (1 - \theta^2)/5$. Consequently, as $\theta \to 0$, the function $\phi(\theta) \to 1$, with $\phi(\theta) < 1$. Further, let us consider the case $\theta \to \pi/2$. For the sake of convenience, we introduce the variable $\omega = \pi/(2 - \theta)$, which approaches zero as $\theta \to \pi/2$. Substituting the variable ω for ϕ in Eq. (2.28) and eliminating θ and then decomposing ϕ in a Taylor series for small values of ω, we derive the estimate $\phi \approx [3\omega(\pi - 4\omega)]/4$. Therefore, ϕ approaches zero as $\theta \to \pi/2$. Using asymptotic series and some simple calculations, we obtain a graph of the function $\phi(\theta)$, which is illustrated in Fig. 2.12. We can see from this graph that as the variable θ varies between zero and $\pi/2$, the function $\phi(\theta)$ monotonically decreases from one to zero. Hence, $P < 1$. In terms of subsequently introduced dimensional quantities, this inequality can be written in the form $P < c/l$. It represents the condition of stability for a perfectly rigid clamped bar (see, for example, Ref. 117). Equation (2.28) gives us a

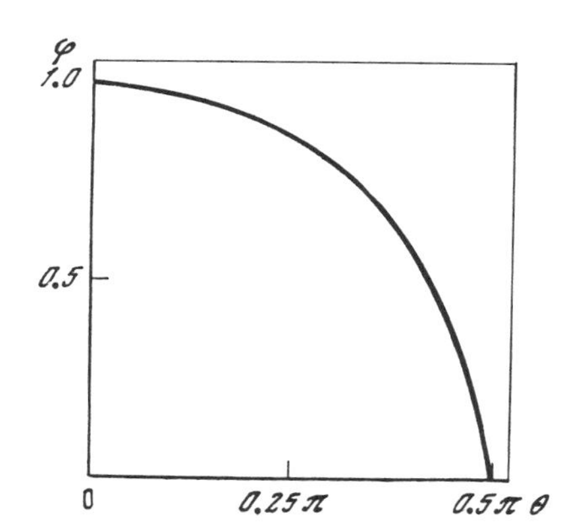

Figure 2.12

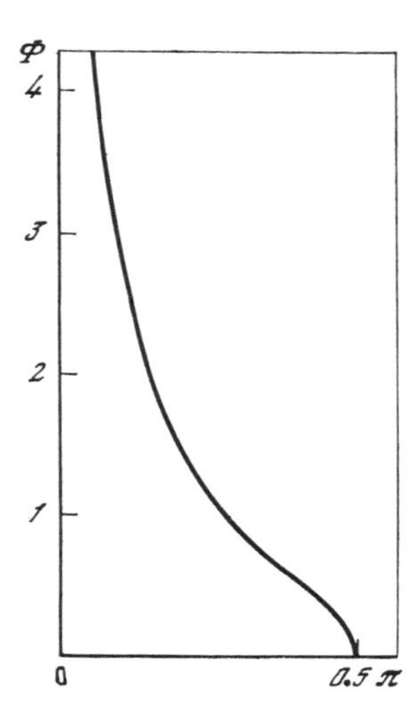

Figure 2.13

relation between the unknown quantities θ_* and P. To obtain a second equation that is necessary for determining θ_* and P, we substitute the solution of Eqs. (2.24) and (2.27) into the isoperimetric condition of Eq. (2.23) and eliminate P from the resulting equation by utilizing the Eq. (2.28).

$$\Phi(\theta_1) = \gamma, \quad \Phi(\theta_1) \equiv \psi(\theta_1)\, \varphi^{-3/2}(\theta_1) \tag{2.29}$$

The graph of the function $\Phi(\theta)$ shown in Fig. 2.13 was drawn with the help of very simple calculations. As θ decreases from $\pi/2$ to zero, the function Φ increases in a monotone fashion from zero to infinity. As $\theta \to 0$, this function has an asymptotic representation $\Phi(\theta) \approx 3\sqrt{3}/(5\theta)$. Consequently, there exist solutions to Eq. (2.29), satisfying the condition $0 < \theta_1 \leqslant \pi/2$ for an arbitrary nonnegative value of the parameter γ.

Thus, for given values of the constants C_2, l, V, and c, finding the optimum distribution of the cross-sectional area $S(x)$ and of the corresponding value of the load P is reduced to the following sequence of computational steps. For

Table 2.1.

γ	θ_1	P	γ	θ_1	P
0.5	1.283	0.488	3	0.339	0.976
1	0.882	0.811	3.5	0.292	0.983
1.5	0.640	0.909	4	0.257	0.987
2	0.496	0.948	4.5	0.229	0.989
2.5	0.404	0.966	5	0.206	0.991

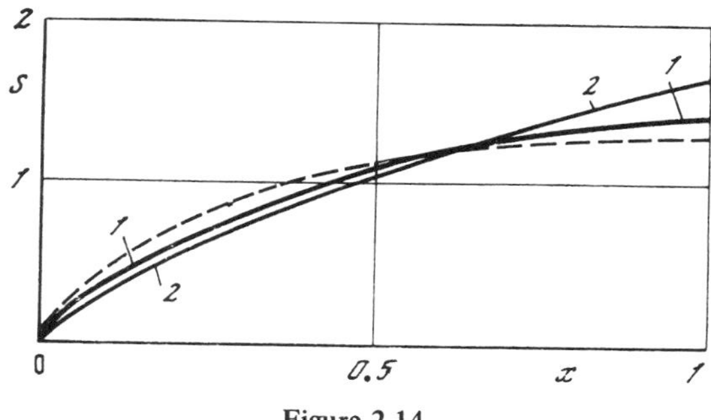

Figure 2.14

given values of C_2, l, V, and c, we compute the value of the dimensionless parameter γ. Using this value of γ, we solve Eq. (2.29) and compute the value of θ_1. Next, we calculate the magnitude of the critical load P, which according to Eq. (2.28) is equal to $P = \phi(\theta_1)$. Computation of the critical shape (of the deformed column) is carried out by using the formulas

$$x = \frac{\theta - \frac{1}{2}\sin 2\theta}{\theta_1 - \frac{1}{2}\sin 2\theta_1}, \qquad S = \frac{2\sqrt{P}\sin^2\theta}{\sqrt{3}\,\gamma\left(\theta_1 - \frac{1}{2}\sin 2\theta_1\right)} \tag{2.30}$$

where $0 \leqslant \theta \leqslant \theta_1$. Equations (2.30) follow directly from Eqs. (2.22), (2.24), and (2.26).

Calculations were performed on a computer in the order given above. Results derived for $\gamma = 0.5i$, $i = 1, 2, \ldots$ and for ten values of the quantities θ_1 and P are illustrated in Table 2.1.

In Fig. 2.14, curves 1 and 2 illustrate the optimum distribution of $S = S(x)$ for corresponding values of $\gamma = 0.5$ and $\gamma = 5$. The dotted curve shows the distribution $S(x)$ for a rigidly held beam ($\gamma = 0$). By comparing these graphs and the computed distributions that correspond to different values of γ, it is clear that as the rigidity of the support c increases ($\gamma \to 0$), the distribution of $S(x)$ approaches the corresponding distribution for a rigidly held beam. As the rigidity c decreases (i.e., γ increases), the material of the beam is "displaced" from the free end toward the built-in end.

2.3. Optimal Configuration of Branched Beams

Closely related to the problems of optimal design in modern technology and to the study of branching in growing organisms is the solution of problems concerning optimal configuration of branching systems of beams, having minimum weight for a given constraint regarding its strength. Below we shall follow Ref. 147 and consider structures loaded by point forces. We shall determine the values of some design parameters such that a branching structure is optimum.

1. We shall analyze a design consisting of three homogeneous straight elastic beams as shown in Figure 2.15. The beam OX lies on the x axis; while the beams XA and XB lie in the horizontal plane x-y. These beams are identical and are situated symmetrically relative to the vertical plane x-z. The points O, X, A, and B are all situated in the x-y plane, with coordinates equal to $[0,0]$, $[x,0]$, $[a,b]$, and $[a,-b]$, where $a \geqslant 0$ and $b \geqslant 0$ are given real numbers and the coordinate X of the branching point lies in the interval $0 \leqslant x \leqslant a$.

We assume that the cross section of each beam is constant along its entire length and is completely determined by a single dimension h_1 for the beams XA and XB and by h_2 for the beam OX.

We shall consider three separate cases with regard to geometry of the cross section: (i) The thickness of all beams in the direction of the z axis is constant and is the same for all three beams. The shape of the cross section depends on the change of its dimension in the direction of the horizontal axis perpendicular to the axis of the beam; (ii) the thickness of all beams in the x-y plane is constant and is the same for all beams. Changes in cross section result from changing the cross-sectional dimension in the direction of the z axis; (iii) the cross-sectional areas of all beams have a similar shape, which is a priori determined. The linear dimensions h_1 and h_2 have the following geometric meaning in each case: In (i) they are the horizontal thickness of the beams; in (ii) they are the vertical height; in (iii) they are dimensions that completely determine the cross section, e.g., for beams of circular cross section, these dimensions are the radii r_1 and r_2.

The beam OX is clamped at the point O. At point X, all beams are rigidly connected to each other. This structure is subjected to loads that act parallel to

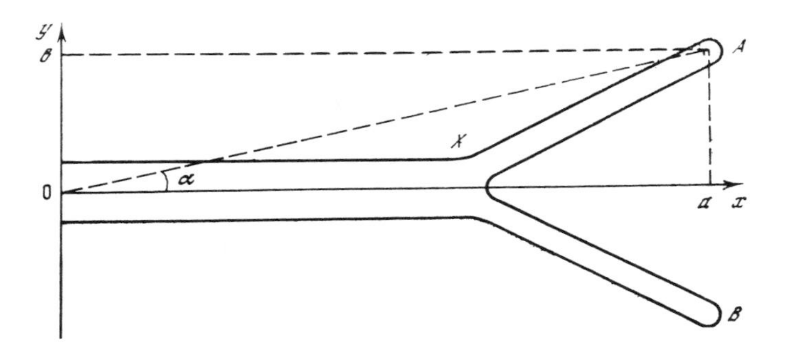

Figure 2.15

the z axis and are symmetrically distributed with respect to the x-z plane. We state the following problem. Let us determine an optimum design of the system of beams $OXAB$, having minimum volume V and satisfying the constraint $\sigma \leqslant \sigma_0$, where σ is the maximum stress arising at some point within the structure for a given distribution of external loads and σ_0 is a given positive constant. The magnitude of σ is determined by the bending moment M and is given by the relation $\sigma = MhI^{-1}$, where I is the moment of inertia of the cross-sectional area with respect to the horizontal axis passing through the center of gravity of the cross section, and h is the maximum distance from this axis to a point of the cross section. The condition $\sigma \leqslant \sigma_0$ acts as a constraint on the magnitude of the variables h_1, h_2, and x. Taking into account the relation $\sigma = MhI^{-1}$, we can restate this constraint in the form

$$\begin{gathered} M_i \leqslant c\sigma_0 h_i^m, \ i = 1, 2, \\ r_1 \geqslant 0, \ r_2 \geqslant 0, \ 0 \leqslant x \leqslant a \end{gathered} \tag{2.31}$$

where M_1 and M_2 are the maximum bending moments for the beams XA and OX, respectively, while the exponents $m = 1, 2$, and 3 correspond to our three separate cases of cross-sectional geometry discussed above. The constant c depends on m and the shape of the cross-sectional area. In particular, for a circular cross section, we have $m = 3$ and $c = \pi/4$.

The volume of the entire structure can be written as

$$V = c_0 \{2h_1^n [(a - x)^2 + b^2]^{1/2} + h_2^n x\} \tag{2.32}$$

where c_0 is a constant depending on the shape of the cross-sectional area, $n = 1$ if $m = 1$ or 2, and $n = 2$ if $m = 3$. Our problem is reduced to the minimization of the function of Eq. (2.32) by selecting parameters X, h_1, and h_2 that satisfy the inequalities of Eq. (2.31).

2. Let our structure be subjected to two equal point loads applied at points A and B. The maximum bending moments acting on beams XA and OX are attained at points X and O, respectively. The inequalities of Eq. (2.31) now assume the form

$$M_1 = P [(a - x)^2 + b^2]^{1/2} \leqslant c\sigma_0 h_1^m, \quad M_2 = 2Pa \leqslant c\sigma_0 h_2^m$$

From these inequalities and the constraints given on the second line of Eq. (2.32), it is clear that for a fixed value of X the minimum of the total volume, regarded as a function of h_1 and h_2, is attained if h_1 and h_2 assume the minimum allowable values. We therefore have:

$$h_1 = \left\{ \frac{P}{c\sigma_0} [(a-x)^2 + b^2]^{1/2} \right\}^{1/m}, \qquad h_2 = \left(\frac{2Pa}{c\sigma_0} \right)^{1/m} \tag{2.33}$$

Substituting these expressions into Eq. (2.32) after some manipulation and introduction of dimensionless quantities, we derive

$$\begin{gathered} V = V_1 f_1(\xi), \qquad V_1 = 2c_0 \left(\frac{P}{c\sigma_0} \right)^{2p-1} (a^2 + b^2)^p \\ \frac{b}{a} = \tan\theta, \qquad \xi = \frac{x}{a}, \qquad p = \frac{1}{2} + \frac{n}{2m} \\ f_1(\xi) \equiv \cos^{2p}\theta \, \{[(1-\xi^2) + \tan^2\theta]^p + 2^{2p-2}\xi\} \end{gathered} \tag{2.34}$$

Here, V_1 denotes the volume of the two beams OA and OB if branching is absent (that is, if $\chi = 0$), θ is the angle AOX (see Fig. 2.15), and the number p is 1, 3/4 and 5/6 when $m = 1, 2, 3$, respectively.

Our original problem is now reduced to finding a minimum for the function f_1 given in Eq. (2.34) for $\xi \in [0, 1]$. Differentiating f_1, we obtain

$$\begin{gathered} f_{1\xi} = \cos^{2p}\theta \, \{-2p \, [(1-\xi)^2 + \tan^2\theta]^{p-1} (1-\xi) + 2^{2p-2}\} \\ f_{1\xi\xi} > 0, \ \xi \geqslant 0, \ 0.75 \leqslant p \leqslant 1 \end{gathered}$$

These formulas indicate that $f_{1\xi} \to -\infty$ as $\xi \to -\infty$, and that $f_{1\xi} > 0$ when $\xi = 1$. Consequently, the function $f_1(\xi)$ is a convex function of ξ on the ray $\xi \in [-\infty, 1]$, and attains a global minimum for the value ξ_0 that is the unique root of the equation $f_{1\xi} = 0$, or

$$p\,(1-\xi_0)\,[(1-\xi_0)^2 + \tan^2\theta]^{p-1} = 2^{2p-3} \tag{2.35}$$

If $\xi_0 \geqslant 0$, then the required minimum of V regarded as a function of $\xi \in [0, 1]$ is attained at the point $\xi = \xi_0$. In the remaining case ($\xi_0 < 0$), function f_1 is monotone increasing on the interval $0 \leqslant \xi \leqslant 1$ and has its minimum at $\xi = 0$. Therefore, the required minimum point ξ_* is given by $\xi_* = \max[0, \xi_0]$. According to this formula, $\xi_* = 0$ corresponds to the absence of branching (i.e., $x = 0$) and the structure consists in fact of two unconnected beams OA and OB. Its dimensionless volume f_1 is equal in the case of Eq. (2.34) to $f_1(0) = 1$. Let us denote the dimensionless volume of the optimum structure by $f_{1*} = f_{1*}(\xi) \leqslant 1$. The magnitude of the quantity $1 - f_{1*}$ describes the relative saving of material used in the optimum structure, as compared to the structure that consists of two unconnected beams. We introduce the ratio of the basic linear dimensions describing the cross-sectional areas of the beams [as given by Eq. (2.33)] for optimal design of the structure,

$$\eta = h_1/h_2 = \left\{\frac{1}{4}[(1-\xi_0)^2 + \tan^2\theta]\right\}^{1/2m} \tag{2.36}$$

By using Eqs. (2.34) to (2.36) and the relation $\xi_* = \max[0, \xi_0]$, we have the following solution for the beam with a constant vertical height (i.e., $m = 1$ and $p = 1$):

$$\begin{aligned} &\xi_* = \xi_0 = 0{,}5, \quad f_{1*}(\theta) = 1 - \frac{1}{4}\cos^2\theta, \quad 0 \leqslant \theta \leqslant \pi/2 \\ &\eta(\theta) = \frac{1}{2}\left(\frac{1}{4} + \tan^2\theta\right)^{1/2} \end{aligned} \tag{2.37}$$

Here the optimal design always contains branching points situated in the interior of the interval $[0, a]$ on the x axis. The maximum relative material saving is $(1 - f_{1*})$, compared to the design consisting of two disjoint beams. According to Eq. (2.37), for $\theta = 0$ this amounts to 0.25. These savings are made by redistributing the material along the length of the beams. In this case $\eta = \frac{1}{4}$, i.e., the width h_1 of the beams XA and XB is four times smaller than the width h_2 of the beam OX. It is also a consequence of Eq. (2.33) that when $b_2 = 0$, $x = a/2$.

For the cases $m = 2$ and 3, $p < 1$, our analysis of Eq. (2.35) indicates that $\xi_* < 0$ if $\theta > \theta_*$ and $\xi_0 > 0$ if $\theta < \theta_*$. Then θ_* denotes the value of θ for which Eq. (2.35) has the root ξ_0. Setting $\xi_0 = 0$ in Eq. (2.35), we find that

$$\begin{aligned} \theta_* &= \arccos[0.5(2p)^{1/(2p-2)}] \\ \theta_* &= \arccos {}^2/_9 \approx 77.16^\circ \ (m = 2,\ p = 0.75) \\ \theta_* &= \arccos 0.108 \approx 83.80^\circ \ (m = 3,\ p = {}^5/_6) \end{aligned} \tag{2.38}$$

Thus, for $\theta \geqslant \theta_*$ the optimum structure consists of two disjoint beams (i.e., $\xi_* = 0$), while for $\theta < \theta_*$ we have a branched design ($\xi_* > 0$). For values of $\theta \geqslant \theta_*$, we deduce from Eqs. (2.34) and (2.36) that

$$\xi_0 = 0, \quad f_{1*} = 1 \tag{2.39}$$

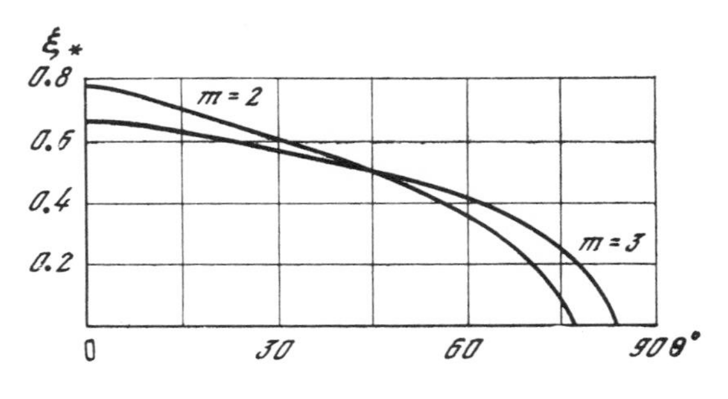

Figure 2.16

In the particular case when $\theta = 0$, let us introduce relations obtained from Eqs. (2.34)-(2.36),

$$
\begin{aligned}
&\xi_* = \xi_0 = 1-2\ (4p)^{1/(1-2p)},\ f_{1*} = (1-\xi_*)^{2p} + 2^{2p-2}\xi_* \\
&\eta = (4p)^{1/(1-m)} (\theta = 0;\ m = 2, 3) \\
&\xi_* = 0.7778,\ f_{1*} = 0.6547,\ \eta = 0.3333\ (m = 2) \\
&\xi = 0.6714,\ f_{1*} = 0.6894,\ \eta = 0.5477\ (m = 3)
\end{aligned}
\tag{2.40}
$$

Figure 2.16 shows the function $\xi_*(\theta)$. It has been established by numerical techniques using Eq. (2.35) and the formula $\xi_* = \max\,[0, \xi_0]$ for values $m = 2$ and $p = 3/4$ (beams of constant thickness in the horizontal direction) and $m = 3$ and $p = 5/6$ (beams of similar cross-sectional shape, e.g., beams with circular cross section).

Figure 2.17 shows the dependence of the dimensionless volume f_{1*} of an optimum structure on the corresponding value of θ. Figure 2.18 shows the values of the dimensionless ratio [as given by Eq. (2.36)] of the basic linear dimensions determining the cross-sectional shape of the beams, as a function

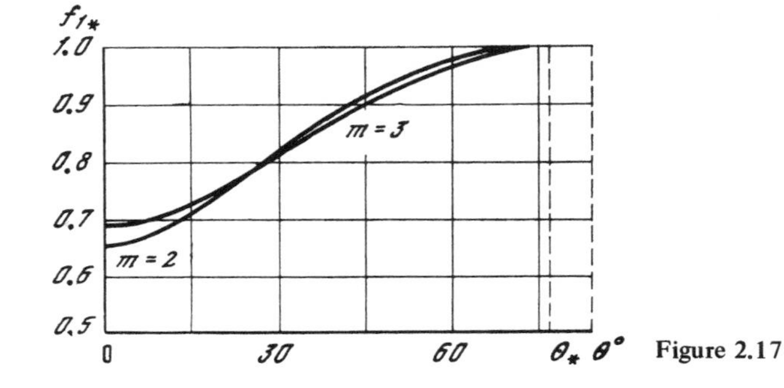

Figure 2.17

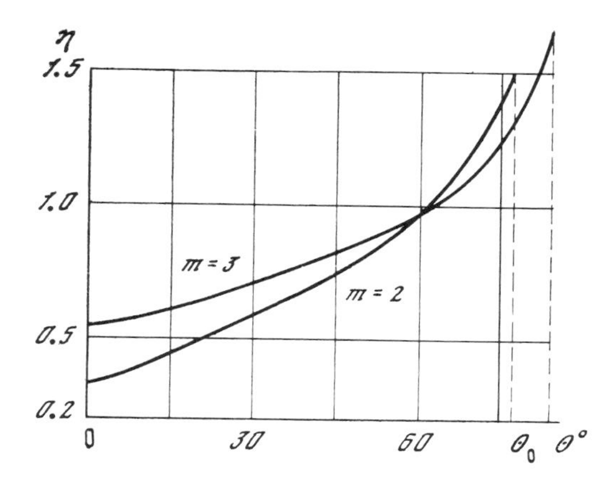

Figure 2.18

of θ. All curves in Figs. 2.16–2.18 are constructed on the interval $0 \leqslant \theta \leqslant \theta_*$, where θ_* is given by Eq. (2.38). The values of the dimensional quantities x, h_1, h_2, and V may be reconstructed by using Eqs. (2.33), (2.34), and (2.36).

Reference 147 deals with the specific case in which the structure (shown in Fig. 2.15) is subjected to the load that is the weight of the structure.

2.4. Design of Optimum Curved Beams

In problems of optimization of beams or columns discussed in Sections 2.1 and 2.2, we regarded the distribution of the thickness as the unknown control variable, while the center line (in the undeformed state) was fixed. In Section 2.3, even the basic configuration of a system of beams was regarded as unknown. However we did assume that the structural elements were straight beams. Aside from other assumptions, it may be of interest to find the original shape of the beam axis (in the undeformed state) from extremum conditions assigned to some stiffness or strength characteristics. A typical problem of this type is the finding of the optimum shape of a curvilinear elastic beam (see Ref. 15).

1. We consider here the problem of determining the optimal shape of an elastic curved beam that is rigidly clamped at one end at the point O (see Fig. 2.19) and is subjected to a static load. We shall restrict our analysis to the case in which the axis of the beam forms a planar curve given by parameterized equations $x = x(s)$, $z = z(s)$, where s is arc length measured from point O. We assume that one of the principal axes of inertia of the cross-sectional area of the beam lies in the plane x–z and that all external forces act in that plane. The deformations are regarded as "small".

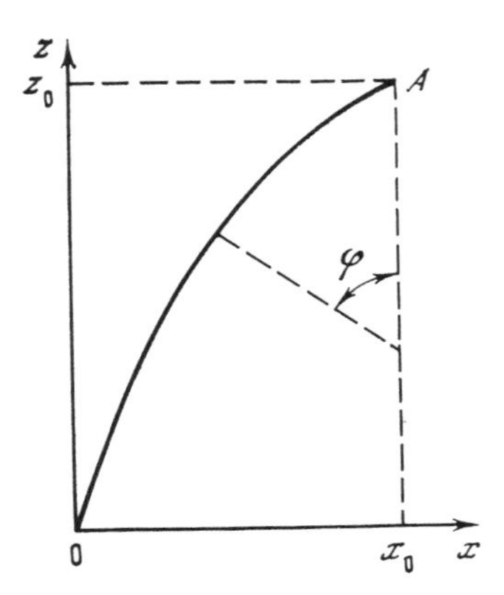

Figure 2.19

The shape of the curved beam is determined by the location of its axis, i.e., by the functions $x(s)$ and $z(s)$ and by the quantity $h = h(s)$ which we shall call the thickness of the beam. We can take as this unknown design variable the height of the rectangular cross section, or the radius if the beam has a circular cross section. The value of h determines the following rigidity or geometric characteristics of the beam, which we shall utilize later: the static moment α, the moment of inertia I, and the cross-sectional area S. We denote by u, w, M, N, and Q, respectively, the components of the displacement vector in the direction of the x and z axes, the bending moment, the tensile (or compressive) force, and the shear force. Assuming that $u = w = 0$ at the point O, we recall the formulas given in Refs. 128 and 132, which describe the displacement of the free end of the beam

$$u(l) = -\frac{1}{E}\int_0^l \Big[\Big(N + \frac{M}{R}\Big)\cos\varphi + \Big(\frac{MS}{Rd} + \frac{N}{R}\Big)(z_0 - z) - \varkappa Q \sin\varphi\Big]\frac{ds}{S}$$

$$w(l) = -\frac{1}{E}\int_0^l \Big[\Big(N + \frac{M}{R}\Big)\sin\varphi - \Big(\frac{MS}{Rd} + \frac{N}{R}\Big)(x_0 - x) + \varkappa Q \cos\varphi\Big]\frac{ds}{S}$$

where x_0 and y_0 are the coordinates of the free end of the beam in the undeformed state and l is its length. The variables R, E, κ, and ϕ denote, respectively, the radius of curvature, Young's modulus, a coefficient depending on the shape of the cross-sectional area [$\kappa = 3(1+\nu)$ for beams with a rectangular cross section, where ν is the Poisson ratio], and the angle between the line normal to the axis of the beam and the z axis (see Fig. 2.19).

The location (x_0, z_0) of the free end of the beam in the undeformed state is considered as known. The variables M, N, Q, R, and ϕ that appear in the integrands in the formulas for u and w are also regarded as known and they depend only on x, z, and their derivatives with respect to s. These are well-known geometric relations for R and ϕ. The corresponding relations for M, N, and Q can be easily derived in each specific case, when the external loads are given by writing the equilibrium conditions since this problem is statically determinate.

The length of the beam l, the beam volume V, and the coordinates x_0 and z_0 are given. We write these assumptions and some necessary purely geometric relations

$$\int_0^l S(h)\,ds = V, \quad x(0) = z(0) = 0, \quad x(l) = x_0 \qquad (2.41)$$
$$z(l) = z_0, \quad x_s^2 + z_s^2 = 1, \quad x_0^2 + z_0^2 \leqslant l^2$$

The functional J that is to be minimized is a function F that depends on the displacements of the free end of the beam,

$$J = F(u(l), w(l)) \to \min \qquad (2.42)$$

The optimization problem consists of finding, on the interval $0 < s < l$, the functions $x(s)$, $z(s)$, and $h(s)$ that satisfy Eq. (2.41) and minimize the functional given by Eq. (2.42).

We comment that if the equality sign is valid in the last relation in Eq. (2.41), then the axis of the beam is a straight line connecting the points [0, 0] and $[x_0, z_0]$ and the optimization process consists in finding a single function $h(s)$.

2. Here we shall assume that the location of the axis is given, i.e., that the functions $x = x(s)$ and $z = z(s)$ are given, and we need only to find the function $h(s)$. By $h = h(s)$ we denote the thickness of a beam having a rectangular cross section, as a function of length, or the thickness of the load-bearing layers of a trilayer beam ($EI = A_\alpha h^\alpha$, $S = bh$). First, let us consider the case of a thin beam, where

$$h_{\max}/l \ll 1, \quad h_{\max}/R_{\min} \ll 1 \qquad (2.43)$$

In this case, $h_{\max} = \max_s h(s)$ and $R_{\min} = \min_s R(s)$. With the assumptions of Eq. (2.43), expressions for the displacement of the end of the beam are given by

$$u(l) = -\frac{1}{A_\alpha}\int_0^l \frac{(z_0 - z)\,M\,ds}{h^\alpha}, \qquad w(l) = \frac{1}{A_\alpha}\int_0^l \frac{(x_0 - x)\,M\,ds}{h^\alpha} \tag{2.44}$$

Euler's equations for the generally nonadditive functionals of Eqs. (2.42) and (2.44), with the isoperimetric volume condition of Eq. (2.41), may be written in the following form (see Section 1.7):

$$Mh^{-(\alpha+1)}[(z_0 - z)F_u - (x_0 - x)F_w] = \lambda$$

where λ is a Lagrange multiplier arising from the isoperimetric condition. The partial derivatives $F_u = F_u(u, w)$ and $F_w = F_w(u, w)$ are computed at values $u(l)$ and $w(l)$ that correspond to the extremum of a variational problem. To find the solution of this problem, we shall use the following device: We denote F_u and F_w in the corresponding extremum of the variational problem by γ_1 and γ_2 and we rewrite the extremality condition in the form

$$h = \left(\frac{M}{\lambda}[(z_0 - z)\gamma_1 - (x_0 - x)\gamma_2]\right)^{\frac{1}{\alpha+1}} \tag{2.45}$$

The constants γ_1 and γ_2 may be found from the solution of the system of equations $F_u \equiv f_1(\gamma_1, \gamma_2, \lambda) = \gamma_1$, $F_w \equiv f_2(\gamma_1, \gamma_2, \lambda) = \gamma_2$, which are obtained by substituting for h in Eq. (2.45) from Eq. (2.44) and computing the values of F_u and F_w. After determining the constants γ_1 and γ_2, the multiplier λ may be found from

$$\lambda = \left(\frac{b}{V}\right)^{\alpha+1}\left\{\int_0^l (M[(z_0 - z)\gamma_1 - (x_0 - x)\gamma_2])^{\frac{1}{\alpha+1}}ds\right\}^{1+\alpha}$$

In the case $F = \beta_1 u(l) + \beta_2 w(l)$ where β_1 and β_2 are given constants, we derive from the above formulas

$$h = \frac{V}{b}[M(z_0 - z)\beta_1 - (x_0 - x)\beta_2]^{\frac{1}{\alpha+1}}\left(\int_0^l (M[(z_0 - z)\beta_1 - (x_0 - x)\beta_2])^{\frac{1}{\alpha+1}}ds\right)^{-1}$$

We now compute the gain in the rigidity of the optimal trilayer beam compared with the beam with constant load bearing layer thickness. In the case considered, the center line of the beam forms a segment of a circle $z = r \sin \phi$, and $x = r(1 - \cos \phi)$, $x_0 = r$, $z_0 = r$, $0 \leqslant \phi \leqslant \pi/2$, where r is the radius and the external load consists of a point force applied at (x_0, z_0), acting in the negative z-direction (Fig. 2.20). As the functional that is to be minimized, we take the magnitude of the vertical displacement w of the beam's free end ($\beta_2 = 1, \beta_1 = 0$). As can be seen from Eqs. (2.41) and (2.42), the magnitude of the displacement w of a trilayer beam of constant width is given by $w = \pi^2 r^4 Pb/(8A_1 EV)$. In the case of a beam having the same volume of reinforcing layers, but the optimum distribution of thickness $h = (V \cos \phi)/rb$, the magnitude of the displacement is given by $w_* = Pr^4 b/(EA_1 V)$. Comparing these displacements, we see that the relative gain in rigidity of the optimum beam, as compared with the beam of constant thickness, amounts to about 18%.

Let us now study a more general case in which no assumptions are made with regard to beam thickness. We consider uniform beams with a rectangular cross section ($I = bh^3/12$ and $S = bh$). We rewrite Eq. (2.44) as

$$u(l) = \int_0^l \left(\frac{\Phi_1}{h} + \frac{\Phi_2}{h^3}\right) ds, \qquad w(l) = \int_0^l \left(\frac{\Phi_3}{h} + \frac{\Phi_4}{h^3}\right) ds \tag{2.46}$$

where the quantities $\Phi_1, \ldots, \Phi_4$ are independent of h and are equal to

$$\Phi_1 = -\frac{1}{Eb}\left[\left(N + \frac{M}{R}\right)\cos\varphi + \frac{N}{R}(z_0 - z) - \varkappa Q \sin\varphi\right]$$

$$\Phi_3 = -\frac{1}{Eb}\left[\left(N + \frac{M}{R}\right)\sin\varphi - \frac{N}{R}(x_0 - x) + \varkappa Q \cos\varphi\right]$$

$$\Phi_2 = -\frac{12M}{Eb}(z_0 - z), \qquad \Phi_4 = \frac{12M}{Eb}(x_0 - x)$$

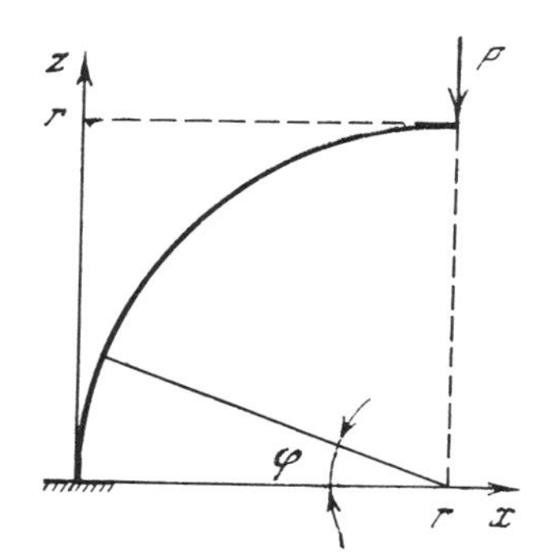

Figure 2.20

Necessary conditions for a minimum of the functional of Eq. (2.42) or (2.46), with an integral constraint of Eq. (2.41) on h, have the following form (see Section 2.7):

$$h^{-2}(\Phi_1 F_u + \Phi_3 F_w) + 3h^{-4}(\Phi_2 F_u + \Phi_4 F_w) = \lambda \tag{2.47}$$

Here λ is a Lagrange multiplier. The quantities F_u and F_w have the same meaning as in the case of a thin beam and therefore Eq. (2.47) can be utilized exactly as before. If the minimized functional F has the form $F = \beta_1 u(l) + \beta_2 w(l)$ and consequently $F_u = \beta_1$ and $F_w = \beta_2$, then Euler's equation is reduced to a biquadratic algebraic equation in h. In this case, the solution of the problem stated above is reduced to a computation of the constant λ. This can be accomplished after substituting the expression for h given in Eq. (2.47) into the isoperimetric condition of Eq. (2.41).

3. In the analysis given in part 2 above the position of the center line of the beam is regarded known and only the single control variable $h(x)$ is to be determined. In this part, we shall regard as unknown not only the distribution of the thickness $h(s)$ but also the location of the beam axis, i.e., the two functions $x(s)$ and $z(s)$. We consider again the optimization problem of Eqs. (2.41) and (2.42) regarding beam thickness [see Eq. (2.43)]. To compute the functions $u(l)$ and $w(l)$, we shall utilize Eqs. (2.44). We shall also restate the problem in more specific terms. The external load acting on the beam is assumed to be a concentrated force P applied to the free end of the beam in a direction parallel to the z axis. The functional to be minimized is the absolute value of the vertical displacement w at the free end, i.e., the magnitude of the displacement in the direction of action of the point load.

We shall now make use of our hypotheses. Taking coordinate x as our independent variable, we list the basic relations of our variational problem,

$$\frac{P}{A_\alpha}\int_0^{x_0} \frac{(x_0 - x)\sqrt{1 + z_x^2}}{h^\alpha}\,dx \to \min \tag{2.48}$$

$$z(0) = 0, \qquad z(x_0) = z_0 \tag{2.49}$$

$$\int_0^{x_0} \sqrt{1 + z_x^2}\,dx = l \tag{2.50}$$

$$b\int_0^{x_0} h(x)\sqrt{1 + z_x^2}\,dx = V \tag{2.51}$$

where $z_x = dz/dx$ and $\alpha = 1, 2, 3$. The case in which $\alpha = 0$ is regarded as an exceptional one, since if $\alpha = 0$ then the beam's rigidity does not depend on h and the shape of the beam axis $z(x)$ is the only function that is considered in the optimization procedure. Equation (2.51) need not be considered in the case $\alpha = 0$. We look for the solution of the variational problem of Eqs. (2.48) to (2.51) in the class of continuously differentiable functions $z(x)$ and continuous functions $h(x)$. After some easy manipulation the necessary conditions for the extremum of the problem of Eqs. (2.48)-(2.51) (see Refs. 39 and 57) be reduced to the form

$$\frac{z_x}{\sqrt{1+z_x^2}}\left[(x_0-x)^{\frac{2}{1+\alpha}}+C_0\right]=C_1 \tag{2.52}$$

$$h=[\lambda_0\,(x_0-x)^2]^{1/1+\alpha} \tag{2.53}$$

where C_0, C_1, and λ_0 are arbitrary constants.

Equation (2.52) determines the optimum shape $z = z(x)$ of the beam axis for the optimum variation of the beam thickness $h(x)$. To compute the values of the constants C_0, C_1, and λ_0 and of the arbitrary constant that arises when we integrate the Eq. (2.52), we can use the boundary conditions of Eq. (2.49) and the isoperimetric conditions of Eqs. (2.50) and (2.51).

It is possible to show (see Ref. 15) that in all cases in which either $z_0 \neq 0$, or $z_0 = 0$ but $l > x_0$, a solution of this problem does not assume the zero value on the entire interval $0 \leqslant x \leqslant x_0$.

In what follows we can assume, without any loss of generality, that $z_0 > 0$. Then, because of the property of solutions noted above, it follows that both the optimal solution $z(x)$ and its derivative z_x are positive (i.e., $z(x) > 0$ for $0 < x \leqslant x_0$, while $z_x > 0$ for $0 \leqslant x \leqslant x_0$). The nonnegative property of the derivative z_x implies that

$$l < x_0 + z_0 \tag{2.54}$$

This is a necessary condition for the existence of a solution to our problem.

If we take Eqs. (2.49) and (2.50) into account the integral of Eq. (2.52) may be written in the form

$$\begin{aligned}
&z = C_1\int_0^x \frac{dt}{\sqrt{[(x_0-t)^{2/(\alpha+1)}+C_0]^2-C_1^2}} \qquad (C_1>0)\\
&C_1\int_0^{x_0} \frac{dt}{\sqrt{[(x_0-t)^{2/(\alpha+1)}+C_0]^2-C_1^2}} = z_0\\
&\int_0^{x_0} \frac{[(x_0-t)^{2/(\alpha+1)}+C_0]\,dt}{\sqrt{[(x_0-t)^{2/(\alpha+1)}+C_0]^2-C_1^2}} = l
\end{aligned} \tag{2.55}$$

The last two relations in Eq. (2.55) can be regarded as equations determining the constants C_0 and C_1. To find the constant λ_0, which enters into Eq. (2.53), we need to substitute into Eq. (2.51) expressions from Eqs. (2.52) and (2.53) for h and z_x and determine λ_0 from the resulting equality. As a result, we obtain

$$\lambda_0 = \left(\frac{V}{b}\right)^{1+\alpha}\left\{\int_0^{x_0}\frac{(x_0-t)^{2/(1+\alpha)}[(x_0-t)^{2/(1+\alpha)}+C_0]\,dt}{\sqrt{[(x_0-t)^{2/(1+\alpha)}+C_0]^2-C_1^2}}\right\}^{-(1+\alpha)} \quad (2.56)$$

which allows us to find λ_0 if we know C_0 and C_1.

The values given in Eqs. (2.55) and (2.56) depend on the value of the parameter α that determines the type of the beam cross section, but they are independent of the coefficient A_α. Therefore, the optimum shape of the beam axis turns out to be the same for different types of beams that may have different exponents in the relation $EI = A_\alpha h^\alpha$.

Below, we offer a brief exposition of research results concerning the optimum shapes of curved beams for the case $\alpha = 1$. A more detailed exposition and a discussion of different cases can be found in Ref. 15.

4. We shall introduce, for the sake of convenience, the following dimensionless variables $z' = z/x_0$, $l' = l/x_0$, and $\delta = z_0/x_0$. We shall omit the primes in future discussion.

We shall now compute the integrals that occur in Eqs. (2.55) and (2.56) and use dimensionless variables. As a result, we obtain an equation that determines the shape of the axis and the relations that must be satisfied by the constants $c_0 = C_0/x_0$ and $c_1 = C_1/x_0$,

$$\begin{aligned} z &= c_1 \ln\left[\frac{1+c_0+\sqrt{(1+c_0)^2-c_1^2}}{1+c_0-x+\sqrt{(1+c_0-x)^2-c_1^2}}\right] \\ c_1 \ln&\left[\frac{1+c_0+\sqrt{(1+c_0)^2-c_1^2}}{c_0+\sqrt{c_0^2-c_1^2}}\right] = \delta \\ \sqrt{(1+c_0)^2-c_1^2}&-\sqrt{c_0^2-c_1^2} = l \end{aligned} \quad (2.57)$$

Necessary conditions for the existence of a solution [i.e., the inequalities in Eqs. (2.41) and (2.54)] can be rewritten in the form $l - 1 < \delta \leqslant (l^2 - 1)^{1/2}$. Following some manipulation, the system of equations that determines the quantities c_0 and c_1 may be transformed into the form

$$c_1 = \frac{1}{2l}\sqrt{l^2-1}\,[(2c_0+1-l)\,(2c_0+1+l)]^{1/2}$$

$$\begin{aligned}&\frac{\sqrt{l^2-1}}{2l}\,[(2c_0+1-l)\,(2c_0+1+l)]^{1/2}\\&\times\ln\left[\frac{1+c_0+\dfrac{1}{2l}\,|2c_0+1+l^2|}{c_0+\dfrac{1}{2l}\,|2c_0+1-l^2|}\right]=\delta\end{aligned}\qquad(2.58)$$

The constant c_0 may be found by solving the second of Eq. (2.58). Then by using the first of Eq. (2.58) and a straightforward computation, we can determine the constant c_1. The constants c_0 and c_1 were found numerically, the computer program making use of the relations stated above. Subsequently, Eq. (2.57) for z was employed to establish the relation $z = z(x)$. Figure 2.21 illustrates the optimum shapes of the axes obtained by such computations for $l = 2$. The numbers 1 to 4 denote curves that correspond to the values of the parameter δ, when δ is equal to 1.060, 1.308, 1.537, and 1.681, respectively.

We shall study the optimum shapes in the special case for which the value of the parameter δ is small compared to unity. We make use of the representation $z = \epsilon Z$, $l = 1 + \epsilon^2 l_*$, and $\delta = \epsilon\delta_*$, where ϵ is a small parameter. The study of this case may be completed using only the relations that we have already derived. However, it is easier to commence with the following equation, boundary con-

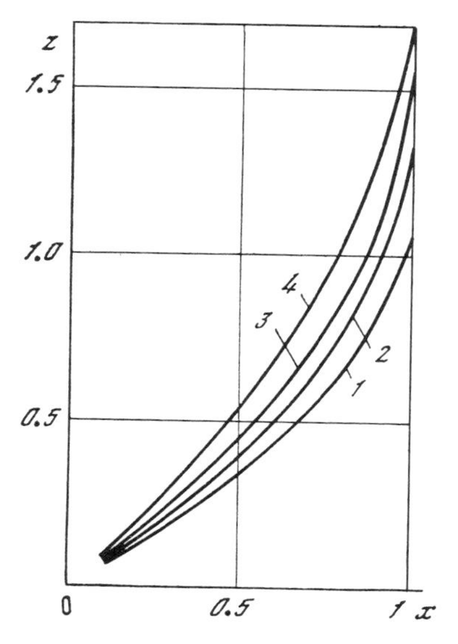

Figure 2.21

dition, and isoperimetric equality

$$\begin{aligned} &(1 + c_0 - x) Z_x = c_1 \\ &Z(0) = 0, \ Z(1) = \delta_* \\ &\int_0^1 Z_x^2 \, dx = 2l_* \end{aligned} \tag{2.59}$$

to which we adjoin Eqs. (2.49), (2.50), and (2.52). We also make use of the equations for z, l, and δ, while we ignore small quantities (i.e., of higher order of smallness). By solving these relations, we obtain an expression for Z as a function of x and t and a transcendental equation for constant values of $t = 1/c_0$

$$\begin{aligned} &Z = \frac{\delta_*}{\ln(1+t)} \ln\left(\frac{1+t}{1+t-xt}\right) \\ &\psi_1 \equiv \frac{(1+t)\ln^2(1+t)}{t^2} = \frac{\delta_*}{2l_*} \end{aligned} \tag{2.60}$$

In this case the condition $l - 1 < \delta \leqslant (l^2 - 1)^{1/2}$ assumes the form $\delta_*^2/(2l_*) \leqslant 1$.

Now, we proceed to study the behavior of the function $\psi_1(t)$ for $t > 0$. For small values of t, the function $\psi_1(t)$ has an asymptotic representation $\psi_1(t) = (1 - t^2)/12$. Consequently in a sufficiently small neighborhood of zero, as t approaches zero the function $\psi_1(t)$ converges to one from below. For large values of t, the function $\psi_1(t)$ behaves like $(\ln^2 t)/t$ and therefore it converges to zero, as t goes to $+\infty$. Numerical results combined with these asymptotic properties imply that as t varies between 0 and $+\infty$ the function $\psi_1(t)$ decreases in a monotone fashion between one and zero. The graph of the function $\psi_1(t)$, obtained by numerical computations, is shown in Fig. 2.22, curve 1. By studying the behavior of the function $\psi_1(t)$, we come to the conclusion that for all admissible values of parameters of our problem, i.e., satisfying the condition $\delta_*^2/(2l_*) \leqslant 1$, a solution of the variational problem of Eq. (2.60) exists and is unique.

Figure 2.23 illustrates the optimum shapes of curved beams derived by a numerical computation. Numbers 1 to 4 denote curves corresponding to the parameter values $t = 0.5$, 2.5, 10, and 50, respectively. The values of the parameter $\delta_*^2/(2l_*)$ corresponding to these values of t are, respectively, 0.986, 0.879, 0.632, and 0.315.

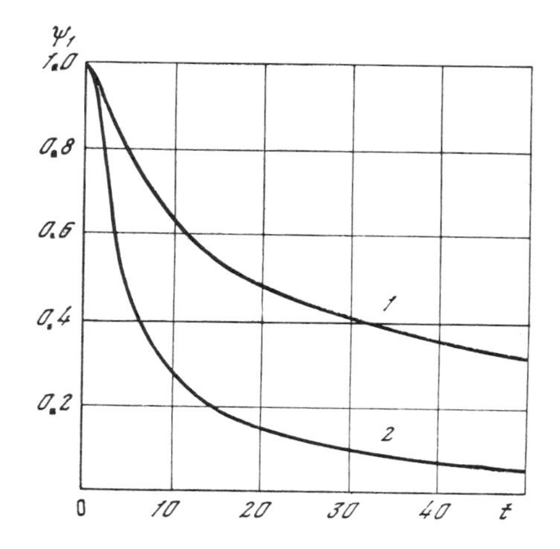

Figure 2.22

For values of the parameter $\delta_*^2/(2l_*)$ that are close to one, we make use of the asymptotic representation for ψ_1 that is valid for small values of t. We obtain a very simple expression for t and the equation for the optimum axis

$$Z = \delta_* x\,[1 + t\,(x - 1)/2], \quad t = 2\,(2 - \delta_*^2/l_*)^{1/2}$$

The discussion contained in this paragraph illustrates the following point. Increasing the number of control variables does not necessarily add to the complexity of an optimization problem. On some occasions, the presence of addi-

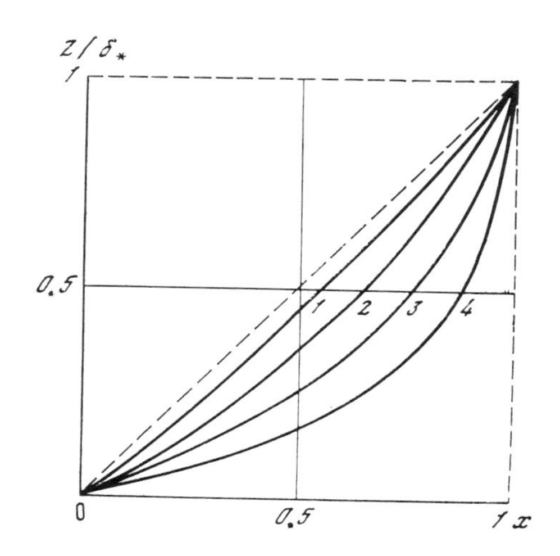

Figure 2.23

tional conditions for optimality caused by a larger number of control variables results in a simpler formulation and may even permit us to derive an analytical solution. Examples of this phenomenon are contained in Ref. 55, where the author discusses a closely related problem concerning simultaneous optimization of the shape of the middle surface and of the thickness distribution for a shell of revolution.

2.5. Optimization of Nonuniformly Heated and Prestressed Beams

The rigidity and strength of structural members is seriously affected by the presence of thermal stresses or residual stresses. A well-known example of decline in rigidity is the phenomenon, which occurs in certain conditions of flight, of decrease in effective torsional rigidity of an unequally heated supersonic wing. The presence of axial thermal stresses caused by aerodynamic heating may lead, for example, to the loss of stability (divergence or flutter) of the wing, loss of load carrying capacity, reversal of ailerons, etc. (see Ref. 38). Aside from thermal effects, in a thin wing (regarded as a thin-walled beam) axial stresses may be caused by other working conditions. There may be numerous causes, such as manufacturing techniques, leading to prestressing of the structure. Since the thermal load is supposed to be known (particularly in the aircraft design stage) techniques of prestressing are widely practiced in order to improve mechanical properties of the structures. It is interesting to reexamine optimization problems, taking into account these factors, and to study the optimal manner of prestressing a structure (see Ref. 203). We shall give an exposition of recent results obtained by V. M. Kartvelishvili and A. A. Mironov, who investigated optimization of torsional rigidity for a nonuniformly heated and prestressed thin-walled beam.

1. To estimate the torsional rigidity of a long cylindrical beam in the presence of a self-equilibrating axial stress σ_x, we make use of the following formula (see Refs. 38 and 136):

$$GC_e = GC_0 + \int_\Omega \sigma_x r^2 \, d\Omega \tag{2.61}$$

where the axial stress satisfies the integral conditions

$$\int_\Omega \sigma_x \, d\Omega = 0, \qquad \int_\Omega r\sigma_x \, d\Omega = 0 \tag{2.62}$$

Here GC_0 denotes the torsional rigidity, according to Saint Venant's definition,[6] G is the modulus of torsional rigidity, $d\Omega$ is a surface element for the cross-sectional area occupied by the material, and r is the distance from $d\Omega$ to the axis of torsion. The axis of the beam coincides with the x axis.

The magnitude of axial stress and the angle θ of beam centerline rotation are constrained by the relation

$$G\theta d \leqslant \sigma_x \leqslant \sigma_0 \tag{2.63}$$

where d is a characteristic linear dimension and σ_0 is a constant that is dependent on the properties of the given structure.

It follows from Eq. (2.61) that for a given distribution of thermal stresses and prestress in axial stresses, the torsional rigidity may turn out to be equal to zero or may even be negative. This behavior, occurring in a wing, is interpreted in Ref. 179 as a loss of stability. However, it is also possible that while these quantities are altered, the effective torsional rigidity is increased. This may be achieved, for example, by distributing the thickness of the wing's skin in a rational manner.

2. We consider the problem of optimal distribution of thickness of beam cross-sectional area satisfying (for given axial stress) the criterion of maximum effective torsional rigidity. This problem arises in maximizing the torsional rigidity for supersonic wings. In fact, the practicing designer of a wing chooses the shape of its contour by purely aerodynamic considerations. However, by redistributing the skin's thickness, it is possible to increase its effective torsional rigidity, without altering its aerodynamic characteristics. Visualizing the wing, to which we intend to apply our results, we make the commonly used assumption (see Ref. 38) that physical properties of the material are not altered by heating.

Let s denote the coordinate measuring distance along the contour Γ of the cross section. Let us assume that the distribution of thickness and temperature are functions of the single independent variable s, i.e., $h = h(s)$ and $T = T(s)$. The function $T(s)$ is supposed to be known, while $h(s)$ is to be determined. Further, let us assume that the cross-sectional area, the distribution of temperature $T(s)$, and the thickness $h(s)$ all have two axes of symmetry. In that case, in the absence of external loads, for a statically determinate heated beam the axial stress $\sigma_x = \sigma_x(s)$ is determined (according to Ref. 98) by

$$\begin{gathered} \sigma_x = E\,[\theta - \beta T(s)], \quad E = 2\,G\,(1 + \nu) \\ \theta = \beta\left(\int_\Gamma T(s)\,h(s)\,ds\right) \Big/ \int_\Gamma h(s)\,ds \end{gathered} \tag{2.64}$$

where β is the coefficient of linear expansion and while θ may be interpreted as the average temperature. Using Bredt's formula (as given in Ref. 6) to determine the quantity GC_0 and substituting Eq. (2.64) into Eq. (2.61), we obtain a simple expression for the effective torsional rigidity of a heated, thin-walled beam

$$GC_e = 4G\Omega^2 \left(\int_\Gamma \frac{ds}{h} \right)^{-1} + E \int_\Gamma hr^2 [\theta - \beta T]\, ds \tag{2.65}$$

This formula for effective torsional rigidity can be written in dimensionless variables

$$\begin{aligned} &s' = s/l, \quad r' = r/l, \quad \Omega' = \Omega/l^2, \quad h' = hl/V, \quad T' = \beta T \\ &C'_e = C_e/2(1+\nu)Vl^2 \end{aligned} \tag{2.66}$$

where

$$l = \int_\Gamma ds, \qquad V = \int_\Gamma h\, ds$$

In terms of the dimensionless variables of Eq. (2.66), after dropping the primes, we have [with $\gamma = 2\Omega^2/(1+\nu)$]

$$C_e = C_0 + C_T, \qquad C_0 = \gamma / \int_\Gamma \frac{ds}{h}, \qquad C_T = \int_\Gamma hr^2 (\theta - T)\, ds \tag{2.67}$$

$$\int_\Gamma h\, ds = 1 \tag{2.68}$$

In Eq. (2.67) the effective torsional rigidity C_e is represented as a sum of two terms C_0 and C_T, where C_0 is torsional rigidity of the beam computed by Bredt's formula, and C_T is the contribution to the torsional rigidity caused by heating and by resulting axial stresses. Equation (2.67) for C_T is transformed, by making use of Eq. (2.64), to

$$C_T = \int_\Gamma Th(r_h^2 - r^2)\, ds, \qquad r_h^2 = \int_\Gamma hr^2\, ds \tag{2.69}$$

It is clear from Eq. (2.69) that $C_T = 0$ for a beam having a circular cross section and constant wall thickness. In other words, the torsional rigidity of a circular thin-walled beam is unchanged by axial temperature distribution, no matter what that temperature distribution is.

Analyzing Eq. (2.69), we keep in mind the following case: The thickness h is constant and the temperature distribution is typical of a supersonic wing (see Refs. 38 and 197). The temperature rises as the distance decreases from the leading or trailing edge of the wing, i.e., the highest temperatures occur at points having the greatest distance from the center of torsion. At such points, the value of r_2 exceeds the average value of r^2, which is denoted by r_h^2, i.e., $r^2 > r_h^2$. Therefore, if $h = \text{const}$, $C_T < 0$. Consequently, such temperature fields cause a decrease in torsional rigidity in wings with constant skin thickness.

Looking at Eq. (2.69), it is not hard to conclude that for arbitrary temperature distribution $T(s)$ it is possible to find a distribution of thickness for which $C_T \geqslant 0$ and a distribution $h(s)$ for which $C_T < 0$.

Keeping these remarks in mind, we formulate the following problem: a given contour Γ and given temperature distribution $T(s)$, we wish to find a distribution of thickness $h(s)$ that satisfies the isoperimetric condition of Eq. (2.68), and maximizes the effective torsional rigidity

$$C_{e*} = \max_h (C_0 + C_T) \tag{2.70}$$

where C_0 and C_T satisfy Eq. (2.67). Formulation of this problem, similar to some problems discussed in the preceding paragraph, is closely related to variational problems with nonadditive functionals. Necessary optimality conditions for the problems of Eqs. (2.67), (2.68), and (2.70) are obtained by recalling the formulas introduced in Section 1.7. After some manipulations, this condition assumes the form

$$\gamma \left(h \int_\Gamma \frac{ds}{h}\right)^{-2} + (T - \theta) \left(\int_\Gamma r^2 h \, ds - r^2\right)$$
$$= \gamma \left(\int_\Gamma \frac{ds}{h}\right)^{-1} + \int_\Gamma hr^2 (\theta - T) \, ds = C_{e*} \tag{2.71}$$

After solving Eq. (2.71) for $h = h(s)$, we arrive at an expression that is convenient for the purposes of quantitative analysis

$$h = \frac{C_0}{\sqrt{\gamma (C_{e*} + \psi)}}, \qquad \psi \equiv (\theta - T)\left(\int_\Gamma r^2 h \, ds - r^2\right) \tag{2.72}$$

Values of the constants θ, C_0 and C_{e*} have an integral form of dependence on h, so Eq. (2.72) represents a nonlinear integral equation in h. With the help of Eq. (2.72), we conduct a quantitative analysis of the function $h = h(s)$ for the optimum beam. Let us first consider Eq. (2.72) for the function ψ. Factors in the formula determine the deviation of T and r^2, at a given point along the contour, from the average values of these quantities on the contour. It follows from Eq. (2.72) that at the points where the signs of the two factors agree (which is the case at the edges and in the center of a supersonic wing), the thickness h is smallest. The thickness h increases on segments of the contour Γ where the two factors either differ in sign or are equal to zero.

Finding the thickness distribution $h(s)$ may be accomplished by numerical techniques, employing a successive optimization technique of the type described in Section 1.10. Computations were carried out for thin-walled beams having a hollow cross section. It was assumed that the temperature is a quadratic function of the coordinate y (as indicated by the dotted line in Fig. 2.24), where y coincides with the chord of the wing's profile. This reflects, to a certain degree, the real temperature distribution in a supersonic wing.[179] The curves 1 and 2 shown in Fig. 2.24 represent the thickness and thermal stress distribution, respectively, for half of the beam profile, where the beam cross section has the shape of a rhombus (ratio of the axes is 1:10). It is clear from these graphs that the thickness of the optimum beam attains its maximum value in a neighborhood of points at which the thermoelastic stress is equal to zero. It follows that in order to increase the torsional rigidity for a supersonic wing we should introduce braces at such points. Computations indicate that for other elongated pro-

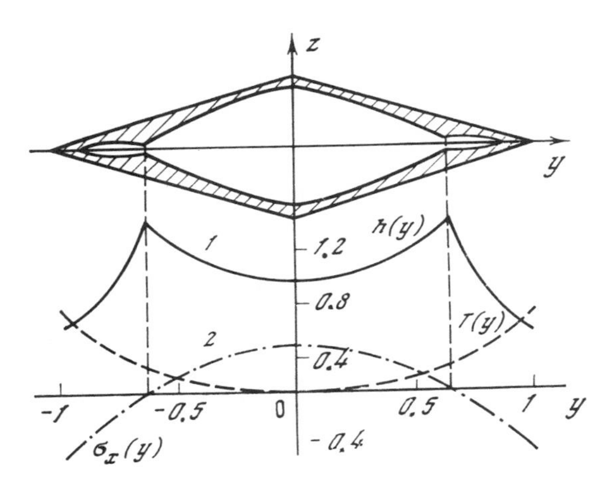

Figure 2.24

files Γ the optimum distribution of thickness basically follows the graph shown in the upper half of Fig. 2.24.

3. As we have indicated above, the axial stress σ_x in a beam may be caused by prestresssing. The effects of prestresss on the magnitude of effective torsional rigidity GC_e are given by the second term in Eq. (2.61). It is clear that depending on the distribution of axial stress, the value of this term may be either negative or positive. Consequently, it is possible to increase GC_e by selecting a prestress satisfying Eqs. (2.62) and (2.63).

We shall pose a problem consisting of maximizing the effective torsional rigidity for a prestressed, thin-walled beam whose cross-sectional contour Γ is a closed curve that is symmetric with respect to the y–z axis (see Fig. 2.25). We need to find the prestress $\sigma_x = \sigma_x(s)$ applied to a quarter Γ_1 of the contour Γ contained between the axes of symmetry, such that the following functional is maximized:

$$J_* = \max_{\sigma_x} J = \max_{\sigma_x} \int h\sigma_x r^2 \, ds \tag{2.73}$$

while the equilibrium condition of Eq. (2.62) and the inequality of Eq. (2.63) are satisfied.

It is not hard to become convinced that the optimal solution must be of the form

$$\sigma_x = \begin{cases} \sigma_0 & \text{on} \quad \Gamma^+ \\ -\sigma_0 & \text{on} \quad \Gamma^- \end{cases} \tag{2.74}$$

Here, Γ^+ and Γ^- are segments of the contour Γ_1 (i.e., $\Gamma^+ \cup \Gamma^- = \Gamma_1$) that satisfy the conditions

$$\begin{gathered} r(s^+) > r(s^-), \quad s^+ \in \Gamma^+, \quad s^- \in \Gamma^- \\ \int_{\Gamma^+} h\sigma_0 \, ds = \int_{\Gamma^-} h\sigma_0 \, ds \end{gathered} \tag{2.75}$$

If $h = \text{const}$ and $r(s)$ is a monotone function of the parameter s, then the arc Γ_1 is decomposed into only two subarcs Γ^+ and Γ^- (see Fig. 2.25), so the points separating the subarcs may be found from Eq. (2.75).

Let us take $s_* = 1/8$ (or, in dimensional variables, $s_* = 1/8\, l$). Then the gain

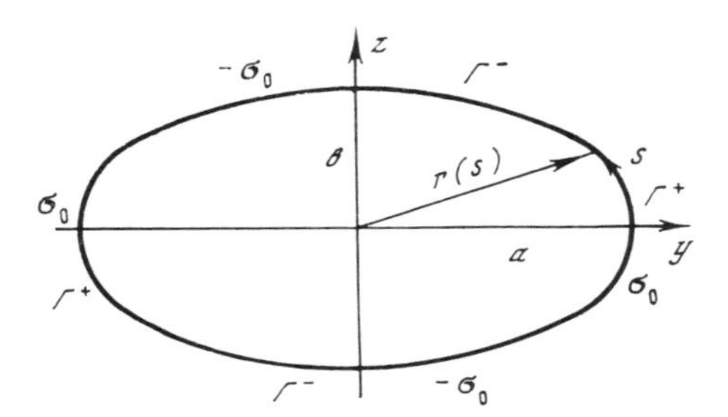

Figure 2.25

attained by optimizing is equal to

$$\varkappa = \frac{C_e - C_0}{C_0} = \frac{\sigma_0}{C_0}\left(\int_{\Gamma^+} r^2 ds - \int_{\Gamma^-} r^2 ds\right) \tag{2.76}$$

It becomes clear by examining Eq. (2.76) that the gain in effective torsional rigidity becomes greater as the difference between the lengths Γ^+ and Γ^- increases; in other words, the more elongated the contour Γ becomes.

Figure 2.26 illustrates graphically the dependence of effective rigidity on the torque C_e (curve 1) and of torsional rigidity, in the sense of Saint Venant (curve 2). It also shows the gain due to optimization (curve 3) as function of the ratio of the axes $\eta = a/b$ for an elliptic cross section of constant thickness. All graphs are drawn for the value $\sigma_0 = 0.2$.

It should be clear from Fig. 2.26 that for a given value of σ_x (say $\sigma_x = 0.2$, while $h = \text{const}$) the effective torsional rigidity of a thin-walled beam of elliptic cross section, with ratio of the axes $1 < \eta < \eta_*$, is greater than the effective torsional rigidity of a thin-walled beam having a circular cross section. We assert

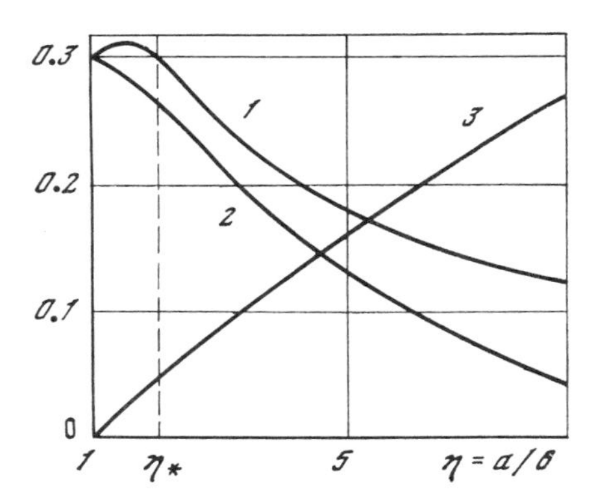

Figure 2.26

that this effect takes place not only in the specific case $\sigma_x = 0.2$, but also for an arbitrary value of $\sigma_x < \sigma_0$. To prove this assertion, we could consider for some (arbitrary) value of $\sigma_x < \sigma_0$ the change δC_e caused by varying the lengths by δa and δb of the axes of the ellipse, while preserving the isoperimetric condition stating that the length of the contour remains unchanged: $\int ds = l = 1$. It is clear that if the quantity $\delta C_e = \delta C_0 + \delta \int \sigma_x r^2 \, ds$ is positive for $\eta = a/b = 1$, then in some neighborhood of $\eta = 1$ it is possible to find an elliptic cross section such that the torsional rigidity of the beam having that cross section is greater than the torsional rigidity of the thin-walled beam having a circular cross section. Because the increase δCe was computed for a circular cross-section, the isoperimetric condition ($l = 2\pi r = 1$, where r is the radius of the circle) implies that the first variation δC_0 is equal to zero. Let us express the equation for the ellipse in a parametrized form $x = a \cos \phi$, $y = b \sin \phi$. We compute the second term in the expression for δC_e as

$$\delta C_e = \frac{\sigma_x}{\pi} \left[\int_{\Gamma+} (\cos^2 \varphi \delta a + \sin^2 \varphi \delta b) \, ds - \int_{\Gamma-} (\cos^2 \varphi \delta a + \sin^2 \varphi \delta b) \, ds \right]$$

Hence, after some rearrangement, using the isoperimetric condition, we derive (for $\delta a > 0, \delta b < 0$)

$$\delta C_e = \frac{\sigma_x}{\pi} (\delta a - \delta b) > 0$$

if $\sigma_x \leqslant \sigma_0$. This proves the validity of the above-stated assertion. Hence, if we consider thin-walled beams prestressed in the axial direction, a twisted beam having a circular cross section is not optimum.

3

Optimal Design of Elastic Plates

Control by Varying Coefficients of the Equations

We examine in this chapter optimization problems for elastic plates subjected to bending. The distribution of thickness is regarded as the "controlling" variable (the coefficients of the state equations depend on the "the control"). In Section 3.1 we formulate the problem of minimizing the maximum deflection of a bent elastic plate. Using relations between norms in the space of continuous functions and norms in the space of p-integrable functions, we reduce the original optimization problem with a local functional to a problem with an integral quality criterion for the study of optimality conditions. In Section 3.2 we determine, by numerical techniques, the optimum shapes of plates under different support conditions at the edges and under different types of loads. In Section 3.3 we study optimum trilayer plates with rigidity as the quality criterion and derive some analytical solutions. In Section 3.4 we derive a solution for the shapes of strongest plates. In this type of problem, we minimize the maximum value of stress (in the region occupied by the plate). In Section 3.5 we study the problem of optimizing the maximum deflection for a circular plate, through location of supports. The results given in this chapter may be found in Refs. 13, 25–28, and 64.

3.1. Plates Having the Greatest Rigidity

1. Let an elastic plate having a variable thickness be supported along its boundary, i.e., along a contour Γ situated in the x–y plane. The plate is bent

due to the action of transverse loads $q = q(x, y)$. The contour Γ forms the boundary of a region Ω whose area is equal to S. The volume of the plate is V. In the undeformed state the middle surface of the plate coincides with Ω. On a part Γ_1 of the boundary Γ the plate is freely supported, while on the remaining part Γ_2 of Γ it is clamped ($\Gamma = \Gamma_1 + \Gamma_2$). We denote by $w = w(x, y)$ and $h = h(x, y)$, respectively, deflection and thickness of the plate, where $h(x, y)$ obeys the constraint

$$\iint_{\Omega} h\, dx\, dy = V \tag{3.1}$$

We introduce the following dimensionless quantities:

$$x' = \frac{x}{\sqrt{S}}, \quad y' = \frac{y}{\sqrt{S}}, \quad w' = \frac{w}{\sqrt{S}}, \quad h' = \frac{hS}{V}$$
$$q' = \frac{12\,(1 - \nu^2)\, S^{9/2}}{EV^3}\, q$$

where, as usual, E stands for Young's modulus and ν for Poisson's ratio of the plate material. Our arguments shall be carried out in terms of these dimensionless qualities, primes being omitted.

In terms of these quantities, the equilibrium and boundary equations may be written in the following form (see, for example, Ref. 99):

$$\begin{aligned} L\,(h)\, w \equiv [h^3\,(w_{xx} + \nu w_{yy})]_{xx} + [h^3\,(w_{yy} + \nu w_{xx})]_{yy} \\ + 2\,(1 - \nu)\,(h^3 w_{xy})_{xy} = q \end{aligned} \tag{3.2}$$

$$(w)_\Gamma = 0, \quad \left(\frac{\partial w}{\partial n}\right)_{\Gamma_2} = 0, \quad \left(h^3 \left[\Delta w - \frac{1 - \nu}{R} \frac{\partial w}{\partial n}\right]\right)_{\Gamma_1} = 0 \tag{3.3}$$

The subscripts x and y denote corresponding partial derivatives, while $\partial w/\partial n$, R, and Δ denote, respectively, the derivative of w in the direction of the exterior normal to the contour Γ, the radius of curvature of the contour Γ, and the Laplace operator.

In the discussion given below, plate rigidity is judged by the absolute value of the maximum deflection $J = \max_{x,y} |w|$. The functional J depends on the distribution of plate thickness, through the function w that is the solution of Eqs. (3.2) and (3.3).

The optimization problem consists in finding in the space of continuous functions satisfying the conditions

$$h_{\min} \leqslant h(x, y) \leqslant h_{\max} \tag{3.4}$$

$$\iint_{\Omega} h \, dx \, dy = 1 \tag{3.5}$$

a function that minimizes the functional J, i.e., that assigns a minimum value to the maximum deflection of the plate

$$J_* = \min_h J = \min_h \max_{xy} | w | \tag{3.6}$$

The problem of Eqs. (3.2)–(3.6) is a minimax problem of optimization, with a local quality criterion. Equations (3.2), (3.5), and (3.4) play, respectively, the part of a differential relation, an isoperimetric condition, and constraints assigned to the smallest and largest value of the control variable ($0 \leqslant h_{\min} \leqslant h_{\max}$ are assigned dimensionless constants). The case differs from the usual case of optimizing integral functionals for elastic bodies, such as the energy of elastic deformation or the fundamental frequency of elastic vibration of a plate (given by the Raleigh quotient). In this case, the functional $J = \max_{x,y} |w|$ to be minimized is determined not by the total behavior of the state function, but only by its values at unknown points where it attains its maximum. The "local" nature of the functional to be optimized causes some basic difficulties in solving the problem of Eqs. (3.2)–(3.6).

Investigations of one-dimensional optimization problems for structures involving local functionals were carried out in Refs. 9, 12, 122, and 180. Some general techniques for dealing with one-dimensional optimization problems were developed in Ref. 58.

2. We shall examine the constraints of Eqs. (3.4) and (3.5) imposed on the control variable. Accounting implicitly for the inequalities of Eq. (3.4) in problems of optimization results in the necessity of determining lines on which $h(x,y)$ intersect the constraints, and "patching" the unknown solution along these lines. This approach results in serious difficulties. However, as we have already seen in Chapter 1, we can utilize the concept of auxiliary control variables, which was introduced in Ref. 239, thus permitting us to exclude constraints of the type of Eq. (3.4) from future considerations. Indeed, by introducing an auxiliary function ϕ and the relations

$$h = \alpha + \beta \sin \varphi, \quad \alpha = \frac{1}{2}(h_{\max} + h_{\min}), \quad \beta = \frac{1}{2}(h_{\max} - h_{\min}) \tag{3.7}$$

and regarding ϕ as a new control variable, we can eliminate from our problem the necessity of dealing with the inequalities of Eq. (3.4). The only constraint imposed on ϕ is necessitated by the isoperimetric condition of Eq. (3.5), which may be written in the form

$$\iint_{\Omega} \sin\varphi \, dx \, dy = \gamma, \qquad \gamma = \beta^{-1}(1 - \alpha) \tag{3.8}$$

We shall use the transformation of Eq. (3.7) and apply a technique described in Chapter 1 to find the optimum shape. This technique permits us to reduce our problem with a local quality criterion to one with an integral functional. We observe that the functional J represents a norm in the space of continuous functions, in the compact region Ω, and that this norm is related to the norm in the space of p-integrable functions. Therefore, in parallel with the original optimization problem of Eqs. (3.2), (3.3), (3.6), (3.7), and (3.8) (i.e., Problem 1), we shall consider the problem of optimizing with respect to ϕ the integral functional

$$J_{p*} = \min_{\varphi} J_p = \min_{\varphi} \left(\frac{1}{S} \iint_{\Omega} | w |^p \, dx \, dy \right)^{1/p} \tag{3.9}$$

with constraints of Eqs. (3.2), (3.3), (3.7), and (3.8). We shall call this Problem 2. In what follows, we shall assume that p is an even integer. In that case we do not have to worry about the sign of the integrand in Eq. (3.9). For sufficiently large values of p, the solutions of Problems 1 and 2 differ by very small amounts, while $J_{p*} < J_*$.

Let us now derive optimality conditions for Problem 2. To do this, we follow the reasoning in Section 1.6 (see also Ref. 27) and we express the first variation of the functional J_p in terms of the variation of the control variable ϕ

$$\begin{gathered} \delta J_p = -3\beta \iint_{\Omega} \cos\varphi \Lambda \delta\varphi \, dx \, dx \\ \Lambda = (\alpha + \beta \sin\varphi)^2 [\Delta w \Delta v - (1 - \nu)(w_{xx} v_{yy} + w_{yy} v_{xx} - 2 w_{xy} v_{xy})] \end{gathered} \tag{3.10}$$

where v denotes the adjoint variable, which is defined as the solution of the boundary-value problem

$$L(h) v = \Phi_p, \qquad \Phi_p \equiv \left(\frac{w}{\| w \|_{L_p}} \right)^{p-1} \tag{3.11}$$

$$(v)_\Gamma = 0, \quad \left(\frac{\partial v}{\partial n}\right)_{\Gamma_2} = 0, \quad \left(h^3\left[\Delta v - \frac{1-\nu}{R}\frac{\partial v}{\partial n}\right]\right)_{\Gamma_1} = 0 \tag{3.12}$$

We observe that w and v obey identical boundary conditions. It is a consequence of Eq. (3.8) that the variation of the control variable $\delta\phi$ that enters into the integrand of Eq. (3.8) must satisfy

$$\iint_\Omega \cos\varphi\delta\varphi\, dx\, dy = 0 \tag{3.13}$$

Taking into account this relation and the equality $\delta J_p = 0$, which is a necessary condition for a minimum of J_p, we derive an optimality condition for Problem 2

$$(\Lambda - \lambda)\cos\varphi = 0 \tag{3.14}$$

where λ is a constant (a Lagrange multiplier).

For a given value of p, Problem 2 reduces to the solution of boundary-value problems of Eqs. (3.2), (3.3), and (3.11) and (3.12) for the function w and the adjoint variable v. The control variable ϕ may be determined from the optimality condition of Eq. (3.14), while the unknown constant λ is determined by making use of the isoperimetric condition of Eq. (3.8). Hence, to solve (in the sense of Problem 2) for the optimum distribution of thickness of the plate, we have a boundary-value problem that is nonlinear, because of nonlinearity in Eqs. (3.11) and (3.14).

3. We obtain now an asymptotic form of the optimality conditions. Dependence of solutions of Problem 2 on p is conditioned by the manner in which this parameter enters into the right-hand side of Eq. (3.10) for the adjoint variable v. The remaining relationships of Eqs. (3.2), (3.3), (3.8), (3.12), and (3.14) do not depend explicitly on p. In particular, the optimality condition of Eq. (3.14) does not depend on p and consequently retains its form in the asymptotic case $p = \infty$.

We study the behavior of the right-hand side of Eq. (3.11) for the adjoint variable v, as $p \to \infty$. In this case, we can assume that the sequences $\{w_p\}$, $\{v_p\}$, $\{h_p\}$, and $\{J_p\}$ have limits. We denote by w_p, v_p, h_p, and J_p the solution of Problem 2 for a given p, and by w, v, h, and J the limits of these quantities. We obtain a preliminary inequality that is satisfied by the quantity $w_p/(\|w\|_{L_p})$. Let the maximum of the function $|w|$ in the region Ω be attained at the point $(\xi, \eta) \in \Omega$, i.e., $\|w\|_c = |w(\xi, \eta)|$, and let $|w(x, y)| < |w(\xi, \eta)|$ for all $(x, y) \in \Omega$

such that $(x,y) \neq (\xi, \eta)$. Let us fix the point $(x,y) \neq (\xi, \eta)$, $[(x,y) \in \Omega]$, and derive an estimate for the unknown quantities at that point. Since $w(x,y) < \|w\|_c$, there exists a small number $\epsilon > 0$ such that $|w(x,y)| \leqslant \|w\|_c - 2\epsilon$.

Furthermore, the assumption that $w_p \to w$ as $p \to \infty$ implies the existence of a number p' such that for all $p > p'$ the inequality $|w_p(x,y) - w(x,y)| < \epsilon$ holds. Using this inequality, we obtain

$$|w_p| < |w| + \varepsilon < \|w\|_C - \varepsilon \tag{3.15}$$

We shall use the equality $\lim \|w\|_{L_p} = \|w\|_c$ as $p \to \infty$, assuming convergence in the norm $\|w\|_{L_p} \to \|w\|_c$ (as $p \to \infty$). It follows from the definition of convergence that there exists a number p'' such that for all $p > p''$

$$\big| \|w_p\|_{L_p} - \|w\|_C \big| < \frac{1}{2}\varepsilon$$

This inequality permits us to form a lower bound estimate

$$\|w_p\|_{L_p} > \|w\|_C - \frac{1}{2}\varepsilon \tag{3.16}$$

Because of Eqs. (3.15) and (3.16), we have, for all $p > \max(p', p'')$

$$\frac{|w_p(x,y)|}{\|w_p\|_{L_p}} \leqslant \frac{\|w\|_C - \varepsilon}{\|w_p\|_{L_p}} \leqslant \frac{\|w\|_C - \varepsilon}{\|w\|_C - \frac{1}{2}\varepsilon} \leqslant 1 - \frac{\varepsilon}{2\|w\|_C} \tag{3.17}$$

As $p \to \infty$, for $(x,y) \neq (\xi, \eta)$, the function $\Phi_p(x,y)$ that appears in the right-hand side of Eq. (3.11) satisfies the relation

$$\lim_{p\to\infty} \Phi_p(x, y) = \lim_{p\to\infty} \left[\frac{w_p(x,y)}{\|w_p\|_{L_p}} \right]^{p-1} = 0 \tag{3.18}$$

Let us compute the limit of $\Phi_p(\xi, \eta)$ as $p \to \infty$. With this purpose in mind, we multiply $\Phi_p(\xi, \eta)$ by $w_p(x,y)$ and integrate over the region Ω. Thus we have

$$\iint_\Omega w_p \Phi_p \, dx\, dy = \iint_\Omega (w_p)^p \, dx\, dy \, (\|w\|_{L_p})^{1-p} = \|w\|_{L_p}$$

Let p approach infinity in this relation. It is a consequence of our assumption concerning the convergence of w_p to w and of $\|w_p\|$ to $\|w\|_c = w(\xi, \eta)$

that

$$\iint_{\Omega} w\Phi \, dx \, dy = \| w \|_C = (w\xi, \ \eta) \tag{3.19}$$

Equations (3.18) and (3.19) imply that the limit of the sequence of functions $\Phi_p(x, y)$, as $p \to \infty$, is the Dirac delta function $\delta(x - \xi, \ y - \eta)$, where (x, y) is an arbitrary point lying in the region Ω. Consequently, in the limit, the equation for the adjoint variable v assumes the form

$$L(h) v = \delta(x - \xi, y - \eta) \tag{3.20}$$

To determine the coordinates (ξ, η) in this equation, we add the maximality condition for $|w|$ in $\Omega + \Gamma$. If the maximum of $|w|$ occurs simultaneously at certain isolated points, the expression on the right-hand side of Eq. (3.20) is replaced by a sum of Dirac delta functions. [TRANSLATION EDITOR'S NOTE: The hypothesis that the load and design are symmetric and the maximum values occur at symmetrically located points should be included; otherwise serious theoretical difficulties arise].

3.2. Numerical Search for Optimal Thickness Distribution of Homogeneous Plates

1. Numerical solution of the boundary-value problem of Eqs. (3.2), (3.3), (3.7), (3.8), (3.12), (3.14), and (3.20) presents well-known difficulties because of the appearance of the Dirac delta function of the right-hand side of Eq. (3.20), and because the coordinates (ξ, η) are not known in advance and have to be determined as a part of the solution process. For these reasons we introduce a technique for computing quasi-optimal solutions. A quasi-optimal solution is a solution of Problem 2 that coincides with a solution of Problem 1 as $p \to \infty$.

To solve Problem 2 we utilize a technique of successive optimizations described in Section 1.10. In the case considered here we use the following expression describing the variation of the control variable ψ:

$$\delta\varphi = \tau\psi, \qquad \psi = \cos\varphi \left[\Lambda - \frac{\iint_{\Omega} \Lambda \cos^2 \varphi \, dx \, dy}{\iint_{\Omega} \cos^2 \varphi \, dx \, dy} \right] \tag{3.21}$$

where τ is some constant. It is easy to see that if $\delta\psi$ is defined according to Eq. (3.21), Eq. (3.13) remains true and $\delta J_p < 0$. We find the functions $w(x,y)$, and $v(x,y)$ [in the $(k+1)$-st approximation] for a fixed thickness distribution by solving the variational problems

$$J_w = \iint_\Omega h^3 [w_{xx}^2 + w_{yy}^2 + 2\nu w_{xx} w_{yy} + 2(1-\nu) w_{xy}^2]\, dx\, dy - 2\iint_\Omega qw\, dx\, dy - 2 \to \min_w \tag{3.22}$$

$$J_v = \iint_\Omega h^3 [v_{xx}^2 + v_{yy}^2 + 2\nu v_{xx} v_{yy} + 2(1-\nu) v_{xy}^2]\, dx\, dy - 2\iint_\Omega \Phi_p v\, dx\, dy \to \min_v \tag{3.23}$$

using the technique of local variation with a variable step size. First, we solve Eq. (3.22) and determine the function $w = w^{k+1}(x,y)$. Next, we solve Eq. (3.23) by finding the function $v^{k+1}(x,y)$, which corresponds to the presently considered approximation. The value of the function Φ_p is given by the $(k+1)$-st approximation, i.e.,

$$\Phi_p = \Phi_p^{k+1} = (w^{k+1}/\|w^{k+1}\|_{L_p})^{p-1}$$

Minimization of the functionals J_w and J_v is accomplished for w and v in a space of functions satisfying the boundary conditions given in Eqs. (3.13) and (3.12). The third boundary condition given in Eqs. (3.3) and (3.12) is a natural condition for the functionals Eqs. (3.22) and (3.23), and therefore need not be satisfied *a priori.*

2. The symmetric case. It is convenient to use the exact conditions of Eqs. (3.2), (3.3), (3.7), (3.8), (3.12), (3.14), and (3.20), corresponding to Problem 1 (with $p = \infty$) in a specific case when the position of the point of maximum deflection is known. Let us consider the problem of Eqs. (3.2), (3.3), (3.6), (3.7), and (3.8) in the case for which the shape of the region Ω and the boundary conditions are symmetric with respect to the origin of the coordinate system. Let us assume that the load q is a concentrated load applied at the point (0, 0), i.e., $q = P\delta(x,y)$, where P denotes the magnitude of the load. The equilibrium equation can be written in the form:

$$L(h)\, w = P\delta(x, y) \tag{3.24}$$

In this case, for an arbitrary symmetric distribution of plate thickness, the maximum deflection occurs at the point (0, 0). Consequently, in Eq. (3.20) we can substitute $\xi = 0$ and $\eta = 0$. Therefore, the adjoint variable v encountered in Problem 1 satisfies

$$L(h)\, v = \delta(x, y) \tag{3.25}$$

with the same boundary conditions as in Eq. (3.12), which are satisfied by the deflection function. Consequently, $w = Pv$, and in each step of the algorithm described above we need to solve only a single boundary-value problem. Let us take w as the principal unknown function and rewrite Eq. (3.10), i.e., the expression for Λ, in the form

$$\Lambda = \frac{1}{P}(\alpha + \beta \sin \varphi)^2 [(\Delta w)^2 - 2(1 - \nu)(w_{xx}w_{yy} - w_{xy}^2)]$$

To solve the boundary-value problem of Eqs. (3.12) and (3.24) we utilize the following device. Assuming that a given magnitude of deflection $w_0 = w(0, 0)$ depends on the load P_0, we may find the distribution of deflections and the magnitude of the force P_0 from the following variational principle

$$P_0 = \frac{1}{w_0} \min_w \iint_\Omega (\alpha + \beta \sin \varphi)^3 [(\Delta w)^2 - 2(1 - \nu)(w_{xx}w_{yy} - w_{xy}^2)]\, dx\, dy \tag{3.26}$$

The minimum in Eq. (3.26) is computed by assuming that $w(x, y)$ belongs to a class of functions satisfying the boundary conditions of Eq. (3.3) and the condition $w(0, 0) = w_0$. Let us denote the function that realizes a minimum of this functional and obeys the above stated conditions by $w'(x, y)$. A solution of the boundary-value problem of Eqs. (3.3) and (3.24) can be computed from the formula

$$w = \frac{P}{P_0} w' \quad \left(J = \| w \|_C = \frac{P}{P_0} w_0\right) \tag{3.27}$$

Finding the deflection function $w(x, y)$ and the minimum value for the functional J_* can be reduced to solving the variational problem of Eqs. (3.3) and (3.26), with the condition $w(0, 0) = w_0$, and the calculation involving the Eqs. (3.27). Otherwise, the technique of numerical computation remains unchanged, i.e., as given in part 1. For homogeneous plates subjected to a point load at the center of symmetry, i.e., $q = \delta(x, y)$ (with $P = 1$), the computation of optimum shapes can proceed with various assumptions concerning the support conditions for the plate and various values of the parameters $h_{\min}$ and $h_{\max}$. Computations were performed for square plates (Ω: $-1/2 \leqslant x \leqslant 1/2$, $-1/2 \leqslant y \leqslant 1/2$), with clamped and simply supported conditions for the plate edges. In each case, we made use of the symmetry and solved the problem in the segment $0 \leqslant x \leqslant 1/2$, $0 \leqslant y \leqslant 1/2$, constituting a quarter of the region Ω. The optimum distributions of plate thickness shown in Figs. 3.1–3.3 correspond to the case $h_{\min} = 0.8$ and $h_{\max} = 1.2$, and to various support conditions. In all cases the initial distribution of thickness was assumed constant, $h^0(x, y) \equiv 1$.

The distribution of thickness shown in Fig. 3.1 corresponds to the case of a clamped boundary. Computational results indicate that the greatest concentration of the material occurs in the middle and along the straight lines connecting the center points of opposite edges. In the vicinity of these lines regions occur in which the thickness attains its upper bound $h_{\max}$. Qualitatively, the behavior of the function h along these straight lines resembles the optimum distribution of thickness in a beam that is clamped at both ends. It is clear from our computations that the material is poorly utilized in the vicinity of plate corners, and here the thickness assumes the lower bound $h = h_{\min}$. Intermediate zones occur between the plateaus $h = h_{\max}$ and $h = h_{\min}$.

We remark that as we increase the value of the parameter $h_{\max}$, the general behavior of the solution is preserved, but the region in which the material is concentrated becomes sharply defined. A true optimum shape begins to emerge. The constraint of Eq. (3.4) is essential in the case for which cylindrical rigidity depends on the control variable h.

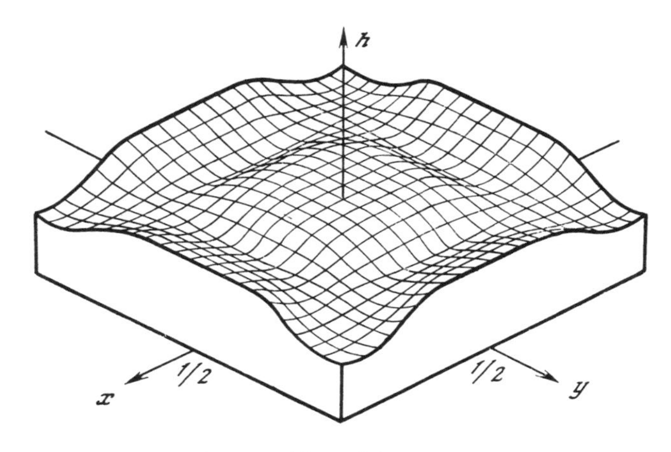

Figure 3.1

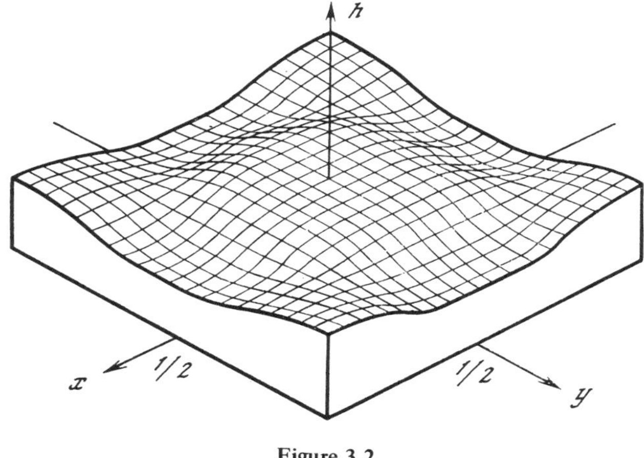

Figure 3.2

Figure 3.2 shows an optimum distribution of thickness for a plate that is simply supported along its boundary. The computed solution indicates that the greatest concentration of material occurs in the middle and along diagonal lines of the plate. Along these lines, the thickness of the plate attains its upper bound $h = h_{max}$. For a simply supported plate the material is added to the corners, where the greatest dependence on the upper bound occurs. In regions close to the middle of the edges of the plate $h = h_{min}$.

Computations were also made for the case in which two opposite edges ($x = \pm 1/2$, $-1/2 \leqslant y \leqslant 1/2$) are clamped, while the remaining two edges are simply supported. The corresponding distribution of thickness is shown in Fig. 3.3.

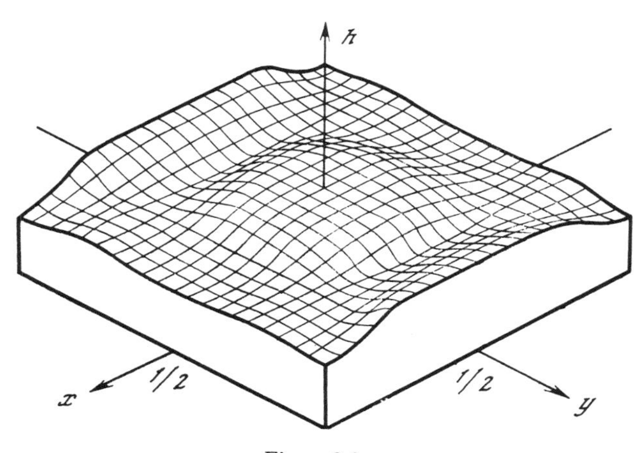

Figure 3.3

Table 3.1

Boundary conditions	$h_{\min}$	$h_{\max}$	$J \cdot 10^3$	$J_* \cdot 10^3$	Improvement %
Clamped edges (see Fig. 3.1)	0.8	1.2	1.39	0.98	30.6
	0	1.6	1.39	0.43	70.0
Simply supported edges (see Fig. 3.2)	0.8	1.2	2.74	1.97	28
	0	1.6	2.74	1.04	63
Mixed supports (see Fig. 3.3)	0.8	1.2	1.69	0.91	46

The values of the functional J computed for the plate of constant thickness $h = 1$ for $P = 1$, and of J_* for the optimum plate, are shown in Table 3.1.

3. *The asymmetric case.* Here we apply the algorithm described in part 1 to compute the optimum shape of plates with asymmetric conditions (the absence of symmetry is caused by the boundary conditions), so that the point of maximum deflection is not known in advance. We find the optimum distribution of thickness for a square plate (Ω: $-1/2 \leqslant x \leqslant 1/2, -1/2 \leqslant y \leqslant 1/2$) subjected to a constant load $q(x, y) \equiv 1$, $[(x, y) \in \Omega]$. We consider the following cases: (i) clamped condition along one edge ($y = -1/2, -1/2 \leqslant x \leqslant 1/2$) and (ii) simply supported along the remaining edges (see Fig. 3.4). Since the problem is symmetric with respect to the axis $y = 0$, we shall carry out our computations only in the region $-1/2 \leqslant x \leqslant 1/2, 0 \leqslant y \leqslant 1/2$. This region is subdivided by a rectangular net containing 100 cells. For this net we find the deflections (for a plate of constant thickness) using the classical techniques given in Ref. 130. We conduct our computations of the thickness by assuming the initial distribution of thickness to be

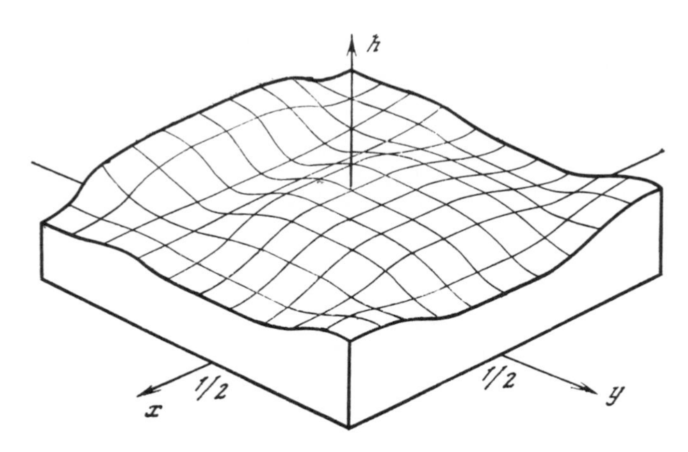

Figure 3.4

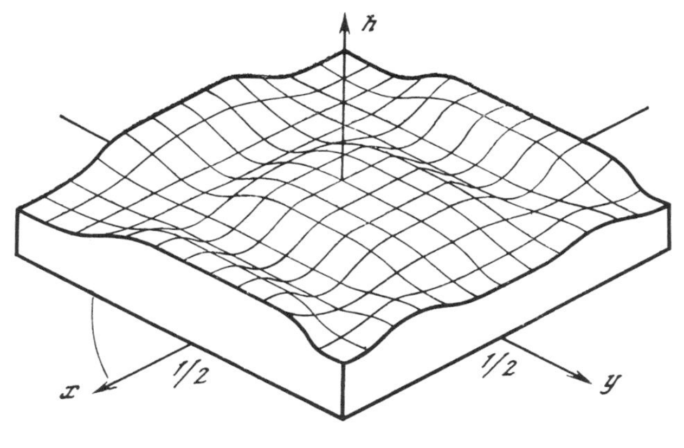

Figure 3.5

constant $h^0(x, y) \equiv 1$, and then determining the initial approximation to the function $\phi^0(x, y)$, utilizing Eq. (3.7).

Figure 3.4 illustrates the computational results, giving the optimum distribution of plate thickness for the case (i). The computed distribution of thickness corresponds to the following values of the parameters: $h_{max} = 1.2$, $h_{min} = 0.8$, $V = 1$, and $p = 100$. These computations indicate that a concentration of material occurs in the middle for the clamped edge and at the corners for the simply supported edges. In these regions the thickness attains the upper bound, i.e., $h = h_{max}$. Near the corners adjacent to the clamped edge and in the center of the freely supported edges, the utilization of the material is poor, and here $h = h_{min}$.

Figure 3.5 illustrates the optimum distribution of thickness for case (ii). Values of the parameters h_{min}, h_{max}, V, and p are the same as for the plate shown in Fig. 3.4. In this case, as can be seen from Fig. 3.5, the most effective use of material takes place at the center of the plate and in the middle of the rigidly held edges. At the corners of the plate, use of the material is least effective.

The values of the functionals J and J_p for plates of constant thickness and of J_* and J_{p*} for the optimum distribution of thickness are given in Table 3.2.

3.3. Optimal Rigidity of Trilayer Plates

We consider the problem of optimizing rigidity of trilayer plates (see Ref. 13). The cylindrical rigidity of such plates is equal to $A_1 h$, where $A_1 = EH^2/$

Table 3.2

Boundary conditions	$h_{\min}$	$h_{\max}$	$J \cdot 10^3$	$J_p \cdot 10^3$	$J_* \cdot 10^3$	$J_{p*} \cdot 10^3$	Improvement %
Clamped at $y = -1/2$, $-1/2 \leqslant x \leqslant 1/2$; free elsewhere	0.8	1.2	2.95	2.80	2.19	2.09	25.8
Free at $y = -1/2$, $-1/2 \leqslant x \leqslant 1/2$; clamped elsewhere	0.8	1.2	1.70	1.61	1.20	1.14	29.3

$[4(1 - \nu^2)]$, H is the constant thickness of the middle layer, and $(1/2)\,h(x,y)$ denotes the thickness of the (outer) reinforcing layers of the plate. Let the plate be supported along the contour Γ that forms the boundary of the region Ω occupied by the plate in the x–y plane. The total volume of material of the outer working layers is assumed to be constant. This leads to the isoperimetric condition of Eq. (3.1) treated as a constraint on the thickness distribution of the reinforcing layers. In this section, our analysis is conducted in terms of dimensionless variables.

1. We investigate the bending of plates assuming that the shape of Ω is symmetric with respect to the origin of the coordinate system, while the load is concentrated and applied at the origin, i.e., $q = P\delta(x,y)$. The optimization problem consists of finding a function $h(x,y)$ satisfying the isoperimetric condition of Eq. (3.1) and providing the minimum of $w(0,0)$, which is the plate deflection at the point $(0,0)$, i.e., the cost functional. Necessary conditions for optimality have the form

$$U \equiv w_{xx}^2 + w_{yy}^2 + 2\nu w_{xx} w_{yy} + 2(1-\nu)\, w_{xy}^2 = \lambda^2 \tag{3.28}$$

where λ^2 is a Lagrange multiplier. This condition can be obtained in a simple manner by making use of known variational principles developed in the bending theory of elastic plates. Indeed, when we use a variational principle, the functional expressing the magnitude of $J = w(0,0)$, i.e., the deflection at the point $(0,0)$ assumes the form

$$J = -\min_w \frac{1}{P}\left\{A_1 \iint_\Omega hU\,(w_{xx}, w_{xy}, w_{yy})\,dx\,dy - 2Pw\,(0,\,0)\right\} \tag{3.29}$$

and consequently

$$J_* = \min_h J\,(h) = -\max_h \min_w \frac{1}{P}\left\{A_1 \iint_\Omega hU\,dx\,dy - 2Pw\,(0,\,0)\right\} \tag{3.30}$$

We can easily show that the optimality condition for this functional in terms of h, with the constraint of Eq. (3.1), coincides with Eq. (3.28), while the condition for the minimum of Eq. (3.30) with respect to w represents the equilibrium equation in terms of the displacements

$$h\Delta^2 w + 2h_x(\Delta w)_x + 2h_y(\Delta w)_y + \Delta h \Delta w - (1-\nu)[h_{xx}w_{yy} - 2h_{xy}w_{xy} + h_{yy}w_{xx}] = P\delta(x, y) \tag{3.31}$$

It is easy to show by straightforward arguments that Eq. (3.28) is not only a necessary but also a sufficient condition for the optimum. To prove this assertion, let us consider the thickness distribution functions h^* and h, satisfying Eq. (3.1). Let w^* and w denote the actual displacement, i.e., solutions of Eq. (3.31) with boundary conditions

$$(w)_\Gamma = 0, \quad \left(h\left[\Delta w - \frac{1-\nu}{R}\frac{\partial w}{\partial n}\right]\right)_L = 0 \tag{3.32}$$

corresponding to the distributions of thickness h^* and h, respectively. Let us assume that the functions h^* and w^* satisfy the optimality condition of Eq. (3.28), while h and w do not satisfy this condition. Since the actual displacement w corresponding to the distribution of thickness h minimizes the potential energy, the following inequality must be true:

$$A_1 \iint_\Omega hU\,dx\,dy - 2Pw(0,0) \leqslant A_1 \iint_\Omega hU^*\,dx\,dy - 2Pw^*(0,0)$$

Using this inequality and introducing the estimate

$$\begin{aligned} J - J_* = w(0,0) - w^*(0,0) = -\frac{1}{P}\Big[A_1 \iint_\Omega hU\,dx\,dy - 2Pw(0,0) \\ - A_1 \iint_\Omega h^*U^*\,dx\,dy + 2Pw^*(0,0)\Big] \geqslant \\ -\frac{A_1}{P}\iint_\Omega U^*(h - h^*)\,dx\,dy = 0 \end{aligned}$$

we obtain $J \geqslant J_*$. Consequently, the cost functional for our problem attains a minimum for the functions h^* and w^* satisfying Eq. (3.28), where the equation of equilibrium, the boundary conditions, and the isoperimetric conditions are assumed to be satisfied. Hence, Eq. (3.29) is not only a necessary, but also a sufficient condition for optimality.

In the problem of trilayer plates, which is now considered, the process of finding the optimum distribution of h and w and the magnitude of $w^*(0,0)$ can be decomposed into two boundary-value problems involving partial differential equations of second order. Theoretically, the optimum distribution of displacements may be found by solving the boundary-value problem given for Eq. (3.28) by the first condition of Eq. (3.32). Introducing a new variable, $u = w/\lambda$, this boundary-value problem may be restated in the form

$$(\Delta u)^2 - 2(1 - \nu)\,(u_{xx}u_{yy} - u_{xy}^2) = 1, \qquad (u)_\Gamma = 0 \tag{3.33}$$

We observe that for the optimum shape of the plate, the following relation exists between the force P and the deflection of the plate at the point $(0,0)$:

$$P = \frac{A_1}{w^*}\iint\limits_{\Omega} h^* U^*\,dx\,dy = \frac{A_1 V}{(u\,(0,\,0))^2}\,w^*\,(0,\,0) \tag{3.34}$$

The following expression is obtained for the constant λ:

$$\lambda = w^*\,(0,\,0)/u\,(0,\,0) = u\,(0,\,0)\,P/A_1 V \tag{3.35}$$

It is not necessary to know the optimum distribution $h(x,y)$ of plate thickness to compute the value of $J_* = w^*(0,0)$. This can be deduced from Eq. (3.34).

We need now the following observation. The solution of Eq. (3.33) does not depend on the location of the point at which the force P is applied, and is completely determined by the shape of the region Ω. Having solved for some shape of the region Ω, solved Eq. (3.33), and having determined the function $u(x,y)$, we can obtain, by simple computation involving u, the optimum distribution of deflections and the value of the minimized quantity $w(0,0)$ for variants involving different locations of the point $(0,0)$ in Ω.

After determining the optimum deflection function $w(x,y)$, we substitute this function into Eqs. (3.31) and (3.32). The function $h(x,y)$ is determined in the region Ω by solving Eq. (3.31) with the second boundary condition of Eq. (3.32).

We investigate asymptotic behavior of the optimum solution in a neighborhood of the contour Γ, assuming that this contour is smooth. We introduce a local system of orthogonal coordinates (ξ, η). The origin of this coordinate system is some arbitrary point 0 situated on the curve Γ. The coordinate η is tangent to the curve Γ, and ξ points towards the interior of the region Ω. In a vicinity of the point 0, Eq. (3.33) has an asymptotic representation $u_{\xi\xi}^2 = 1$, while the deflection function [which satisfies the boundary condition of Eq. (3.33)] is given by a quadratic function $u = {}^1/_2\, \xi^2 + a_0\, \xi$, where a_0 is an arbitrary constant. Making further use of the asymptotic formula $u_{\xi\xi}^2 = 1$ and of the second condition of Eq. (3.32) we obtain

$$h = 0, \qquad (x,\ y) \in \Gamma \tag{3.36}$$

To determine the behavior of h in a neighborhood of Γ, we make use of the derived asymptotic function u and write the asymptotic representation of Eq. (3.31) for small values of ξ: $\eta_{\xi\xi} = 0$, from which we derive, by using the boundary condition of Eq. (3.36)

$$h = a_1 \xi \tag{3.37}$$

where a_1 is the asymptotic constant. Hence, it is necessary to make use of the boundary condition of Eq. (3.36) to determine the function h, by solving the boundary-value problem of Eq. (3.31).

2. We now find the optimum shape of a circular plate. Consider the optimal design problem for a circular plate with radius R, subjected to a point load P. We solve this problem in the polar coordinate system (r, θ), assigning the origin $r = 0$ to the center of the plate. Attempting to find the thickness and deflection functions, we make use of the axial symmetry of this problem, assuming dependence of these functions on a single variable r, i.e., $w = w(r)$ and $h = h(r)$. Using this symmetry, we note that Eqs. (3.33) and (3.35), determining the deflection $w = \lambda u$ for the optimum plate design, assume the form

$$\left(u_{rr} + \frac{u_r}{r}\right)^2 - \frac{2\,(1-\nu)}{r} u_r u_{rr} = 1 \qquad (0 \leqslant r \leqslant R) \tag{3.38}$$

$$u\,(R) = 0, \qquad u_r\,(0) = 0, \qquad \lambda = w\,(0)/u\,(0) \tag{3.39}$$

We have formulated a boundary-value problem for the function $u(r)$. The condition $u_r(0) = 0$ follows directly from Eq. (3.38). Indeed, assuming to the contrary that $u_r(0) \neq 0$, and rewriting Eq. (3.38) in the form

$$\left(u_{rr} + \frac{\nu u_r}{r}\right)^2 + \frac{u_r^2}{r^2}\,(1 - \nu^2) = 1 \tag{3.40}$$

we conclude that as r tends to zero the left-hand side of Eq. (3.40) grows without bound, and therefore that Eq. (3.40) cannot be satisfied by $u_r(0) \neq 0$.

The solution of Eq. (3.11), satisfying boundary conditions of Eq. (3.39), has the form

$$u = (R^2 - r^2)/2\sqrt{2(1+\nu)}$$

while $\lambda = PR^2/\{2[2(1+\nu)]^{1/2}A_1V\}$. Hence, the optimum deflection function is described by a function that is quadratic in r

$$w^* = PR^2(R^2 - r^2)/8(1+\nu)A_1V$$

Substituting $r = 0$ into this formula, we have

$$J_* = (w^*)_{r=0} = PR^4/8(1+\nu)A_1V \qquad (3.41)$$

We note that for a circular plate whose reinforcing layers have constant thickness, $w(0) = PR^4(3+\nu)/[16(1+\nu)A_1V)]$. Comparing the maximum deflection of the optimum plate with that of a plate having constant thickness, we find that the gain in the value of the cost functional obtained by redistribution of the thickness is given by $(w - w^*)/w = 1 - [2/(3+\nu)]$ and varies between 33% and 56%, depending on the value of the Poisson ratio ν (where $0 < \nu < 0.5$).

To find the optimum distribution of thickness, we reformulate Eq. (3.31) in polar coordinates

$$\left[h_r(rw_{rr} + \nu w_r) + h\left(rw_{rrr} + w_{rr} - \frac{w_r}{r}\right)\right]_r = P\delta(x, y) \qquad (3.42)$$

To determine the constants of integration for Eq. (3.42), we use the conditions

$$h(R) = 0, \qquad 2\pi\int_0^R hr\,dr = V \qquad (3.43)$$

Substituting the computed deflection function $w(r)$ into Eq. (3.42) and solving Eqs. (3.42) and (3.43) for $h(r)$, we have

$$h = -(4V/R^2)\ln(r/R) \qquad (3.44)$$

The optimum thickness function given by Eq. (3.44) has a singularity at the point of application of the force. An identical singularity was discovered in Ref.

212, where the authors investigated the optimum shape $h(r)$ of a circular plate, within the framework of plastic limit design.

This optimal solution cannot be directly applied to practical projects. However, it can be applied to assess theoretical chances of optimizing a structure, or to compute quasi-optimum solutions, which are close to the optimum solution but do not have any singularities.

3. The case of an elliptic plate. We now consider a two-dimensional bending problem for an elliptic plate with a point load P applied at the point $(0, 0)$. The plate is simply supported along its boundary $(x^2/a^2) + (y^2/b^2) = 1$, to check that for this shape of the region Ω, the solution of the boundary-value problem of Eq. (3.33) has the form

$$u = \frac{a^2b^2}{2\sqrt{a^4 + b^4 + 2\nu a^2b^2}}\left(1 - \frac{x^2}{a^2} - \frac{y^2}{b^2}\right)$$

$$\lambda = \frac{a^2b^2P}{2A_1V\sqrt{a^4 + b^4 + 2\nu a^2b^2}} \tag{3.45}$$

while the optimum deflection function is given by

$$w^* = \frac{a^4b^4P}{4A_1V(a^4 + b^4 + 2\nu a^2b^2)}\left(1 - \frac{x^2}{a^2} - \frac{y^2}{b^2}\right) \tag{3.46}$$

The value of the cost functional corresponding to the optimum solution is equal to

$$J_* = w^*(0, 0) = Pa^4b^4/4A_1V(a^4 + b^4 + 2\nu a^2b^2) \tag{3.47}$$

3.4. Strongest Plates

In preceding sections we have found the plate thickness function that minimizes the maximum deflection. Another problem that is as important as maximizing the rigidity of plates, in technical applications, is that of minimizing the maximum level of stress "intensity." We shall formulate this problem in a specific case of bending of homogeneous plates. Let $\sigma_x, \sigma_y, \ldots, \tau_{yz}$ be components of the stress tensor. We characterize the stress condition of the material at each of its points (x, y, ζ) [with $(x, y) \in \Omega$, $-h/2 \leqslant \zeta \leqslant h/2$] by the magnitude of the second invariant of the stress tensor deviator part

$$g \equiv \frac{1}{3}[\sigma_x^2 + \sigma_y^2 - \sigma_x\sigma_y + 3\tau_{xy}^2 + 3(\tau_{xz}^2 + \tau_{yz}^2)] \tag{3.48}$$

For the sake of convenience we introduce the dimensionless quantities introduced in Section 3.1 and also substitute

$$\sigma_x' = \frac{\sigma_x}{E}, \qquad \sigma_y' = \frac{\sigma_y}{E}, \ldots, \tau_{yz} = \frac{\tau_{yz}}{E}, \qquad g' = \frac{12(1-\nu^2) S^{3/2} g}{VE^2}$$

where primes will be omitted in the formulas given below.

The bending equilibrium equation and the boundary conditions expressed in terms of the dimensionless variables are given in Eqs. (3.2) and (3.3). The equations expressing components of the stress tensor in terms of deflections are as follows:

$$\begin{aligned}
\sigma_x &= -\frac{\zeta V}{(1-\nu^2) S^{3/2}} (w_{xx} + \nu w_{yy}) \\
\sigma_y &= -\frac{\zeta V}{(1-\nu^2) S^{3/2}} (w_{yy} + \nu w_{xx}) \\
\tau_{xy} &= -\frac{\zeta V}{(1+\nu) S^{3/2}} w_{xy} \\
\tau_{xz} &= -\frac{V^2 (h^2 - 4\zeta^2)}{8S^3 (1-\nu^2) h^3} [(1-\nu)(h^3 w_{xy})_y + (h^3 w_{xx})_y + \nu (h^3 w_{yy})_x] \\
\tau_{yz} &= -\frac{V^2 (h^2 - 4\zeta^2)}{8S^3 (1-\nu^2) h^3} [(1-\nu)(h^3 w_{xy})_x + (h^3 w_{yy})_x + \nu (h^3 w_{xx})_y]
\end{aligned} \tag{3.49}$$

After substitution of Eq. (3.49) into Eq. (3.48), we have the following expression for g:

$$\begin{aligned}
g &= 4\zeta^2 g_1 + \frac{V^2 (h^2 - 4\zeta^2)^2}{S^3 h^6} g_2 \\
g_1 &= (w_{xx} + \nu w_{yy})^2 + (w_{yy} + \nu w_{xx})^2 \\
&\quad - (w_{xx} + \nu w_{yy})(w_{yy} + \nu w_{xx}) + 3(1-\nu)^2 w_{xy}^2 \\
g_2 &= \frac{3}{16} \{[(1-\nu)(h^3 w_{xy})_y + (h^3 (w_{xx} + \nu w_{yy}))_x]^2 \\
&\quad + [(1-\nu)(h^3 w_{xy})_x + (h^3 (w_{yy} + \nu w_{xx}))_y]^2\}
\end{aligned} \tag{3.50}$$

We adopt as our cost functional the quantity

$$J = \max_{xy\zeta} g, \qquad (x, y) \in \Omega, \qquad \zeta \in [-h/2, h/2] \tag{3.51}$$

and formulate the following optimization problem. We wish to find the thickness function for a plate, such that in the region $\Omega_h = \Omega \times (-h/2, h/2)$ the maximum value of the quantity g is minimized, i.e.,

$$J_* = \min_h J = \min_h \max_{xy\zeta} g \tag{3.52}$$

A minimum with respect to h of the functional given in Eq. (3.52) needs to be computed, subject to constraints on admissible thickness given by Eq. (3.3) for the isoperimetric condition of Eq. (3.5).

Let us first offer some explanations concerning the formulation of this problem, as given by Eqs. (3.48)-(3.52). Let us suppose that for some plate and for a given (sufficiently small) load $q^0(x,y)$ the boundary-value problem corresponding to pure bending of this plate has been solved, and the deflection function $w^0(x,y)$ has been found. Then, using Eq. (3.49), we can determine the stress components σ_x^0, σ_y^0, τ_{xy}^0, τ_{xz}^0, and τ_{yz}^0, and we can determine the point set $\Omega_h^0 \subset \Omega_h$ where the function g attains its maximum

$$J^0 = (g)_{\Omega_h^0} = \max_{xy\zeta} g$$

Let us increase the load proportionally to a parameter t, i.e., let us take $q = tq^0(x,y)$, so that for some value of this parameter $t^0 = k/\sqrt{J^0}$ (where k is the constant determining onset of plasticity), at the point $(x,y,\zeta) \in \Omega_h^0$ where the plastic condition is first initiated. We have assumed the validity of the von Miess criterion of plasticity, which asserts that the plastic condition is initiated when $g = k^2$. Clearly, the smaller the value of J^0, the higher the loads (that is the value of t^0) for which the plastic condition first occurs in the plate. Hence, we can attain a greater diversity of loads that do not cause the limiting state $g = k^2$ to occur, i.e., that do not cause the existence of a plastic region if the quantity J^0 is at its minimum.

We observe that the dependence of g on ζ is very simple. Therefore, we can immediately compute the maximum of g with respect to ζ. It is easy to show that

$$\max_\zeta g = \max \{h^2 g_1, V^2 g_2/S^3 h^2\} \tag{3.53}$$

Here we consider only thin plates, where the coefficient V^2/S^3 is small and $h_{\min} \neq 0$. Therefore, the maximum of g is equal to the first value inside the curly brackets in Eq. (3.53). We have, therefore,

$$J = \max_{xy} h^2 g_1$$

Following the technique explained in Section 1.5, we replace the functional J by the functional J_p, which is defined by

$$J_p = \Big[\iint_\Omega (h^2 g_1)^p \, dx \, dy \Big]^{1/p}$$

and proceed with the usual necessary computations. We have the following expression for the variation of this functional:

$$\begin{aligned} \delta J_p &= \iint_\Omega \Lambda \delta h \, dx dy, \\ \Lambda &= 2h\chi \, [(1 - \nu + \nu^2)(w_{xx}^2 + w_{yy}^2) + (4\nu - 1 - \nu^2) w_{xx} w_{yy} \\ &\quad + 3(1 - \nu^2) w_{xy}^2] - [w_{xx} v_{xx} + w_{yy} v_{yy} \\ &\quad + \nu (w_{xx} v_{yy} + w_{yy} v_{xx}) + 2(1 - \nu) w_{xy} v_{xy}] \\ \chi &= (h^2 g_1 / \| h^2 g_1 \|_{L_p})^{p-1} \end{aligned} \tag{3.54}$$

where v denotes the adjoint variable that satisfies the boundary-value problem

$$\begin{aligned} & L(h)\, v = \Phi_p \\ & (v)_\Gamma = 0, \quad \left(\frac{\partial v}{\partial n} \right)_{\Gamma_2} = 0, \quad \left(h^3 \left[\Delta v - \frac{1-\nu}{R} \frac{\partial v}{\partial n} \right] \right)_{\Gamma_1} = 0 \\ & \Phi_p \equiv 2(1 - \nu + \nu^2)\, [(\chi h^2 w_{xx})_{xx} + (\chi h^2 w_{yy})_{yy}] + (4\nu - 1 - \nu^2) \\ & \qquad \times [(\chi h^2 w_{xx})_{yy} + (\chi h^2 w_{yy})_{xx}] + 6(1 - \nu^2)(\chi h^2 w_{xy})_{xy} \end{aligned} \tag{3.55}$$

The difference between this boundary-value problem for the adjoint variable v and the corresponding boundary-value problem for v given in Eqs. (3.10) and (3.11), which arose in finding the shape of a plate having the greatest rigidity, consists in the different definition of the functions Φ_p occurring in the right-hand sides of these equations.

These relations permit us to apply, in our problem of optimizing strength, the same algorithm that we used in the problem of minimizing the maximum deflection. As an example, we offer a computation of the optimum shape for a square plate ($-1/2 \leqslant x \leqslant 1/2$, $-1/2 \leqslant y \leqslant 1/2$) that is simply supported along the edge $y = -1/2$, $-1/2 \leqslant x \leqslant 1/2$ and is clamped along the remaining edges. Data for this plate were $h_{\min} = 0.8$, $h_{\max} = 1.2$ and $q \equiv 1$. The computed optimum thickness function is illustrated in Fig. 3.6. Comparing Figs. 3.5 and 3.6, we observe that the computed optimum distributions of thickness do not substantially differ

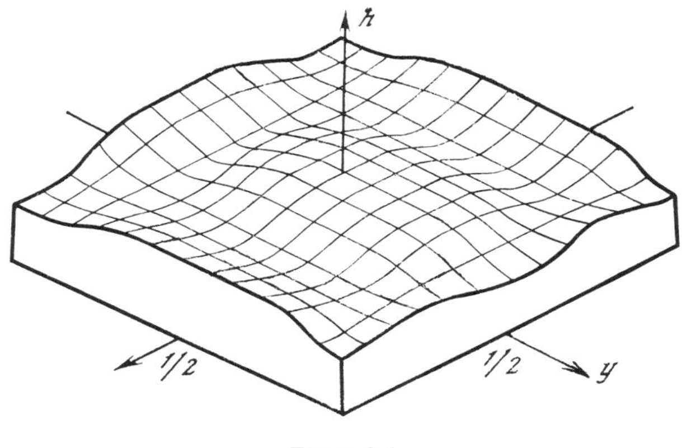

Figure 3.6

from each other, i.e., the thickness function of the strongest plate does not substantially differ from that of the most rigid plate (as shown in Fig. 3.5), obeying the same boundary conditions and having the same design criteria.

3.5. Optimum Support Conditions for Thin Plates

We consider minimization of the maximum deflection of a load-carrying circular plate by optimizing location of the supports (see Ref. 64). Let the plate by supported at n-point supports (perfectly rigid columns) located at the points O_k. There are no friction forces between the plate and the columns at the points of contact. We locate the origin of the polar coordinate system (r, θ) at the center of the plate. The z axis points in the direction normal to the plane of the undeformed plate. Let us assume that the plate has a constant thickness and is clamped along the boundary $r = R$. The normal load has a constant intensity q and acts in the direction of the positive z axis. For a given number n and load q, we need to find in the region $\Omega = \{r \leqslant 1, 0 \leqslant \theta \leqslant 2\pi\}$ the location of the support points $O_k = (r_k, \theta_k)$, $k = 1, 2, \ldots, n$, that minimizes the maximum deflection of the plate.

We shall restrict our investigation of the optimum location of the supporting columns by considering only cases for which the number of support points does not exceed four. We shall consider (as most sensible) a symmetric distribution of plate support points. Therefore, the coordinates of these support points are

$$O_k = \{\rho, 2\pi k/n\}, \quad 0 \leqslant \rho \leqslant R, \; k = 0, 1, \ldots, n-1 \qquad (3.56)$$

In addition, when $n \geqslant 3$, we consider cases for which one of the supports is located at the center of the plate, while the remaining supports are located along radial lines

$$\theta_k = 2\pi k/(n-1), \qquad k = 0, \ldots, n-2 \tag{3.57}$$

at a distance ρ from the origin of the coordinate system.

For a given number and a given location of the supports, i.e., for given n, θ_k, and ρ, the determination of the deflection w in terms of dimensionless variables $r' = r/R$, $\rho' = \rho/R$, and $w' = wEh^3/[12(1-\nu^2)qR^4]$ (primes will be omitted henceforth) reduces to minimization of the functional

$$J = \int_0^{2\pi}\int_0^{1} \left\{\left[\frac{1}{r}(rw_{,r})_{,r} + \frac{1}{r^2} w_{\theta\theta}\right] - 2w\right\} r\, dr\, d\theta \tag{3.58}$$

with boundary conditions

$$(w)_{r=1} = (w_{,r})_{r=1} = 0 \tag{3.59}$$

and constraints

$$(w)_{O_k} \leqslant 0, \qquad k = 0, 1, \ldots, n-1 \tag{3.60}$$

The unknown optimum location of the supports can be determined from the relation

$$\min_\rho \max_\Omega |w(r,\theta)| \qquad (0 \leqslant \rho \leqslant 1) \tag{3.61}$$

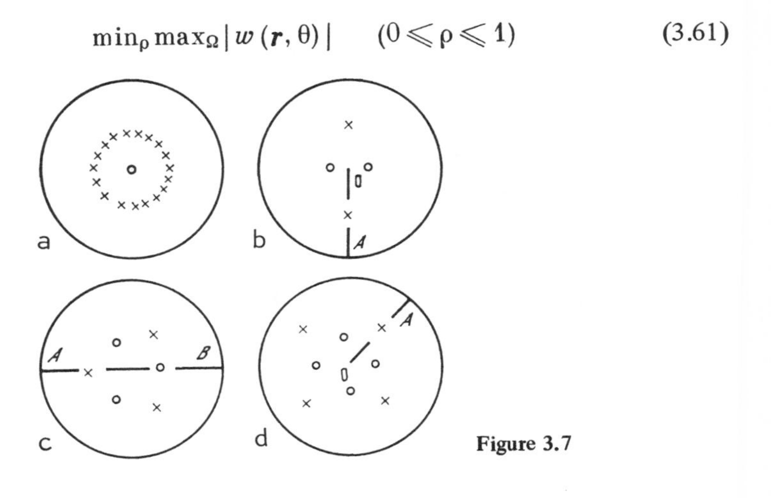

Figure 3.7

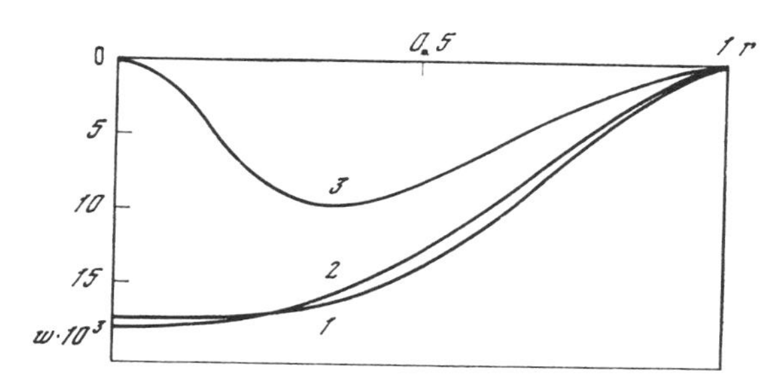

Figure 3.8

where the deflection $w(r, \theta)$ corresponds to the computed value of $0 \leqslant \rho \leqslant 1$, found from the variational relations of Eqs. (3.58)-(3.60).

To solve numerically the problem of Eqs. (3.56)-(3.61), we subdivide the region $\Omega = \{0 \leqslant r \leqslant 1, 0 \leqslant \theta \leqslant 2\pi\}$ by means of a net $r_i = i\Delta r$, $i = 0, 1, \ldots, i_0$, $\Delta r = 1/i_0$, $\theta_j = j\Delta\theta$, $j = 0, 1, \ldots, j_0 - 1$, and $\Delta\theta = 2\pi/j_0$. According to Eq. (5.1), to accommodate the location of the supports the integer j_0 is chosen to be a multiple of n and, if Eq. (3.57) is valid, a multiple of $(n - 1)$. The function $w(r, \theta)$ is replaced by a discrete function $w_{ij} = w(r_i, \theta_j)$, which is defined only at the lattice points. Making use of the symmetry of supports in Ω, we choose a finite difference approximation to the functional of Eq. (3.58), which assures the zero of the derivative w_r at the center of the plate.

The boundary conditions of Eq. (3.59) are approximated up to the second order of accuracy ($w_{i_0 j} = 0$, $w_{i_0,j} = (1/4)\, w_{i_0-2,j}$, $j = 0, 1, \ldots, j_0 - 1$). The constraints of Eq. (3.60) are assigned to the lattice points with indices (i_*, k), where k is determined by Eqs. (3.56) and (3.57) and the integer i_* (for a given value of k) assumes all values between 0 and $i_0 - 1$.

For each fixed value of i_*, $0 \leqslant i_* \leqslant i_0 - 1$, the variational problem of Eqs. (3.58)-(3.60) is solved numerically, using the technique of local variations with a locally optimum variation step, and the maximum value of the displacement $\max_{ij} |w_{ij}| = J$ is established.

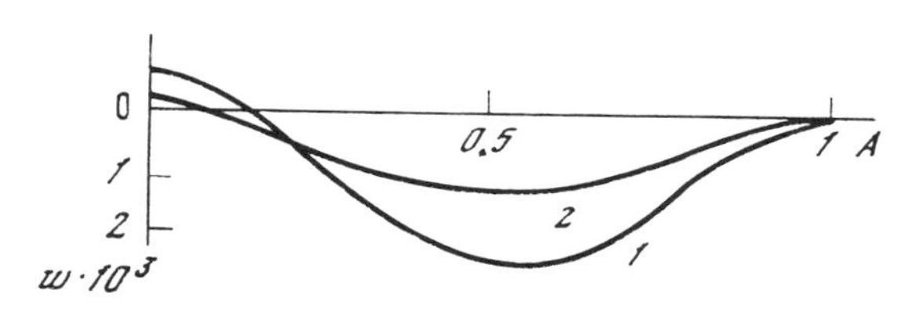

Figure 3.9

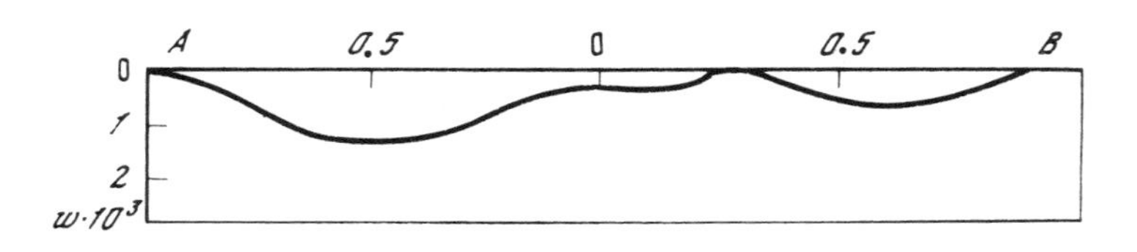

Figure 3.10

In this manner, for a given n we obtain a table of the value of maximum displacements, which depends on the index i_*. Choosing among the integers i_0 the one for which J has the smallest value, we also determine the index i_* for which the location of supports is optimum.

In our numerical computations we have taken $i_0 = 10$ and $j_0 = 24$. The minimum value of the variational step chosen in the technique of local variations was 10^{-9}. Circles marked in Fig. 3.7 show the optimum location of supports obtained by numerical techniques. Figures 3.7a–d correspond respectively to $n = 1$–4. The crosses indicate the location of maximum deflection points.

Figure 3.8 illustrates the behavior of the deflection function in the absence of supports, computed from the explicit solution given in Ref. 130 and by the technique of local variations (curves 1 and 2, respectively). The values of maximum deflection agree with accuracy of 0.5%. Curve 3 illustrates the deflection of a plate supported at its center.

Figure 3.9 shows the deflection of the plate in the plane cross section OA (of Figs. 3.7b and d) for $n = 2$ and 4 (curves 1 and 2, respectively), while Fig. 3.10 illustrates the deflection along the section AB (of Fig. 3.7c) for $n = 3$.

Aside from supporting a plate at some interior points, we can in practice improve the rigidity by reinforcing the plate with stiffening ribs. Hence, the problem arises of optimizing the location of the ribs. In Ref. 120 we can find some necessary conditions for optimality of the location of a circular rib attached to a circular, uniformly loaded plate of constant thickness.

4

Optimization Problems with Unknown Boundaries in the Theory of Elasticity

Control by Varying the Boundary of the Domain

This chapter is devoted to the study of optimization problems for systems of partial differential equations, specifically of finding the optimum shapes of elastic bodies. We consider the boundary of the domain in which our equations are defined as the control variable. In the first three sections, the quality criteria are integral functionals. In Section 4.1, we formulate the problem of optimizing rigidity of a shaft in torsion and derive some optimality criteria. Using these criteria in Section 4.2, we derive the shape of the cross section such that the bar has the greatest torsional rigidity. An optimization problem for a thin-walled bar is solved analytically, using a perturbation technique. For the case in which the walls are not thin, we employ a technique utilizing the theory of functions of complex variables. In Section 4.3 we study some problems of optimization of piecewise-homogeneous bars subjected to torsion and consider the problem of optimal reinforcement.

In Sections 4.4 to 4.6 we consider problems involving local functionals. In Section 4.4, we introduce a formulation and initiate the study of finding the shapes of holes in elastic plates that cause the minimum stress concentration. We prove that holes whose boundaries have uniform stress distribution are optimum for plates subjected to tension. In Section 4.5 we use techniques based on mapping the domain bounded by the contours of the holes into a canonical domain of the complex plane, to determine the shapes of holes with uniform distribution of stress on their boundaries.

Optimization problems for plates with holes subject to bending moments are considered in Section 4.6. The contents of this chapter are based on the results of Refs. 14, 18-21, 46, 78, 79, 144, and 162.

4.1. Maximizing the Torsional Rigidity of a Bar

We consider the problem of torsion of a cylindrical elastic shaft, posed in the Cartesian x-y-z coordinate system (see Fig. 4.1). We denote by Ω the region occupied by the bar cross section, i.e., intersected by the x-y plane, which for the sake of simplicity of our exposition will be assumed to be doubly connected.

We denote by Γ_0 and Γ, respectively, the exterior and interior boundary of the region Ω (see Fig. 4.2). The direction of the torque couple coincides with the z axis. We assume that the bar material is homogeneous and isotropic. We express the nonzero components of the stress tensor in terms of the stress function

$$\tau_{xz} = G\theta\varphi_y, \quad \tau_{yz} = -G\theta\varphi_x \tag{4.1}$$

where G is the shear modulus, θ is the angle of twist per unit length of the bar, $\phi(x, y)$ is the stress function satisfying $\phi_x = \partial\phi/\partial x$, and $\phi_y = \partial\phi/\partial y$.

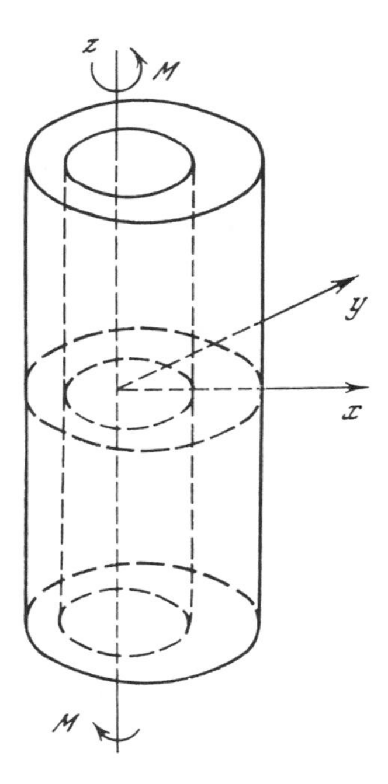

Figure 4.1

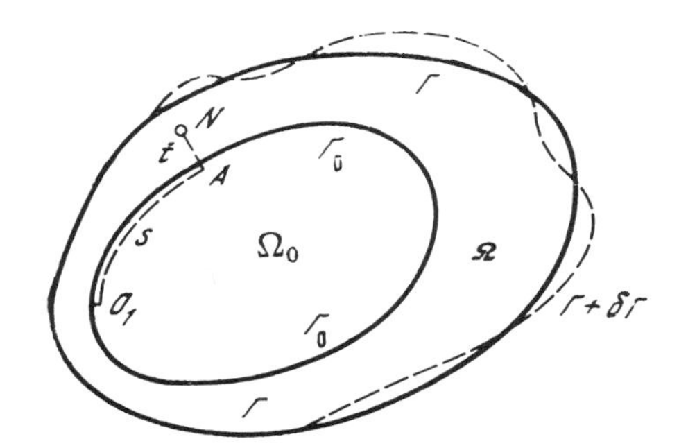

Figure 4.2

It is well known that for given boundaries Γ_0 and Γ, the torsion problem is reduced to finding the stress function $\phi(x, y)$ that satisfies the relations given below (see, for example, Ref. 6 or 94):

$$\varphi_{xx} + \varphi_{yy} = -2, \ (x, y) \in \Omega \tag{4.2}$$

$$(\varphi)_\Gamma = 0, \qquad (\varphi)_{\Gamma_0} = C \tag{4.3}$$

$$\int_{\Gamma_0} \frac{\partial \varphi}{\partial n} \, ds = 2S_0 \tag{4.4}$$

where S_0 is the area of the region bounded by the curve Γ_0. The constant C occurring in Eq. (4.3) is an unknown quantity. To determine its value, we can use the condition of Eq. (4.4). The torsional rigidity K is given by

$$K = 2\left(\iint_\Omega \varphi \, dx \, dy + CS_0 \right) \tag{4.5}$$

The curve Γ_0 and therefore the area S_0 of the region Ω_0 bounded by Γ_0 is assumed to be known. We also assume that the length of the bar and its volume are given. This assumption implies the isoperimetric condition

$$\text{mes } \Omega = S \tag{4.6}$$

where S is a given constant. The shape of the boundary Γ is not fixed and is to be determined.

The optimization problem consists of finding the shape of the curve Γ that maximizes the functional of Eq. (4.5), subject to the condition of Eqs. (4.2) to (4.4) and (4.6).

To derive optimality conditions, it is necessary to restate the optimization problem by introducing a torsion function ψ that is related to ϕ by the equations $\psi_x = \phi_y + y$, $\psi_y = -\phi_x - x$. The torsion function ψ may be found by solving the boundary-value problem of Neumann type

$$\begin{aligned} \psi_{xx} + \psi_{yy} &= 0, \qquad (x, y) \in \Omega \\ \frac{\partial \psi}{\partial n} &= y n_x + x n_y \qquad (x, y) \in \Gamma_0 + \Gamma \end{aligned} \tag{4.7}$$

where n_x and n_y are projections on the x and y axes, respectively, of the unit vector normal to the boundary of the region Ω. The bar rigidity K is given in terms of ψ by

$$K = I_0 + \iint_\Omega (x\psi_y - y\psi_x)\, dx\, dy$$

$$I_0 = \iint_\Omega (x^2 + y^2)\, dx\, dy$$

For the given boundaries of the region Ω, the function ψ may be found by solving the following variational problem[94]:

$$\Pi = \iint_\Omega [(\psi_x - y)^2 + (\psi_y + x)^2]\, dx\, dy \to \min_\psi \tag{4.8}$$

We note that it is not necessary for the function ψ to satisfy the boundary condition of Eq. (4.7) because this condition is a "natural" condition for the functional of Eq. (4.8). For a function ψ that minimizes the integral of Eq. (4.8) (we cannot choose just any arbitrary function that satisfies the equality below), we have $K = -\pi$ (see, for example, Ref. 94). This relation permits us to state our variational problem in the following manner:

$$K_* = -\max_\Gamma \min_\psi \Pi \tag{4.9}$$

The maximum with respect to Γ in Eq. (4.9) must be computed while considering the isoperimetric condition of Eq. (4.6). We take care of this constraint by constructing a Lagrangian functional $J = K - \lambda^2 \int_\Omega \int dx\, dy$, where λ^2 is a Lagrange multiplier. Making use of the formulas given in Section 1.8 of Chapter 1, we obtain necessary conditions for an extremum of the functional J. As is well-known, these conditions are also necessary conditions for the existence of an extremum of the functional K, with the constraint of Eq. (4.6).

The equations and boundary conditions of Eq. (4.7) are necessary conditions for the extremum of the functional J, regarded as a function of ψ. The necessary conditions for optimality of J with respect to Γ are written in the form

$$(\psi_x - y)^2 + (\psi_y + x)^2 = \lambda^2, \qquad (x, y) \in \Gamma \tag{4.10}$$

We restate Eq. (4.10) in terms of the stress function ϕ by substituting $\psi_x = \phi_y + y$ and $\psi_y = -\phi_x - x$ (see Ref. 14 or 78)

$$(\nabla\varphi)^2_\Gamma = \lambda^2 \tag{4.11}$$

where $(\nabla\phi)^2 = \phi_x^2 + \phi_y^2$. This condition may be rewritten in a slightly different form (which will be used in our discussion below) if we replace the derivatives of ϕ with respect to x and y by derivatives in the directions s and n, which are, respectively, tangential and normal to the curve Γ.

According to Eq. (4.3), the derivative tangential to the boundary is identically equal to zero. Therefore, along the optimum boundary, the square of the normal derivative is equal to a constant, i.e.,

$$(\partial\varphi/\partial n)^2_\Gamma = \lambda^2 \tag{4.12}$$

Let us investigate some properties of the optimum solutions. First, we observe that on the optimum boundary the tangential stress τ satisfies a simple relationship $\tau^2 = \tau_{xz}^2 + \tau_{yz}^2 = G^2\theta^2(\nabla\phi)^2 = \lambda^2 G^2\theta^2$.

We can now derive a relation between the constant λ, the length l of the optimum boundary, and the area of the region bounded by that curve. To accomplish this, we apply Bredt's theorem to the curve Γ and use the optimality condition of Eq. (4.12). We have

$$2(S_0 + S) = -\int_\Gamma \frac{\partial\varphi}{\partial n}\, ds = \lambda l$$

This relation implies, that $\lambda \neq 0$ and $\lambda \neq \infty$.

We can show that the optimum curve is smooth and does not have either external (pointing outward) or interior (pointing inward) sharp corners. Assuming the contrary, let us first assume that the boundary curve Γ has an exterior corner. As we approach this corner point we must have $\tau = \partial\phi/\partial n \to 0$ (see Ref. 94 for a rigorous proof). Consequently, the condition $\lambda \neq 0$ is violated in this case. Let us now assume that the boundary curve Γ has an interior corner point. As this point is approached, along Γ, the quantity $\tau = \partial\phi/\partial n$ approaches infinity. Consequently, the condition $\lambda \neq \infty$ is violated. In this case, when Ω is simply connected (no holes exist in Ω), the extremal boundary curve Γ enclosing Ω and satisfying Eq. (4.11) turns out to be a circle. A rigorous proof of this optimal property of a circular cross section, based on symmetrization arguments, is presented in Ref. 110.

4.2. Finding Optimum Shapes of Cross-Sectional Areas for Bars in Torsion

1. *Application of perturbation techniques.* We derive an analytical solution to the problem considered in Section 4.1 for the specific case of a thin-walled bar. For the sake of convenience, we introduce a new st coordinate system, determined by the properties of the fixed curve Γ_0. For a point $P \in \Omega$, the coordinate s measures the distance along Γ_0 from some fixed point $0_1 \in \Gamma_0$ to the point 0_2 of the intersection of Γ_0 with the normal to Γ_0 drawn from the point P. The coordinate t is equal to the length of the interval $0_2 P$. Let $R = R(s)$ and $h = h(s)$ denote, respectively, the radius of curvature of the curve Γ_0 and the equation of the curve Γ, while l_0 denotes length of Γ_0. Equations (4.2) to (4.6) and (4.12) assume the following form, in terms of coordinates s and t:

$$
\begin{gathered}
(T\varphi_t)_t + (T^{-1}\varphi_s)_s = -2T, \qquad T = 1 + t/R \\
\varphi(s, 0) = C, \qquad \varphi(s, h) = 0, \qquad \varphi_\eta(s, h) = -\lambda \\
\int_0^{l_0} \varphi_t(s, 0)\, ds = -2S_0, \qquad \int_0^{l_0}\int_0^{h} T\, dt\, ds = S \\
K = 2\left(\int_0^{l_0}\int_0^{h} T\varphi\, dt\, ds + CS_0\right)
\end{gathered}
\tag{4.13}
$$

The assumption concerning thinness of the bar walls implies that

$$
\max_s h(s) = H \ll l_0 \qquad (0 \leqslant s \leqslant l_0) \tag{4.14}
$$

that is, $H/l_0 = \epsilon$ is a small number ($\epsilon \ll 1$).

Let us assume that the curve Γ_0 does not have crooked segments, i.e.,

$$\min_s R(s) \sim l_0 \qquad (4.15)$$

We introduce new variables denoted by primes (the primes will subsequently be omitted)

$$\begin{aligned} &s = l_0 s', \quad t = Ht', \quad h = Hh', \quad \varphi = Hl_0\varphi', \quad S_0 = l_0^2 S_0' \\ &S = Hl_0 S', \quad R = l_0 R', \quad K = Hl_0^3 K', \quad C = Hl_0 C', \quad \lambda = l_0\lambda \end{aligned} \qquad (4.16)$$

We substitute these variables into Eq. (4.13) to obtain

$$\begin{aligned} &(T\varphi_t)_t + \varepsilon^2 (T^{-1}\varphi_s)_s = -2\varepsilon T, \quad T = 1 + \varepsilon t/R \\ &[\nabla\varphi(s, h)]^2 = \lambda^2, \quad \varphi(s, 0) = C, \quad \varphi(s, h) = 0 \\ &\int_0^1 \varphi_t(s, 0)\, ds = -2S_0, \quad \int_0^1 \left(h + \frac{\varepsilon h^2}{2R}\right) ds = S \\ &K = 2\left(CS_0 + \varepsilon \int_0^1\int_0^h T\varphi\, dt\, ds\right) \end{aligned} \qquad (4.17)$$

To solve this problem using a perturbation technique, we seek a representation of the variables ϕ and h, and of the unknown constants C, λ, and K in infinite power series in the parameter ϵ,

$$\begin{aligned} \varphi &= \varphi^0 + \varepsilon\varphi^1 + \varepsilon^2\varphi^2 + \ldots \\ h &= h^0 + \varepsilon h^1 + \varepsilon^2 h^2 + \ldots \\ K &= K^0 + \varepsilon K^1 + \varepsilon^2 K^2 + \ldots \end{aligned} \qquad (4.18)$$

Similar expansion is carried out for the constants C and λ.

Let us write the equations determining the zeroth, first, and second order approximations. To do this, we substitute the series of Eq. (4.18) into Eq. (4.17) and equate terms that contain identical powers of ϵ. As a result, we derive a system of boundary-value problems, whose solutions give us all the unknown quantities.

To determine these quantities in the zeroth approximation, we have the following boundary-value problem:

$$\varphi^0_{tt}=0, \qquad \varphi^0(s,0)=C^0, \qquad \varphi^0(s,h^0)=0$$
$$\varphi^0_t(s,h^0)=-\lambda^0 \tag{4.19}$$
$$\int_0^1 \varphi^0_t(s,0)\,ds=-2S_0, \qquad \int_0^1 h^0\,ds=S$$

Taking into consideration the values of unknown quantities in the zeroth approximation, we formulate the boundary-value problem for the first order approximation

$$\varphi^1_{tt}=-2-\varphi^0_t/R, \qquad \varphi^1(s,0)=C^1, \qquad \varphi^1(s,h^0)=\lambda^0 h^1$$
$$\varphi^1_t(s,h^0)=-\lambda^1 \tag{4.20}$$
$$\int_0^1 \varphi^1_t(s,0)\,ds=0, \qquad \int_0^1 h^1\,ds=-\frac{1}{2}\int_0^1 \frac{(h^0)^2}{R}\,ds$$

Taking into account Eqs. (4.19) and (4.20), we formulate the second order approximation boundary-value problem

$$\varphi^2_{tt}=-\frac{1}{R}\left[(t\varphi^1_t)_t+2t\right]-\varphi^0_{ss}, \qquad \varphi^2(s,0)=C^2$$
$$\varphi^2(s,h^0)=\lambda^1h^1+\lambda^0h^2, \qquad \varphi^2_t(s,h^0)=(2-\lambda^0/R)\,h^1-\lambda^2 \tag{4.21}$$
$$\int_0^1 \varphi^2_t(s,0)\,ds=0, \qquad \int_0^1 h^2\,ds=-\int_0^1 \frac{h^0h^1}{R}\,ds$$

If we determine the solutions of these boundary-value problems, the torsional rigidity of the bar may be computed, up to a suitable power to ϵ^3, using the following formula:

$$K=K^0+\varepsilon K^1+\varepsilon^2K^2+\ldots=2C^0S_0+2\varepsilon\left(\int_0^1\int_0^{h^0}\varphi^0\,dt\,ds+C^1S_0\right)$$
$$+2\varepsilon^2\left(\int_0^1\int_0^{h^0}\varphi^1\,dt\,ds+C^2S_0\right)+O(\varepsilon^3) \tag{4.22}$$

Let us now find the solution of our optimization problem in the zeroth approximation, i.e., let us solve the problem stated in Eq. (4.19). The function ϕ^0 may be found from the basic equation and the boundary conditions of Eq. (4.19) [i.e., those stated in the first line of Eq. (4.19)]. Using the formula for ϕ^0, which has been now established, and the optimality condition given in the second line of Eq. (4.19), we determine h^0. We have $\phi^0 = C^0 [(1 - t)/h^0]$, with $h^0 = C^0/\lambda^0$. Substituting ϕ^0 and h^0 into the isoperimetric condition of Eq. (4.19) we find the values of the constants λ^0 and C^0. We thus arrive at the following expressions determining the unknown quantities in the zeroth order approximation:

$$\begin{aligned} &h^0 = S, \quad \varphi^0 = 2SS_0\,(1 - t/S), \quad K^0 = 4SS_0^2 \\ &\lambda^0 = 2S_0, \quad C^0 = 2SS_0 \end{aligned} \tag{4.23}$$

It is clear from Eq. (4.23) that if the curvature of the curve Γ_0 is small [see the assumption of Eq. (4.15)], then the optimal distribution of thickness, within the zeroth approximation, is constant.

Let us now find the unknown quantities in the first order approximation. To accomplish this, we integrate Eq. (4.20) for ϕ^1 and determine the integration constants from the boundary conditions. The constants λ^1, C^1, and the function h^1 are found by using the optimality conditions and the isoperimetric condition of Eq. (4.20). As a result, we find the following expressions for the unknown quantities in the first order approximation:

$$\begin{aligned} &\varphi^1 = t^2\left(\frac{S_0}{R} - 1\right) + 2SS_0\left(\int_0^1 \frac{ds}{R} - \frac{1}{R}\right) + S^2 - 2S_0S^2\int_0^1 \frac{ds}{R} \\ &h^1 = -\frac{S^2}{2R}, \quad K = 4S_0S^2\left(1 - S_0\int_0^1 \frac{ds}{R}\right) \\ &C^1 = S^2\left(1 - 2S_0\int_0^1 \frac{ds}{R}\right), \quad \lambda^1 = 2S\left(1 - S_0\int_0^1 \frac{ds}{R}\right) \end{aligned} \tag{4.24}$$

In the expression for h^1 of Eq. (4.24), the curvature of the inner curve Γ_0 influences the optimum shape of the exterior boundary curve Γ. The formula for the optimal distribution of the bar thickness is

$$h = h^0 + \varepsilon h^1 = S\,(1 - \varepsilon S/2R) \tag{4.25}$$

In terms of a dimensional variable, $h = Sl_0^{-1}[1 - S/(2Rl_0)]$. This formula indicates that as the curvature of Γ_0 increases, the quantity h decreases at corresponding points of the curve Γ_0.

In an identical manner, we determine all unknown quantities in the second order of approximation by solving the boundary-value problem of Eq. (4.21). We give the resulting formulas for the function h^2 and the constant C^2;

$$h^2 = -\frac{S^3}{2R^2}\left(3 + \frac{R}{S_0} - 4R^2\int_0^1 \frac{ds}{R^2} - \frac{R^2}{S_0}\int_0^1 \frac{ds}{R}\right)$$

$$C^2 = \frac{2S^3}{3}\left(S_0\int_0^1 \frac{ds}{R^2} - 2\int_0^1 \frac{ds}{R}\right) + 2S^3S_0\left(\int_0^1 \frac{ds}{R}\right)^2 \tag{4.26}$$

Using the formula for C^2 and corresponding quantities given by the zeroth and first order approximations, which occur in Eq. (4.22), we find the correction K^2. Using the formulas we have found and substituting new variables given by Eq. (4.16), we derive the following formula for rigidity of the optimum bar (using the dimensional variables):

$$K = \frac{4SS_0^2}{l_0^2} + \frac{4S^2S_0}{l_0^2}\left(1 - \frac{S_0}{l_0^2}\int_0^{l_0} \frac{ds}{R}\right) + \frac{4S^3}{3l_0^6}\left[3S_0^2\left(\int_0^{l_0} \frac{ds}{R}\right)^2\right.$$
$$\left. + S_0^2 l_0 \int_0^{l_0} \frac{ds}{FR^2} - 4S_0 l_0^2 \int_0^{l_0} \frac{ds}{R} + l_0^4\right] \tag{4.27}$$

Let us estimate the gain achieved by optimizing the shape. We first find the solution to the torsion problem for a bar having a constant thickness h along its inner contour Γ_0. Without getting into details, which are basically the same as given above, we compute an expression for the difference $\Delta K = K - K_C$ between the rigidity of the optimum bar and the rigidity of the bar of constant thickness,

$$\Delta K = \frac{S^3S_0^2}{l_0^6}\left[l_0\int_0^{l_0} \frac{ds}{R^2} - \left(\int_0^{l_0} \frac{ds}{R}\right)^2\right]$$

We apply the Cauchy-Schwartz-Bunyakovski inequality to the expression enclosed in the square brackets and conclude that the quantity inside the square brackets is always positive. Therefore, for any shape of the interior boundary curve Γ_0 for which $\min_s R(s) \sim l_0$, $\Delta K \geqslant 0$. The equality sign occurs only for the circular boundary curve Γ_0, as can be easily checked. In that case, R is the radius of the circle and the thickness is constant.

The solution given in Eqs. (4.23) to (4.27) was computed using the assumption of Eq. (4.15). Let us now investigate a different possibility, namely $\min_s R(s) \sim H$, i.e., in the presence of sectors of the boundary contour having large curvature. The optimization problem is posed again in the coordinates of Eq. (4.16), with the difference that now $R = HR'$. Basic relations of the problem are still obtained from Eq. (4.17) by replacing all expressions containing ϵ/R by $1/R$. We use a perturbation technique to obtain a solution of the form of Eq. (4.18). We shall limit our analysis to finding of the zeroth order approximation, which satisfies the following relations:

$$[(1 + t/R)\varphi_t^0]_t = 0, \quad \varphi^0(s, 0) = C^0, \quad \varphi^0(s, h^0) = 0$$

$$\varphi_t^0(s, h^0) = -\lambda^0, \quad \int_0^1 \varphi_t^0(s, 0)\, ds = -2S_0, \quad \int_0^1 \left(h^0 + \frac{(h^0)^2}{2R}\right) ds = S$$

Solving these relations, we obtain the following expressions for the unknown quantities:

$$\begin{aligned} \varphi^0(s, t) &= 2S_0 \left[\int_0^1 \frac{ds}{R \ln(1 + h^0/R)} \right]^{-1} \left(1 - \frac{\ln(1 + t/R)}{\ln(1 + h^0/R)}\right) \\ R(1 + h^0/R) \ln(1 + h^0/R) &= \text{const} \end{aligned} \tag{4.28}$$

It is clear from the second line of Eq. (4.28) that the thickness of the optimum bar is not constant, even in the zeroth order approximation. If the curvature $1/R$ increases as we vary s $(0 \leqslant s \leqslant 1)$, then according to Eq. (4.28), the function $h^0(s)$ must decrease. It is also clear from Eq. (4.28) that on the sectors with small curvature the bar thickness, computed with a sufficient degree of accuracy, will be almost constant. This is in complete agreement with the foregoing results.

In a similar manner, we can investigate the dependence of the variation of thickness on the curvature of Γ_0, in sectors for which $R \sim \epsilon^p H (p > 1)$. Without going into the details, which are identical with the arguments given above, we shall state the final result. This unknown dependence has the following asymptotic representation: $R = h^0 \exp(-\gamma/h^0)$, where γ is an arbitrary constant.

2. *Application of mappings into a canonical domain.* If the torsional bar is not thin-walled, the optimum shape of the cross section may be computed by use of techniques of the theory of functions of complex variables. The unknown region, which represents the bar cross section, is mapped into a designated canonical domain. Effectiveness of this technique in optimization problems was demonstrated in Refs. 78 and 79.

First, following the ideas of Ref. 78, let us consider a simply-connected region Ω, bounded by a curve Γ. To find the optimum shape of the contour Γ, we introduce a function $f(z)$ that is analytic in the region Ω (i.e., a complex torsion function as given in Ref. 101), such that

$$\operatorname{Re} f(z) = \varphi(x, y) + 0{,}5\,(x^2 + y^2) \tag{4.29}$$

where $z = x + iy$. Boundary conditions for the function f, which take into account the optimality conditions $\phi = -\lambda$, are given in the form

$$f'(t) = \bar{t} - i\lambda\, \overline{dt}/|dt|$$

If $z = \omega(\zeta)$ represents a conformal map of the unit disc $|\zeta| < 1$ into the unknown domain Ω, then the expression $\overline{\omega(\tau)} - \overline{\lambda\tau\, \omega'(\tau)}$ (where $\tau = \exp i\theta$, $0 < \theta < 2\pi$) can represent the boundary value of the function, which is regular in the interior of the unit disc. It is easy to see that the function $\omega(\zeta) = -C\zeta$ satisfies this condition. Consequently, a shaft having a circular cross section satisfies the optimality condition. As we have already commented, the assertion concerning the optimality of a circular cross section was formulated by Saint Venant,[126] and was proved by Polya and Szegö.[110]

Let us return to the problem of determining the exterior boundary curve Γ of a doubly-connected region Ω with a given interior boundary curve Γ_0, using the assumption that in the region Ω we can define a function satisfying Eq. (4.2) and the conditions of Eqs. (4.3), (4.4), and (4.12), i.e., we can introduce a complex torsion function. This problem is reduced to computation of the exterior boundary curve for a doubly-connected region in which there exists a univalent analytic function $f(z)$ that satisfies, on the given curve Γ_0, the condition

$$f(t) + \overline{f(t)} = 2C + t\bar{t} \tag{4.30}$$

while on the unknown curve Γ it satisfies the condition

$$f(t) = \frac{1}{2}\, t\bar{t} + \frac{1}{2i} \int_{t_0}^{t} (\bar{t}\, dt - t\, d\bar{t}) - i\lambda \int_{t_0}^{t} |dt| \tag{4.31}$$

To obtain a solution, we use the mapping

$$z = \omega_1(\zeta) = B\zeta \sum_{k=0}^{n} b_k \zeta^{-k}$$
$$z = \omega_2(\xi) = A\xi \sum_{k=0}^{n} a_k \xi^{-k}, \qquad a_0 = b_0 = 1 \tag{4.32}$$

which maps the exterior of the circle $|\xi| = 1$ (i.e., $|\xi| > 1$) into the exterior of the unknown curve Γ. Coefficients a_k in Eq. (4.32) are assumed to be known, while the quantities b_k are to be determined. The real variables A and B determine the scale (the quantity $\nu = A/B$ describes the relative size of the cross section).

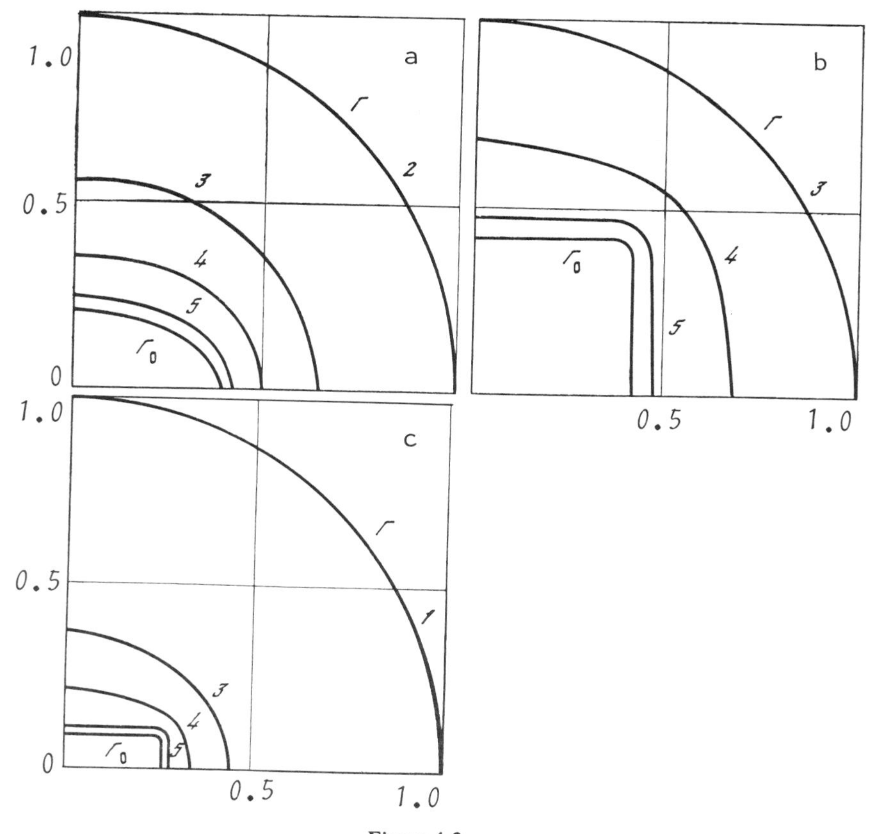

Figure 4.3

Using the techniques of the theory of functions (as given in Ref. 154) to find the values of the constants b_k, we obtain a system of n nonlinear equations. In Ref. 79 this system is solved by Newton's method and the coefficients b_j are determined for a given set of values of the a_j and ν. The solution was computed in the interval $0 < \nu < 1$. In these computations the inner curve Γ_0 was assumed to be an ellipse ($a_1 = 0.1, 0.2, 0.3, \ldots, 0.9$, while $a_j = 0$ if $j = 2, \ldots, n$) or a rectangle with various values of the ratio δ of the sides ($\delta = 1, 2, \ldots, 10$).

Sketches of the shape of the exterior boundary curve for the cross section of the bar are shown in Fig. 4.3 for the specific cases in which the inner curve Γ_0 is an ellipse and $a_1 = 0.3$ (see Fig. 4.3a), a square (Fig. 4.3b), and a rectangle (Fig. 4.3c). Curves 1 to 5 represent the shapes of Γ for values of ν equal, respectively, to 0.2, 0.3, 0.5, 0.7, and 0.9.

4.3. Torsion of Piecewise Homogeneous Bars and Problems of Optimal Reinforcement

1. *Piecewise homogeneous bar.* We follow the ideas of Ref. 18 in considering an optimization problem for a piecewise homogeneous elastic bar of constant cross section. The region Ω occupied by the cross section of the bar in the xy plane has the boundary Γ (see Fig. 4.4.). We assume that the bar is made from two materials having shear moduli G_0 and G_1. These materials occupy, respectively, regions Ω_0 and Ω_1 ($\Omega_0 \cup \Omega_1 = \Omega$) of the cross section. The inner region Ω_0 is convex and is separated from the outer region Ω_1 by a smooth curve Γ_1. The curves Γ and Γ_1 do not have any points in common. We assume that $G(x,y) = G_0$ if $(x,y) \in \Omega_0$ and $G(x,y) = G_1$ if $(x,y) \in \Omega_1$. To find a stress function $\phi(x,y)$ defined in all of Ω, which is related to the components of the stress tensor in the usual manner $\tau_{xy} = \theta\phi_y$, $\tau_{yz} = -\theta\phi_x$, we need to solve the

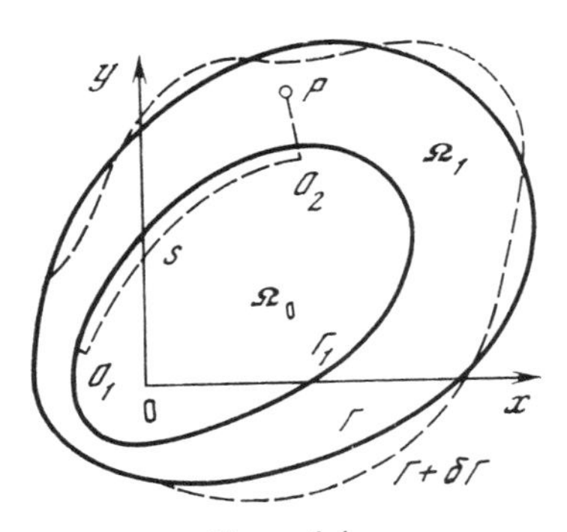

Figure 4.4

following boundary value problem (6):

$$\begin{aligned} &\Delta\varphi = -2G, \quad (\varphi)_\Gamma = 0 \\ &(\varphi)^+_{\Gamma_1} = (\varphi)^-_{\Gamma_1}, \quad \left(\frac{1}{G}\frac{\partial\varphi}{\partial n}\right)^+_{\Gamma_1} - \left(\frac{1}{G}\frac{\partial\varphi}{\partial n}\right)_{\Gamma_1} = 0 \end{aligned} \tag{4.33}$$

where Δ is Laplace's operator, $\partial\varphi/\partial n$ denotes the derivative in the direction normal to the boundary, and θ is the angle of twist per unit length. The upper indices (+) and (−) in Eq. (4.33) indicate the limits of the corresponding quantities on Γ_1 as we approach Γ_1, respectively, from the interior or the exterior region bounded by Γ_1.

The relations given above between the stress function and the components of the stress tensor differ from the corresponding relations given in Eq. (4.1), because of the presence of the factor G in the right-hand terms of Eq. (4.1).

We regard the curve Γ_1 as fixed, and denote the area of the region Ω_0 bounded by this curve by S_0. We also assume that the area of the region Ω_1 is given and is equal to S_1, i.e., meas $\Omega_1 = S_1$. We study the optimization problem of finding the curve Γ that satisfies the isoperimetric condition meas $\Omega_1 = S_1$, such that the torsional rigidity K (the payoff functional) is maximized, where

$$K = 2\iint_\Omega \varphi(x, y)\, dx\, dy \to \max_\Gamma \tag{4.34}$$

The fundamental difference between this problem and the problems that were considered in Sections 4.1 and 4.2 is that the function ϕ is given by an equation in the domain Ω that has a discontinuous right-hand side. However, it is easy to show that the necessary condition for optimality of the curve Γ is unchanged and is still of the form of Eq. (4.12).

If we regard the curve Γ as fixed and the location of the discontinuity curve Γ_1 as unknown, and we maximize the functional of Eq. (4.34) by choosing an optimal curve Γ_1, we arrive at a problem of finding an optimum distribution of material in a given cross section Ω of the bar.

In case $G_0 = 0$, we use Eqs. (4.33) and (4.34) and some elementary transformations to arrive at a corresponding problem for a thin-walled cylindrical bar whose inner boundary curve is fixed and whose outer boundary curve is unknown. The solution of this problem was already discussed in Section 4.2.

2. *Optimal reinforcement of a bar in torsion.* We use Eq. (4.12) to study the possibility of covering the torsional bar in an optimal manner with a thin

reinforcing layer, made of a material having greater rigidity (i.e., greater shear modulus). The region Ω_1 shown in the Fig. 4.4 represents the reinforcing layer. In addition to the original Cartesian system of coordinates, we introduce a system of curvilinear coordinates (s, t) that are related to the shape of the curve Γ_1. The coordinate s of a point $P \in \Omega$ is the distance measured along Γ_1, starting from some fixed point $0_1 \in \Gamma_1$ to a point 0_2 (which is an intersection point of Γ_1 with the line normal to Γ_1, drawn from the point P). Let us denote by $h = h(s)$ the function describing the position of the unknown boundary curve Γ, in this curvilinear coordinate system. Let l denote the total length of the curve Γ. To solve this problem, we make the assumptions $l^{-1} \max_s h(s) \sim \epsilon$, $G_0/C_1 \sim \epsilon$, and $\min_s R(s) < l$. Here ϵ denotes a small parameter and $R(s)$ is the radius of curvature of the curve Γ_1 at a point having a coordinate s. We also make assumptions that are commonly made in problems concerning reinforcement,[6] and find only the first few terms of the series in powers of the parameter ϵ. We denote the leading terms of the series by the same symbols that designate the quantities expanded into these series. In the approximation considered here, the function ϕ defined in the region Ω_1 satisfies Eq. (4.33), the boundary condition $\phi = 0$ on Γ, and the assumption that ϕ is continuous on Γ_1 and ϕ is determined by the formula $\phi(s, t) = \phi(s, 0)\ [(1 - t)/h(s)]$ (see Ref. 6). Here $\phi(s, 0)$ denotes the value of the function ϕ at points lying on the curve Γ_1. Using this expression for ϕ in the domain Ω_1, computation of the reinforced bar is reduced to the finding of a function ϕ in the region Ω_0 as the solution of the boundary-value problem

$$\triangle \varphi = -2G_0, \quad (x, y) \in \Omega_0, \quad \varphi_t = -G_0\varphi/G_1 h, \ (x, y) \in \Gamma_1 \quad (4.35)$$

The symbol ϕ_t in Eq. (4.35) stands for the quantity $(\phi_t)^+$. The boundary condition in Eq. (4.35) may be obtained by using for ϕ in the domain Ω_1 the above expression and substituting it into the relation $(G^{-1}\ \partial\phi/\partial n)^+ = (G^{-1}\ \partial\phi/\partial n)^-$. This relation is supposed to hold on Γ_1. If the function $h = h(s)$ is given, the boundary-value problem of Eq. (4.35) is solvable.

Let us now consider the optimization problem. We transfer the optimization condition from the curve Γ to Γ_1. With accuracy up to terms of higher order of smallness, we obtain $(\phi_t)_{\Gamma_1} = -\lambda$. It follows from the formula for the function ϕ in the region Ω that

$$h(s) = \frac{1}{\lambda} (\varphi)_{\Gamma_1} \quad (4.36)$$

We conclude that for the optimally reinforced bar the thickness distribution function $h(s)$ and the function ϕ evaluated on the curve Γ are directly proportional to each other. Equation (4.36) allows us to reduce the third (mixed) boundary-value problem for the function $\phi(x,y)$ (in the region Ω_0) to the Neumann problem

$$\triangle \varphi = -2G_0, \quad (x,\, y) \in \Omega_0, \quad \varphi_t = -\lambda G_0/G_1, \quad (x,\, y) \in \Gamma_1 \quad (4.37)$$

Let us denote the position vector for the point whose coordinates are (x,y) by $r(r^2 = x^2 + y^2)$ and introduce an auxiliary function $\psi = \phi/G_0 + r^2/2$. We substitute ψ in Eq. (4.37) for ϕ and utilize the equation $\partial r^2/\partial t = (\nabla r^2, n) = 2(r, n)$. We obtain for the function ψ the following boundary-value problem (which is Neumann's problem for the Laplace equation):

$$\begin{aligned} &\triangle \psi = 0, \quad (x,\, y) \in \Omega_0 \\ &\psi_t = -\frac{\lambda}{G_1} + (r,\, n), \quad (x,\, y) \in \Gamma_1 \end{aligned} \tag{4.38}$$

A necessary condition for the existence of a solution to Eq. (4.38) is $\int_{\Gamma_1} \psi_t \, ds = 0$. This condition may be satisfied by a suitable choice of the constants

$$\lambda = \frac{G_1}{l} \int\limits_{\Gamma_1} (r,\, n)\, ds = \frac{2G_1 S_0}{l} \tag{4.39}$$

The constant λ was chosen to satisfy the necessary condition of solvability for the boundary-value problem, instead of the commonly used isoperimetric condition. To satisfy the isoperimetric condition, we make use of one more arbitrary constant that may be chosen in Eq. (4.38). This arbitrary condition occurs in the transition from the problem of Eq. (4.35) to the problem of Eqs. (4.37) and (4.38), and is caused by the fact that using the condition of Eq. (4.36) changes the type of the boundary-value problem. As a result, instead of the third boundary-value problem of Eq. (4.35), we have the Neumann problem. The solution of this problem can be determined only up to an arbitrary constant term.

Let γ denote an arbitrary constant, and let us write the function ψ as a sum

$$\psi = \Psi + \gamma, \quad \Psi\,(x_0,\, y_0) = 1, \quad (x_0,\, y_0) \in \Omega$$

In addition to this condition, the function Ψ satisfies the equation and the boundary condition given in Eq. (4.38) (with $\lambda = 2GS_0/l$). In this approximation the isoperimetric condition can be written in the form

$$\int_{\Gamma_1} h\,ds = S_1 \tag{4.40}$$

The following constraint applied to the function ϕ can be deduced from Eq. (4.40) and Eqs. (4.36) and (4.39):

$$\int_{\Gamma_1} \varphi\,ds = \frac{2G_1S_0S_1}{l}$$

Substituting the expression $\phi = G[(\Psi + \gamma - r^2)/2]$ for ϕ into the formula for computing the constant γ gives

$$\gamma = \frac{1}{l}\left[\frac{2G_1S_0S_1}{lG_0} - \int_{\Gamma_1}\left(\Psi - \frac{r^2}{2}\right)ds\right]$$

Using the values found for the constants γ and λ, we finally arrive at the following expressions for the unknown quantities:

$$\begin{aligned}
\varphi &= G_0\left[\Psi - \frac{r^2}{2} - \frac{1}{l}\int_{\Gamma_1}\left(\Psi - \frac{r^2}{2}\right)ds\right] + \frac{2S_0S_1G_1}{l^2} \\
h &= \frac{l}{2G_1S_0}(\varphi)_{\Gamma_1} \\
K &= 2\iint_{\Omega_0}\varphi\,dx\,dy = 2G_0\iint_{\Omega_0}\left(\Psi - \frac{r^2}{2}\right)dx\,dy \\
&\quad - \frac{2G_0S_0}{l}\int_{\Gamma_1}\left(\Psi - \frac{r^2}{2}\right)ds + \frac{4G_1S_0^2S_1}{l^2}
\end{aligned} \tag{4.41}$$

Thus, determining an optimum design is reduced to finding a function satisfying the boundary-value problem of Eqs. (4.38) and (4.39) and an auxiliary condition $\Psi(x_0, y_0) = 1$. The solution of this problem is independent of the physical constants G_0 and G_1 and the parameter S_1. It is completely determined by the geometry of the region Ω_0.

Substituting $G_0 = 0$ into Eq. (4.41), we obtain the optimum design for a thin-walled cylinder subjected to torsion: $h = S_1/l$, $K = 4G_1S_0^2S_1/l^2$. These formulas were derived in Section 4.2. Constant thickness turns out to be optimal. However, as was shown in Section 4.2, in the approximation of the next order, the optimum thickness distribution function is not constant and does depend on the curvature of the boundary $\{h = S_1 l^{-1} [(1 - S_1)/2lR]\}$.

Let the region Ω be a circular disc of radius R_0. In this case $(r, n) = R_0$, $\lambda = G_1 R_0$, and the boundary condition of Eq. (4.38) assumes the form $\partial\psi/\partial t = 0$, $(x, y) \in \Gamma_1$. The function ψ does not depend on x and y, and therefore $\Psi(x, y) \equiv 1\,[(x, y) \in \Omega_0]$. Making use of this fact and Eq. (4.41), we have

$$\varphi = \frac{1}{2} G_0 (R_0^2 - r^2) + \frac{1}{2\pi} S_1 G_1, \qquad h = \frac{S_1}{2\pi R_0}$$
$$K = \frac{\pi}{2} G_0 R_0^4 + G_1 S_1 R_0^2$$

Thus, for a circular region Ω_0 a reinforcing layer of constant thickness is optimum. It is easy to check that this property remains true even for thick layers.

3. *Sufficient conditions for optimality.* We used the necessary conditions for optimality of Eq. (4.12) to construct solutions given in part 2 of this section. Here we show that in problems of optimal reinforcement, this condition turns out to be not only necessary but also sufficient for the existence of a local optimum. Let us consider two sufficiently close thickness functions h^* and h that satisfy the isoperimetric condition of Eq. (4.40). The functions ϕ^* and ϕ satisfying Eq. (4.35) correspond, respectively, to h^* and h.

The smallness of the quantity $\delta h = h^* - h$ is understood in the sense of the sup norm. The small quantity $\delta\phi = \phi^* - \phi$ is denoted by ω (i.e., $\omega = \delta\phi$). We assume that ϕ^* and h^* are related to each other by the condition of optimality, but that ϕ and h are not required to satisfy this condition. Since h^* and ϕ^* satisfy the optimality condition, the function ϕ^* is a solution of Eq. (4.37). As a consequence of the Eq. (4.40), when either h or h^* is substituted into that condition, $\int_{\Gamma_1} \delta h\, ds = 0$.

It remains to be shown that $\Delta K = K^* - K \geqslant 0$. This would indicate that the condition of Eq. (4.12) is not only necessary, but is also a sufficient condition for a local maximum. Before we attempt to estimate directly the magnitude of ΔK, we carry out some preliminary substitutions. Since ϕ and ϕ^* solve, respectively, the boundary-value problems of Eqs. (4.35) and (4.37), the function $\omega = \phi^* - \phi$ solves the Laplace equation with the boundary condition $\partial\omega/\partial t = G_0 G_1^{-1} (\phi - \lambda h)/h$. Decomposing the right-hand side of this condition in power series of δh and ω, and retaining only terms of first order of smallness, we obtain the following boundary-value problem for ω:

$$\Delta\omega = 0, \qquad (x, y) \in \Omega_0, \qquad \omega_t = \frac{G_0(\lambda\delta h - \omega)}{G_1 h^*}, \qquad (x, y) \in \Gamma_1$$

Applying the first form of Green's lemma, we have

$$Q \equiv \iint_{\Omega_0} (\nabla\omega)^2\, dx\, dy = \frac{G_0}{G_1} \int_{\Gamma_1} \frac{\omega}{h^*} (\lambda\delta h - \omega)\, ds$$

Now, let us derive some estimates for the magnitude of ΔK. Using Eqs. (4.35), (4.37), and the second form of Green's formula, we derive the following relation:

$$\Delta K = 2 \iint_{\Omega_0} (\varphi^* - \varphi)\, dx\, dy = -\frac{1}{G_0} \iint_{\Omega_0} (\varphi^* \Delta\varphi - \varphi\Delta\varphi^*)\, dx\, dy$$
$$= \frac{1}{G_0} \int_{\Gamma_1} (\varphi\varphi_t^* - \varphi^*\varphi_t)\, ds$$

We replace, in the boundary integral the functions ϕ_t^* and ϕ_t by their boundary values, as given by Eqs. (4.35) and (4.37). Using Eqs. (4.26), (4.37), and some elementary manipulations, we have

$$\Delta K = \frac{2S_0}{l} \int_{\Gamma_1} \frac{\varphi}{h} (h^* - h)\, ds$$

Our assumptions concerning the smallness of δh and ω, and the fact that the integral of δh along the contour Γ_1 is equal to zero, permit us (with accuracy up to terms of high order of smallness) to express ΔK in the form

$$\Delta K = \frac{2S_0}{l} \int_{\Gamma_1} \frac{\delta h}{h} (\lambda\delta h - \omega)\, ds$$

Subtracting from both sides of this equality the positive constant Q/G_0, we find that

$$\Delta K - \frac{Q}{G_0} = \frac{1}{G_1} \int_{\Gamma_1} \frac{(\lambda\delta h - \omega)^2}{h^*}\, ds \geqslant 0$$

which establishes the required inequality $\Delta K \geqslant 0$. The sufficiency condition for optimality has thus been established.

4. *Optimum reinforcement of an elliptical bar.* We offer a solution for the optimal reinforcement problem when the region Ω_0 is an ellipse (see Fig. 4.5).

Let the center of the ellipse coincide with the origin of the Cartesian coordinates (x, y). We assign to the origin the value $\Psi = 1$ [(x_0, y_0) coincides with 0]. For the sake of convenience, we express our solution in terms of elliptical coordinates (α, β), related to the Cartesian coordinates by the formulas $x = c \cosh \alpha \sin \beta$, $y = c \sinh \alpha \cos \beta$, where $2c$ is the focal distance. In elliptic coordinates, the boundary ellipse Γ_1 corresponds to the interval $\alpha = \alpha_0, 0 \leqslant \beta \leqslant 2$, while Ω_0 is the rectangle $0 \leqslant \alpha < \alpha_0, 0 \leqslant \beta \leqslant 2\pi$. We make use of the symmetry of our problem and seek the solution in the region $0 \leqslant \alpha \leqslant \alpha_0, 0 \leqslant \beta \leqslant \pi$, i.e., in half of the ellipse. We substitute the following well-known relations:

$$S_0 = \frac{1}{2}\pi c^2 \sinh 2\alpha_0, \qquad \frac{\partial}{\partial n} = \frac{1}{g}\frac{\partial}{\partial \alpha}, \qquad \frac{\partial}{\partial s} = -\frac{1}{g}\frac{\partial}{\partial \beta}$$

$$(r, n)_{\alpha = \alpha_0} = \frac{c^2}{2g} \sinh 2\alpha_0, \qquad g = c\sqrt{1/2(\cosh 2\alpha + \cosh 2\beta)} \tag{4.42}$$

to express, in terms of elliptical coordinates, the equations and boundary conditions determining the function Ψ

$$\Psi_{\alpha\alpha} + \Psi_{\beta\beta} = 0 \qquad (0 \leqslant \alpha \leqslant \alpha_0, \qquad 0 \leqslant \beta \leqslant \pi) \tag{4.43}$$

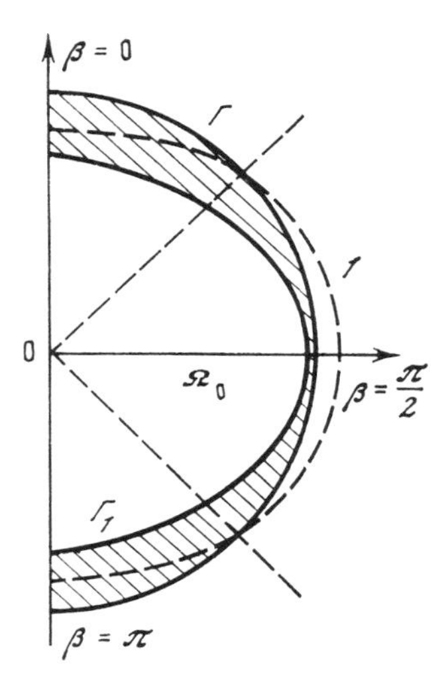

Figure 4.5

$$\Psi = 1 \qquad (\alpha = 0, \beta = 0) \tag{4.44}$$

$$\Psi_\alpha = {}^1\!/_2 c^2 \sinh 2\alpha_0 (1 - 2\pi g/l) \qquad (\alpha = \alpha_0, 0 \leqslant \beta \leqslant \pi) \tag{4.45}$$

$$\Psi_\alpha = 0 \qquad (\alpha = 0, 0 \leqslant \beta \leqslant \pi) \tag{4.46}$$

$$\Psi_\beta = 0 \qquad (0 \leqslant \alpha \leqslant \alpha_0, \beta = 0)\, (0 \leqslant \alpha \leqslant \alpha_0, \beta = \pi) \tag{4.47}$$

We expand the function Ψ in Fourier series

$$\Psi = \sum_{n=0}^{\infty} a_n \cos n\beta \cosh n\alpha \tag{4.48}$$

The series of Eq. (4.48) satisfies the boundary conditions of Eqs. (4.46) (4.47). The coefficients of the Fourier series can be determined from Eqs. (4.44) and (4.45). We have

$$a_0 = 1 - \sum_{n=1}^{\infty} a_n, \qquad a_n = -\frac{2c^2 \sinh 2\alpha_0}{nl \sinh n\alpha_0} \int_0^\pi g \cos n\beta \, d\beta \tag{4.49}$$

Substituting Eqs. (4.48) and (4.49) into Eq. (4.41), using the basic formula $dx\, dy = g^2\, d\alpha\, d\beta$, with $ds = g\, d\beta$ on Γ_1, and Eq. (4.42), we derive

$$\begin{aligned}
\varphi &= G_0 \left[\frac{a_2}{2} + \frac{c^2}{4} (\cosh 2\alpha_0 - \cosh 2\alpha + \cos 2\beta) \right.\\
&\quad \left. + \sum_{n=1}^{\infty} a_n \left(\cos n\beta \cosh n\alpha + \frac{n a_n \sinh 2n\alpha_0}{2c^2 \sinh 2\alpha_0} \right) \right] + \frac{2G_1 S_0 S_1}{l^2} \\
h &= \frac{G_0 l}{2G_1 S_0} \left[\frac{a_2}{2} + \frac{c^2}{4} \cos 2\beta + \sum_{n=1}^{\infty} a_n \left(\cos n\beta \cosh n\alpha_0 \right.\right. \\
&\quad \left.\left. + \frac{n a_n \sinh 2n\alpha_0}{2c^2 \sinh 2\alpha_0} \right) \right] + \frac{S_1}{l} \\
K &= \pi c^2 G_0 \sinh 2\alpha_0 \left[a_2 + \frac{c^2}{8} \cosh 2\alpha_0 + \sum_{n=1}^{\infty} \frac{n a_n^2 \sinh 2n\alpha_0}{2c^2 \sinh 2\alpha_0} \right] + \frac{4G_1 S_1 S_0^2}{l^2}
\end{aligned} \tag{4.50}$$

The expressions for the unknown quantities ϕ, h, and K given in Eq. (4.50) may be simplified if we make an additional assumption, which was introduced in part 2, that $\min_s R(s) \sim l$. This permits us to regard the quantity $1/\cosh 2\alpha_0$ as small when compared with unity. Decomposing the integrand in Eq. (4.49) in series of powers of the small parameter $1/\cosh 2\alpha_0$, it is easy to show that with accuracy up to the order of $(1/\cosh 2\alpha_0)^2$, the following equalities are true

$$a_0 = 1 - a_2, \quad a_2 = -\pi c^3/4\sqrt{2l}\,\sqrt{\cosh 2\alpha_0} = -c^2/8\cosh 2\alpha_0$$
$$a_1 = a_3 = \cdots = 0 \tag{4.51}$$

Making use of Eq. (4.51), after some obvious manipulations, the expressions for h and K given in Eq. (4.50) now assume the form

$$h = \frac{S_1}{l} + \frac{cG_0\sqrt{\cosh 2\alpha_0}}{4\sqrt{2}\,G_1 \sinh 2\alpha_0}\cos 2\beta$$
$$K = {}^1\!/_{16}\,\pi c^4 G_0 \sinh 4\alpha_0 + 4G_1 S_1 S_0^2 l^{-2} \tag{4.52}$$

where $l \approx c\pi(2\cosh 2\alpha_0)^{1/2}$. It is clear from Eq. (4.52) that if the curvature increases as we move along the boundary Γ_1, the thickness h decreases. This is illustrated in Fig. 4.5, which shows the optimum distribution of thickness of the reinforcing layer Ω_1. This layer is shaded in Fig. 4.5, while the dotted line indicates the boundary Γ of the region Ω having a constant thickness $h = S_1/l$. This thickness function turns out to be optimum if the region Ω_0 is a circular disc.

4.4. Optimization of Stress Concentration for Elastic Plates with Holes

1. *Formulation.* In Sections 4.1–4.3 we investigated problems of optimization with unknown boundaries and integral functionals. Optimization problems with local quality criteria are much harder to solve. In particular, these problems arise when we wish to determine optimum shapes of elastic bodies having the smallest concentration of stresses (see Section 1.2 of Chapter 1, and also Section 3.4 of Chapter 3). Here we consider the stressed state of an elastic body with a hole, subjected to the action of external forces. The domain occupied by the elastic medium is denoted by Ω, while Γ denotes the boundary of the hole. The quantity that measures the stress intensity is $J = (f)_{\Omega_0} = \max_P f$, $P \in \Omega$, where Ω_0 is the set of points at which f assumes its maximum. The function f is a function of the stress tensor invariants I_1, I_2, and I_3. If f is written as a function of the stress components, it is assumed to be a homogeneous function

with a homogeneity index equal to two. Throughout this section, we investigate the problem of determining the boundary curve Γ, for which we attain the minimum of the maximum of f in the region $\Omega + \Gamma$, i.e.,

$$J_* = \min_\Gamma J = \min_\Gamma \max_{P \in \Omega} f \tag{4.53}$$

Because of the local nature of the cost functional, the problem described by Eq. (4.53) belongs to the class of optimization problems with local quality criteria. In finding the minimum in Eq. (4.53) with respect to Γ, we assume that Γ cannot be contracted to a single point, i.e., the region does have a hole. The shape of the boundary curve Γ replaces "the control variable," while the equations of equilibrium and of compatibility of deformations, which are used in determining the stresses, enter into the optimization problem as differential constraints.

We shall study this problem in specific cases of application, namely for bending and tension of elastic plates containing holes. First, we need to note a certain property of harmonic functions that will be utilized in our discussion below.

Let us consider the x-y plane with n holes, having closed simple curves Γ_i $(i = 1, 2, \ldots, n)$ as their respective boundaries. The region Ω is bounded by $\Gamma = \Sigma\Gamma_i$ and contains the point at infinity (i.e., Ω is the region exterior to the holes). We consider a family of harmonic functions that are continuous in $\Omega + \Gamma$ and tend to the limit A at infinity, where A is a given positive constant. For an arbitrary function χ, which belongs to this family, if $g(\xi, \eta)$ denotes the values of χ on the boundary Γ, then according to the maximum principle,[48] the following inequality is true:

$$|\chi(x, y)| \leqslant \max_{\xi\eta} |g(\xi, \eta)|$$

where $(x, y) \in \Omega$ and $(\xi, \eta) \in \Gamma$. In particular, it must be true that

$$A \leqslant \max_{\xi\eta} |g| \tag{4.54}$$

If the equality sign is realized in Eq. (4.54), then the function χ must be identically equal to the constant A. Therefore, the minimum with respect to g of the functional $\max_{\xi\eta} |g|$ $[(\xi, \eta) \in \Gamma]$ is realized for a unique function $g(\xi, \eta) \equiv A$, (i.e., $\chi(x, y) \equiv A$), which is a member of our family of harmonic functions and is identically equal to A, i.e.,

$$\min_g \max_{\xi\eta} |g| = A \tag{4.55}$$

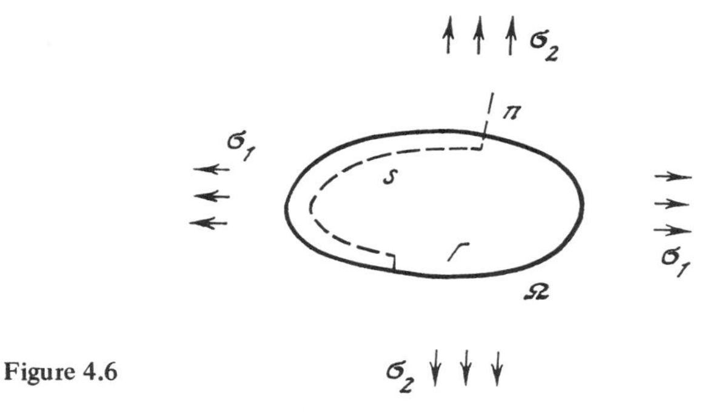

Figure 4.6

We shall use this family of harmonic functions to estimate stresses on the boundary of holes contained in a plate.

2. *Von Mises criteria for a single hole.* We consider a two-dimensional problem in the theory of elasticity. It concerns the stressed state of an infinite plate weakened by a hole. Let Ω denote a region of the x–y plane occupied by the material, and let Γ be the boundary of the hole (see Fig. 4.6).

We use a commonly accepted notation for components of the stress tensor: σ_x, σ_y, and τ_{xy}. We assume that the plate is subjected to tensile forces applied at infinity and that the boundary Γ of the hole has no external loads applied to it. Thus, boundary values on Γ and at the point at infinity can be written as

$$\begin{aligned} \sigma_n &= 0,\ \tau_n = 0,\ (x,\ y) \in \Gamma \\ (\sigma_x)_\infty &= \sigma_1,\ (\sigma_y)_\infty = \sigma_2 \\ (\tau_{xy})_\infty &= 0 \end{aligned} \tag{4.56}$$

where σ_1 and σ_2 are given positive constants and n and s denote, respectively, the directions normal and tangent to the curve Γ. For a given shape of the curve Γ the distribution of stresses is fully determined in the region $\Omega + \Gamma$ by the boundary conditions of Eq. (4.56). As is well known, we can introduce a stress function ϕ related to the stress components by the relations $\sigma_x = \phi_{yy}$, $\sigma_y = \phi_{xx}$, and $\tau_{xy} = -\phi_{xy}$. The problem is then reduced to solution of the biharmonic equation

$$\Delta^2\varphi \equiv \varphi_{xxxx} + 2\varphi_{xxyy} + \varphi_{yyyy} = 0 \tag{4.57}$$

in the domain Ω, where ϕ satisfies the boundary conditions of Eq. (4.56). The subscripts in Eq. (4.57) denote partial differentiation with respect to the designated variables.

For tension in a plate, the third invariant of the stress tensor is equal to zero, hence the function f may be written in the form $f = f(I_1, I_2)$, where $I_1 = \sigma_x + \sigma_y$, and $I_2 = \tau_{xy}^2 - \sigma_x \sigma_y$. The following expression may be assigned to f: $f = I_1^2 + 3I_2$, corresponding to the Von Mises criterion. Aside from this expression for f, we shall consider here some more general forms of dependence of f on the invariants I_1 and I_2. To begin with, we consider the following expression for f:

$$f = I_1^2 + 3I_2 = \sigma_x^2 + \sigma_y^2 - \sigma_x\sigma_y + 3\tau_{xy}^2 \tag{4.58}$$

corresponding to the Von Mises plasticity criterion. The expression of Eq. (4.58) corresponds to the square of the intensity, or concentration of the tangential stress, within the accuracy determined by the value of the coefficients. We introduce an auxiliary function $\chi = \sigma_x + \sigma_y$, which, as is well known,[101] is a harmonic function. Taking into account the boundary conditions of Eq. (4.56) and the equality $\sigma_x + \sigma_y = \sigma_n + \sigma_t$, we have

$$\triangle\chi = 0, \quad (x, y) \in \Omega, \; (\chi)_\Gamma = \sigma_s, \; (\chi)_\infty = \sigma_1 + \sigma_2 \tag{4.59}$$

We can make use of the invariance of Eq. (4.58), under transformation of coordinates from the x–y axes to the local coordinates in the directions n and s, and utilize the boundary values of Eq. (4.56) to arrive at the following formula describing the boundary values of f:

$$(f)_\Gamma = \sigma_s^2 \tag{4.60}$$

The function χ determined by Eq. (4.59) satisfies Eq. (4.55) for a family of harmonic functions described above. Substituting Eqs. (4.59) and (4.60) for the boundary values of f and χ, we assert that the minimum of the maximum value of $|\chi|$, and of f on the curve Γ, is attained if and only if the stress component σ_s is constant along the entire length of this curve, and is equal to

$$(\sigma_s)_\Gamma = \sigma_1 + \sigma_2 \tag{4.61}$$

We now assume the existence of uniformly stressed holes, satisfying Eq. (4.61). In Section 4.5 of this chapter we shall describe techniques (given in Refs. 46 and 144) for finding the shapes of holes having a uniform stress distribution on their boundaries. For the case of a single hole, it was shown in Refs. 143 and 144 that the contours Γ satisfying Eq. (4.61) form a one-parameter family of ellipses $x^2\sigma_1^{-2} + y^2\sigma_2^{-2} = \lambda^2$, where λ^2 is the parameter.

We can now prove that Eq. (4.61) is a necessary and sufficient condition for optimality of the boundary curve Γ. To prove this assertion, it is sufficient to show that for any contour Γ satisfying Eq. (4.61), the maximum of the function f in the domain $\Omega + \Gamma$ is attained on the curve Γ, and therefore

$$\max_{(x,\ y)\in\Omega+\Gamma} f = \max_{(x,\ y)\in\Gamma} f \tag{4.62}$$

Indeed, it is easy to see that the assumption that Eq. (4.62) is true and the obvious relation

$$\max_{\Gamma} \max_{(x,\ y)\in\Omega+\Gamma} f \geqslant \min_{\Gamma} \max_{(x,\ y)\in\Gamma} f \tag{4.63}$$

implies that the necessary and sufficient conditions for the minimum of the maximum value of f on Γ are also the necessary and sufficient conditions for the minimum of the maximum value of f on $(\Omega + \Gamma)$. Therefore,

$$J_* = \min_{\Gamma} \max_{(x,\ y)\in\Gamma} f$$

We shall employ the complex representation of components of the stress tensor in terms of the potentials $\Phi(z)$ and $\Psi(z)$ of Kolosov-Muskhelishvili (see the basic Ref. 101)

$$\sigma_x + \sigma_y = 4 \operatorname{Re} \Phi(z), \qquad \sigma_y - \sigma_x + 2i\tau_{xy} = 2[\bar{z}\Phi'(z) + \Psi(z)] \tag{4.64}$$

where $z = x + iy$, $\bar{z} = x - iy$, $(i^2 = -1)$. If Eq. (4.61) is satisfied, the function χ satisfies the relation $\chi = \sigma_x + \sigma_y = \sigma_1 + \sigma_2$ in $\Omega + \Gamma$, consequently $\Phi'(z) = 0$. Equation (4.64) assumes the form

$$\begin{gathered} \sigma_x + \sigma_y = \sigma_1 + \sigma_2, \qquad \sigma_y - \sigma_x + 2i\tau_{xy} = 2\Psi(z) \\ \left(\Phi(z) = \frac{1}{4}(\sigma_1 + \sigma_2)\right) \end{gathered} \tag{4.65}$$

The second equation in Eq. (4.65) is multiplied (i.e., both sides of the equation) by the expression $\sigma_y - \sigma_x - 2i\tau_{xy} = 2\overline{\Psi}(z)$. Taking into account the first equation in Eq. (4.65), we have

$$f_0 \equiv (\sigma_1 + \sigma_2)^2 + 4(\tau_{xy}^2 - \sigma_x\sigma_y) = 4\Psi\overline{\Psi} \tag{4.66}$$

We observe that the function f_0 differs from f only in the value of the coefficient of the second term. Let us express f as a function of Ψ and $\overline{\Psi}$. Using Eqs.

(4.58) and (4.66), we obtain

$$f = \frac{3}{4} f_0 + \frac{1}{4} (\sigma_1 + \sigma_2)^2 = 3\Psi\overline{\Psi} + \frac{1}{4} (\sigma_1 + \sigma_2)^2 \tag{4.67}$$

Furthermore, we represent the functions Ψ and $\overline{\Psi}$ in the form $\Psi = \psi_1 + i\psi_2$, $\overline{\Psi} = \psi_1 - i\psi_2$, and restate Eq. (4.67) as $f = 3(\psi_1^2 + \psi_2^2) + 1/4\,(\sigma_1 + \sigma_2)^2$. Since the functions ψ_1 and ψ_2 satisfy the Cauchy–Riemann equations, and consequently $\Delta\psi_1 = \Delta\psi_2 = 0$, it is easy to show that

$$\Delta f = 12(\nabla\psi_1)^2 \geqslant 0 \tag{4.68}$$

Therefore, f cannot attain its maximum value at any interior point of $(\Omega + \Gamma)$. Comparing the values of f on the boundary of this region, $(f)_\Gamma = (\sigma_1 + \sigma_2)^2$ and $(f)_\infty = \sigma_1^2 + \sigma_2^2 - \sigma_1\sigma_2$, we arrive at the conclusion that the maximum of f is attained on Γ.

Since the minimum of the maximum value of f is attained on Γ for uniformly stressed [in the sense of Eq. (4.61)] boundary curves Γ, while in the same case of a uniformly stressed curve Γ the maximum value of f in the domain $(\Omega + \Gamma)$ is also attained on Γ, it can be deduced [observe Eqs. (4.62) and (4.63)] that the uniformly stressed contour Γ is optimum. We make the observation that Eq. (4.61) is the necessary and sufficient condition for a global optimum.

3. *Multiple holes.* Now let us assume that the plate contains n holes whose boundaries are the curves $\Gamma_i (\Gamma = \Sigma\Gamma_i,\ i = 1, 2, \ldots, n)$. In this case, if we apply the condition of Eq. (4.55), we can show that the minimum of the maximum value of f is attained on Γ if and only if Eq. (4.61) is satisfied.

We observe that the shapes of uniformly stressed contours that satisfy Eq. (4.61) were found in Ref. 144 for a plate having two holes and also for some periodic arrangements of holes. Also, we note that the proof of the inequality of Eq. (4.68) did not depend on connectivity of the region $(\Omega + \Gamma)$. Therefore, if the region $(\Omega + \Gamma)$ has n equally stressed holes, the maximum of f (in that region) is attained on Γ. Consequently, in this case uniformly stressed holes are optimum also.

Deflection of plates having uniformly stressed holes takes place in such a manner that all points of the region $(\Omega + \Gamma)$ have the same dilation. In fact, since the relation $\sigma_x + \sigma_y = \sigma_1 + \sigma_2$ is satisfied in the region $(\Omega + \Gamma)$, we can use the well-known relations between displacements and stresses to assert that the following equality is true:

$$\operatorname{div} U = \frac{1 - \nu}{E} (\sigma_1 + \sigma_2) \tag{4.69}$$

where U is the displacement vector and ν is Poisson's ratio.

It follows from some properties of the function f, which were discussed above, that in the process of increasing the loads on a plate having optimally shaped holes, plastic flow cannot be originated at any interior point of the region $(\Omega + \Gamma)$. A plastic state of the material first occurs on Γ. Moreover it occurs simultaneously at every point of the contour Γ, where $\Gamma = \Sigma\Gamma_i$.

4. *Tresca failure criterion.* In parts 2 and 3 of this section we have selected as f in Eq. (4.58) the square of the stress intensity, which is the second invariant of the stress deviator corresponding to the Von Mises plasticity criterion. Let us now examine the case in which f is given by

$$f = I_1^2 + 4I_2 = (\sigma_x - \sigma_y)^2 + 4\tau_{xy}^2 \tag{4.70}$$

The quantity f given by Eq. (4.70) is equal to the square of the maximum tangential stress τ_{max} (plane deformation with $\nu = 1/2$). When it attains a certain value, a plastic zone is originated according to Tresca's criterion. Stress boundary values and conditions at infinity are assumed to be known *a priori.* In our considerations of the quality criterion functional it is possible to show that the boundary values assigned to f on Γ should be determined by exactly the same formula $(f)_\Gamma = \sigma_s^2$, i.e., by the formula given in part 2. Hence, our previous assertion that uniformly stressed boundaries of the holes correspond to the minimum of the maximum value of f on Γ still remains valid. The proof of the inequality similar to Eq. (4.68) is shorter, since in this case we have $f = f_0$. We have $\Delta f = \Delta f_0 = 16(\nabla\psi_1)^2 \geqslant 0$. Hence, uniformly stressed holes turn out to be optimum, in the sense of the criterion of Eq. (4.70).

5. *A more general failure criterion.* Let us now consider a more general dependence of f on the stress invariants I_1 and I_2, assuming non-negative properties of the first and second derivatives of f with respect to I_2, i.e.,

$$f = f(I_1, I_2), \qquad \partial f/\partial I_2 \geqslant 0, \qquad \partial^2 f/\partial I_2^2 \geqslant 0 \tag{4.71}$$

We assume also that the assumptions made in part 1 of this section, concerning the dependence of the magnitude of f on the components of the stress tensor and the condition of positive property of f, are still valid. Using homogeneity and positive properties of f, and the equalities $(I_1)_\Gamma = \sigma_s$ and $(I_2)_\Gamma = 0$, we arrive at the following expression for f, restricted to the contour Γ:

$$f = f(I_1, 0) = aI_1^2 = a\sigma_s^2 \qquad (a > 0)$$

Making use of the above stated properties of $(f)_\Gamma$ and of the properties of the class of harmonic functions that were explained in part 1 of this section, we arrive directly at the following conclusion: For a uniformly stressed contour Γ, the minimum of the maximum of f occurs on Γ.

Further, let us make use of Eq. (4.71) and recall the fact that for a uniformly stressed contour Γ, the invariant I_1 has a constant magnitude in the domain $(\Omega + \Gamma)$ (we recall that $I_1 = \sigma_1 + \sigma_2$). Making some substitutions that are analogous to those in part 2, we are able to show that for a plate with holes, which satisfies Eq. (4.61), the following inequality must be true:

$$\Delta f = \frac{d^2 f}{dI_2^2} (\nabla I_2)^2 + 4 \frac{df}{dI_2} (\nabla \psi_1)^2 \geqslant 0, \qquad (x, y) \in \Omega + \Gamma$$

The function f, obeying our assumptions, turns out to be superharmonic. Consequently, it cannot attain its maximum value at any interior point of the domain $(\Omega + \Gamma)$. Comparing the values of f on Γ and at the point at infinity, we compare $(f)_\Gamma = f(\sigma_1 + \sigma_2, 0)$ and $(f)_\infty = f(\sigma_1 + \sigma_2, -\sigma_1 \sigma_2)$. Observing that $(I_2)_\infty = -\sigma_1 \sigma_2 \leqslant (I_2)_\Gamma = 0$, and the first inequality in Eq. (4.71), we conclude that $(f)_\Gamma > (f)_\infty$. Hence, in our case, the inequality of Eq. (4.63) remains valid. Consequently, only uniformly stressed holes are optimum.

6. *Uniform pressure on hole boundaries.* The discussion presented in part 3 may be generalized to the case in which a uniform pressure $\sigma_0 (\sigma_0 > 0)$ is applied to the boundaries of the holes, i.e., $(\sigma_n)_\Gamma = -\sigma_0$. The second boundary condition given on Γ and the conditions at the point at infinity are given by the relations of Eq. (4.56). Making use of the assigned boundary values and of Eq. (4.58), we obtain an expression for the function f restricted to the contour Γ: $(f)_\Gamma = \sigma_s^2 + \sigma_0 \sigma_s + \sigma_0^2$. With this formula and Eq. (4.55), we can show that the minimum of the maximum value of $(f)_\Gamma$ is attained if and only if

$$(\sigma_s)_\Gamma = \sigma_1 + \sigma_2 + \sigma_0 \tag{4.72}$$

It is easy to see that the inequalities $(f)_\Gamma > (f)_\infty$ and $\Delta f \geqslant 0$ are true, and therefore Eq. (4.62) is valid in this case. Hence, the holes satisfying the condition $(\sigma_s)_\Gamma = \sigma_1 + \sigma_2 + \sigma_0$ are optimum with the quality criterion given by Eqs. (4.70) and (4.71).

Without offering details of the proof, we simply state the fact that the optimality conditions remain unaltered in the more general case of two-dimensional deformation of an elastic body with holes.

4.5. Determining the Shape of Uniformly Stressed Holes

In this paragraph we describe two techniques available for finding the shape of uniformly stressed holes, within the framework of two-dimensional theory of elasticity.

1. *Complex variable method.* We follow the ideas of Ref. 144 to outline a technique for determining uniformly stressed contours by using the theory of functions of a complex variable, reducing the original inverse boundary-value problem to a Dirichlet problem for the region exterior to a family of parallel slits.

Finding the stressed state of a plane region with holes satisfying Eq. (4.72) can be reduced to finding one of the two Kolosov–Muskhelishvili potentials, namely $\Psi(z)$, where $z = x + iy$. The function $\Psi(z)$ behaves like $\beta + O(z^{-2})$ as $z \to \infty$, where $\beta = \frac{1}{2}(\sigma_2 - \sigma_1)$. Taking into account Eqs. (4.65) and (4.72), $(\sigma_n)_\Gamma = -\sigma_0$ (see part 6 of Section 4.4), and the well-known relation

$$\sigma_s - \sigma_n + 2i\tau_{sn} = e^{2i\theta}(\sigma_y - \sigma_x + 2i\tau_{xy})$$

we express the boundary conditions for the function Ψ in the form

$$e^{2i\theta}\Psi(z) = \mu, \qquad z \in \Gamma \tag{4.73}$$

where $\mu = (\sigma_1 + \sigma_2 + 2\sigma_0)/2$ and θ is the angle between the exterior normal direction to the contour Γ and the x axis.

We introduce a conformal map $z = \omega(\zeta)$, mapping the exterior of n slits in the complex plane of ζ, where the slits are parallel to the real axis, into the region exterior to the equal number of stressed holes in the z plane (see Fig. 4.7). We make the following observation (see Ref. 65). For any n-tuply connected region that contains the point at infinity, there exists a conformal map that takes the exterior of n slits parallel to the real axis into this region and maps the point at infinity into itself. For $n \geqslant 3$, this map is unique if we describe the behavior of $\omega(\zeta)$ near infinity: $\omega(\zeta) = \zeta + O(1)$ as $\zeta \to \infty$.

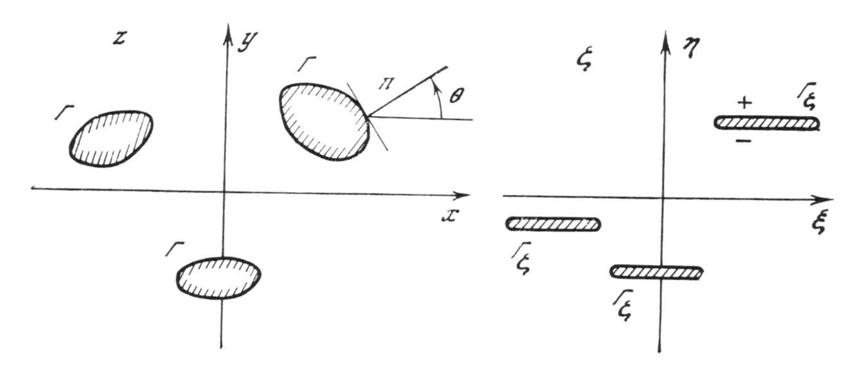

Figure 4.7

We compute the value of $e^{2i\theta}$, which appears in Eq. (4.73). Let us move a point in the z plane in the direction normal to the contour Γ: $dz = e^{i\theta}\,|dz|$. Because of the conformal property of the map ω, the corresponding point in the ζ plane will move in the direction normal to a slit, i.e., normal to the real axis. Therefore, $d\zeta = \pm i\, d\zeta$. Using these expressions for dz and $d\zeta$, we find $e^{i\theta} = dz/|dz| = \pm i\omega'(\zeta)/|\omega'(\zeta)|$, where primes denote differentiation. Substituting this expression for $e^{2i\theta}$ into Eq. (4.73), we have

$$-\psi(\zeta)\,\omega'(\zeta) = \mu\overline{\omega'(\zeta)}, \quad \zeta \in \Gamma_\zeta \tag{4.74}$$

where $\psi(\zeta) = \Psi[\omega(\zeta)]$ and Γ_ζ is the boundary of the straight line slits in the ζ plane. The functions $\psi(\zeta)$ and $\omega(\zeta)$ satisfy the boundary-value problem of Eq. (4.74). Let us consider separately the real and imaginary parts in Eq. (4.74). We have

$$\operatorname{Re} T'(\zeta) = 0, \quad \zeta \in \Gamma_\zeta, \quad T'(\zeta) \equiv (\psi(\zeta) + \mu)\,\omega'(\zeta) \tag{4.75}$$

$$\operatorname{Im} N'(\zeta) = 0, \quad \zeta \in \Gamma_\zeta, \quad N'(\zeta) \equiv (\psi(\zeta) - \mu)\,\omega'(\zeta) \tag{4.76}$$

The functions $T'(\zeta)$ and $N'(\zeta)$ are analytic everywhere outside the slits Γ_ζ. In a neighborhood of the point at infinity these functions are bounded, since the functions $\psi(\zeta)$ and $\omega'(\zeta)$ remain bounded as $\zeta \to \infty$. We assume that the boundaries Γ of the holes in the z plane are smooth. It follows that $\psi(\zeta)$ is bounded everywhere in a neighborhood of the slits Γ_ζ, while the function $\omega'(\zeta)$ is bounded everywhere, with the exception of neighborhoods of the end points of the slits Γ_ζ. In each such neighborhood, the function $\omega'(\zeta)$ has a singularity of the order of power $1/2$. Therefore, the functions $T'(\zeta)$ and $N'(\zeta)$ are bounded in the entire ζ plane, with the exception of the end points of the slits Γ_ζ, where they have a singularity of the order $\frac{1}{2}$, i.e.,

$$T'(\zeta) = O\left(\frac{1}{\sqrt{\zeta - \zeta_k}}\right), \qquad N'(\zeta) = O\left(\frac{1}{\sqrt{\zeta - \zeta_k}}\right) \tag{4.77}$$

The boundary-value problems of Eqs. (4.75) and (4.76) represents the classical Dirichlet problem for the exterior of the slits. The solution of this problem is sought in the class of functions that are bounded at infinity and have a singularity of the type in Eq. (4.77) at the end points of each slit.

If the functions $T(\zeta)$ and $N(\zeta)$ are found, the unknown functions $\psi(\zeta)$ and $\omega(\zeta)$ can be found, according to Eqs. (4.75) and (4.76), by the use of formulas

$$\omega(\zeta) = \frac{1}{2\mu}[T(\zeta) - N(\zeta)] + c_0$$
$$\psi(\zeta) = (T'(\zeta) - N'(\zeta))/2\omega'(\zeta) \tag{4.78}$$

where c_0 denotes an arbitrary constant.

In case only a single hole exists in the z plane, we have only a single slot in the ζ-plane, which we may consider, without any loss of generality, to be situated on the real axis and connecting the points $(-1, +1)$. As $\zeta \to \infty$, the function $\omega(\zeta)$ behaves as $\omega(\zeta) = c_1\zeta + O(\zeta_1^{-1})$, where c_1 is a real constant. This condition, together with the assignment of the length of the slit, eliminates the arbitrariness in the definition of the conformal map between the two domains. According to Eqs. (4.75) and (4.76), as $\zeta \to \infty$, we have

$$T'(\zeta) = (\beta + \mu)c_1 + O(\zeta^{-2}), \quad N'(\zeta) = (\beta - \mu)c_1 + O(\zeta^{-2}) \tag{4.79}$$

Solution of the boundary-value problem of Eqs. (4.75) and (4.76), with conditions of Eqs. (4.77) and (4.79), have the following general form:

$$T'(\zeta) = (c_1\zeta(\beta + \mu) + d_1)/\sqrt{\zeta^2 - 1}$$
$$N'(\zeta) = c_1(\beta - \mu) + id_2/\sqrt{\zeta^2 - 1} \tag{4.80}$$

The real valued constants d_1 and d_2 are arbitrary. Integrating Eq. (4.80), we derive

$$T(\zeta) = c_1\sqrt{\zeta^2 - 1}(\beta + \mu) + d_1 \ln(\zeta + \sqrt{\zeta^2 - 1})$$
$$N(\zeta) = c_1\zeta(\beta - \mu) + id_2 \ln(\zeta + \sqrt{\zeta^2 - 1})$$

Making further use of Eq. (4.78), and substituting $c_0 = 0$, we obtain

$$\omega(\zeta) = \frac{c_1}{2}(m_1\zeta + m_2\sqrt{\zeta^2 - 1}) + \frac{d_1 - id_2}{2\mu}\ln(\zeta + \sqrt{\zeta^2 - 1})$$
$$m_1 = 1 - \beta/\mu, \quad m_2 = 1 + \beta/\mu \tag{4.81}$$

The condition of the univalent property of the map $\omega(\zeta)$ in the region exterior to $(-1, +1)$ is satisfied if we set $d_1 = d_2 = 0$. We can then derive from Eq. (4.81) the following parametric representation of the shape of the boundary of the hole.

$$x = c_1 t m_1/2, \qquad y = \pm c_1 m_2\sqrt{1 - t^2}/2 \quad (-1 < t < 1) \tag{4.82}$$

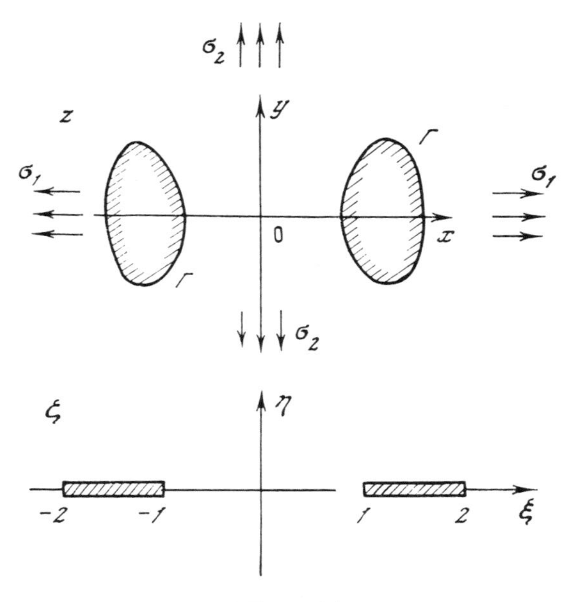

Figure 4.8

where t is a parameter. Eliminating the parameter t in these formulas, we arrive at the equation of an ellipse $x^2/m_1^2 + y^2/m_2^2 = \lambda^2$. If $\sigma_0 = 0$, then the equation of a uniformly stressed boundary of the hole has the form

$$x^2/\sigma_1^2 + y^2/\sigma_2^2 = \lambda^2 \tag{4.83}$$

In the case when two holes exist, the ζ plane has two slits, (λ_1, λ_2) and (λ_3, λ_4), which lie on the real axis ζ (see Fig. 4.8). We can assume that the coefficient c_1 in Eq. (4.79) is real and positive. The solution of the Dirichlet problem of Eqs. (4.75) and (4.76), with the condition of Eq. (4.79) at infinity, is of the form

$$\begin{aligned} T'(\zeta) &= \frac{c_1\zeta^2(\mu+\beta)+d_1\zeta+d_2}{\sqrt{(\zeta-\lambda_1)(\zeta-\lambda_2)(\zeta-\lambda_3)(\zeta-\lambda_4)}} \\ N'(\zeta) &= c_1(\beta-\mu)+\frac{i(d_3\zeta+d_4)}{\sqrt{(\zeta-\lambda_1)(\zeta-\lambda_2)(\zeta-\lambda_3)(\zeta-\lambda_4)}} \end{aligned} \tag{4.84}$$

Here d_1, d_2, d_3, and d_4 are arbitrary real numbers. Using the symmetry of our problem, relative to the x and y axes, we asmume that one hole lies entirely in the left-half plane ($x < 0$) and the other in the right-half plane ($x > 0$). Without any loss of generality we may substitute in Eq. (4.84) $\lambda_1 = -2$, $\lambda_2 = -1$, $\lambda_3 = 1$, $\lambda_4 = 2$, and $d_1 = d_3 = d_4 = 0$. Making use of Eq. (4.78) we compute the function $\omega(\zeta)$. The constant c_0 that appears in Eq. (4.78) can be computed from the condition $\omega(0) = 0$. As a result we have

$$\omega(\zeta) = \frac{c_1}{2\mu}\Big\{(\mu - \beta)\zeta + 2(\mu + \beta)E\left(\arcsin\zeta, \frac{1}{2}\right) - \left[(\mu + \beta) - 2 + \frac{1}{2}d_2\right]F\left(\arcsin\zeta, \frac{1}{2}\right)\Big\} \tag{4.85}$$

Here, E and F are elliptic integrals of the first and second type, respectively. Now the shape of the contour may be found directly from Eqs. (4.78) and (4.85) (for $x > 0$ and $y > 0$)

$$x = \frac{1}{2}c_1\left(1 - \frac{\beta}{\mu}\right)(\xi - 1) + \frac{c_1}{2\mu}\Big\{\mu - \beta + 2(\mu + \beta)E\left(\frac{1}{2}\right) - k\left(\frac{1}{2}\right)\left[(\mu + \beta)2 + \frac{1}{2}d_2\right]\Big\}$$

$$y = -\frac{1}{2}c_1\left(1 + \frac{\beta}{\mu}\right)\int_1^{\xi}\frac{(\xi^2 + d_2)\,d\xi}{\sqrt{(\xi^2 - 1)(4 - \xi^2)}} = -\frac{1}{2}c_1\left(1 + \frac{\beta}{\mu}\right) \times \left[2E\left(\chi, \frac{\sqrt{3}}{2}\right) + \frac{1}{2}d_2F\left(\chi, \frac{\sqrt{3}}{2}\right) - \frac{1}{\xi}\sqrt{(4 - \xi^2)(\xi^2 - 1)}\right] \tag{4.86}$$

$$\left(\chi = \arcsin\left(2\sqrt{\xi^2 - 1}/\sqrt{3}\,\xi\right), \quad 1 < \xi < 2\right)$$

A detailed analysis of the solution in Eq. (4.86) and also of the determination of the shape of the contours of holes arranged in a periodic or doubly periodic manner, with a uniform stress distribution on the boundary of these holes, has been carried out in Ref. 144.

2. *Integral equation method.* A different technique of determining the shape of uniformly stressed holes (in finitely connected domains) consists of reducing the problem to an integral equation. An exposition of this technique is given in Ref. 46.

As is well known,[76] any n-tuply connected region of the complex z plane that contains the point at infinity can be mapped conformally into a canonical domain that consists of the ζ plane with n circles deleted. If $n > 2$, the map $\omega(\zeta)$ that has the form $\omega(\zeta) = c\zeta + \omega_0(\zeta)$, where $\omega_0(\zeta)$ is bounded at infinity, depends on $3n$ real parameters. Six of these parameters (for example, one circumference, a point on that circumference, and a center of another circle) can be assigned quite arbitrarily. The constant c only changes the scale. Consequently, if a system of uniformly stressed holes exists, it determines a $(3n - 6)$ parameter family of curves. As the parameters change, the boundaries can be determined from geometric considerations. The number of parameters is decreased with existence of symmetries.

To find the stresses on the boundary Γ of the holes, we have the relations

$$\sigma_r + \sigma_\theta = 4\mathrm{Re}\, \Phi(\zeta),$$
$$\sigma_\theta - \sigma_r + 2i\tau_{r\theta} = \frac{2(\zeta - \zeta_k)^2}{r_k^2 \omega'(\zeta)} \left(\omega(\zeta)\Phi'(\zeta) + \omega'(\zeta)\Psi(\zeta)\right) \tag{4.87}$$

where σ_r, σ_θ, and $\tau_{r\theta}$ are the normal stresses and the tangential stress component in polar coordinates, with the origin at the point ζ_k of the center of the circle with radius r_k. The circumference of the circle is denoted by Γ_k, $k = 1, 2, \ldots, n$. If the value of the stress tensor, i.e., of its components $\sigma_x = \sigma_1$, $\sigma_y = \sigma_2$, and $\tau_{xy} = 0$, is given at the point at infinity, then $\Phi(\zeta)$ and $\Psi(\zeta)$ must have the form $\Phi(\zeta) = 1/4\,(\sigma_1 + \sigma_2) + \Phi_0(\zeta)$, and $\psi(\zeta) = 1/2\,(\sigma_2 - \sigma_1) + \Psi_0(\zeta)$, where $\Phi_0(\zeta)$ and $\Psi_0(\zeta_0)$ are holomorphic functions with asymptotic behavior $O(z^{-2})$ as $z \to \infty$.

Observing that $(\sigma_r)_\Gamma = -\sigma_0$ on the boundary contour of the holes, we have $\sigma_\theta = \sigma_1 + \sigma_2 + \sigma_0$. Then, as we have already observed, $\Phi(\zeta) = \frac{1}{4}(\sigma_1 + \sigma_2)$, while the second relation in Eq. (4.87) assumes the form

$$\frac{\mu c r_k^2}{(\zeta - \zeta_k)^2} + \frac{\mu r_k^2 \omega_0'(\zeta)}{(\zeta - \zeta_k)^2} = c\beta + \omega_0'(\zeta)\Psi_0(\zeta) \tag{4.88}$$

Let us look at the second term on the left-hand side of Eq. (4.88). We note that $r_k^2/(\zeta - \zeta_k)^2 = -d\bar{\zeta}/d\zeta$, with $\zeta \in \Gamma_k$, which we rewrite in the form

$$\frac{r_k^2 \omega_0'(\zeta)}{(\zeta - \zeta_k)^2} = -\lim_{\Delta\zeta \to 0} \frac{\overline{\Delta\zeta}}{\Delta\zeta} \frac{\overline{\omega_0(\zeta + \Delta\zeta) - \omega_0(\zeta)}}{\overline{\Delta\zeta}} = -\frac{d\overline{\omega_0}}{d\zeta}$$

Substituting this expression into Eq. (4.88) and integrating, we obtain

$$\Lambda(\zeta) + \overline{\omega_0(\zeta)} = -c\left(\frac{\beta}{\mu}\zeta + \frac{r_k^2}{\zeta - \zeta_k}\right) + d_k \tag{4.89}$$
$$\zeta \in \Gamma_k, \quad k = 1, 2, \ldots, n$$

The function $\Lambda(\zeta)$ is holomorphic. $\Lambda'(\zeta) = \omega_0'(\zeta)\,\Psi_0(\zeta)/\mu$, while d_k is an arbitrary constant. By changing the value of d_k, we come to the conclusion that the bounded functions $\Lambda(\zeta)$ and $\omega_0(\zeta)$ decrease as $\zeta \to \infty$. To solve our boundary-value problem, we can follow the ideas of D. I. Sherman.[155] The functions $\Lambda(\zeta)$ and $\omega_0(\zeta)$ are defined by integrals of the Cauchy type

$$\Lambda(\zeta) = \frac{1}{2\pi i}\int_\Gamma \frac{u(t)}{t-\zeta}\,dt, \qquad \omega_0(\zeta) = \frac{1}{2\pi i}\int_\Gamma \frac{\overline{u(t)}}{t-\zeta}\,dt \tag{4.90}$$

The conditions at infinity are defined by Eq. (4.90). Substituting Eq. (4.90) into Eq. (4.89), we derive

$$u(\zeta) + \frac{1}{2\pi i}\int_\Gamma u(t)\,d\ln\frac{t-\zeta}{\bar{t}-\bar{\zeta}} - d_k = -c\left(\frac{\beta}{\mu}\zeta + \frac{r_k^2}{\zeta-\zeta_k}\right) \tag{4.91}$$

The constants d_k are determined as follows

$$d_k = -\frac{1}{2\pi r_k}\int_{\Gamma_k} u(t)\,ds, \qquad ds = |dt| \tag{4.92}$$

Equation (4.92) is an equation of the Fredholm type, with a real symmetric kernel. Separating in this equation the real and imaginary parts, we obtain a pair of integral equations with respect to a potential of a double layer in a transformed Dirichlet problem, for a class of bounded continuous functions. The equations have a unique solution. Hence, Eq. (4.91) with conditions of Eq. (4.74) also has a solution for an arbitrary function on the right-hand side of Eq. (4.91). In case $n = 1$, Eq. (4.91) has the obvious solution: $\Lambda(\zeta) = \zeta$, $\omega_0(\zeta) = \beta/\mu\zeta$. The function $\omega(\zeta) = c[(\zeta + \beta)/\mu\zeta]$ coincides with the function $\omega(\zeta)$ given by Eq. (4.81), if we assign an additional map of the exterior of the unit circle into the exterior of the interval $(-1, +1)$ defined by the Zhukovskii function.

For $n > 1$, Eq. (4.91) was solved in Ref. 46 by a purely numerical technique. The author of that article computed the optimum shapes for a cyclically symmetric distribution of n holes. He assumed that a uniform pressure is applied to the contours Γ of the boundaries, i.e., $(\sigma_r)_\Gamma = -\sigma_0$, $(\tau_{r\theta})_\Gamma = 0$, and that the stresses vanish at infinity, i.e., $\sigma_1 = \sigma_2 = 0$.

Figure 4.9 illustrates the shapes of optimum holes determined by these computations. Curves 1 to 4 illustrate the shapes for values of the only independent parameter, $\lambda = r^{-1}H\sin(\tau/2)$, $\lambda > 0$, when it is chosen to be equal to 1.01,

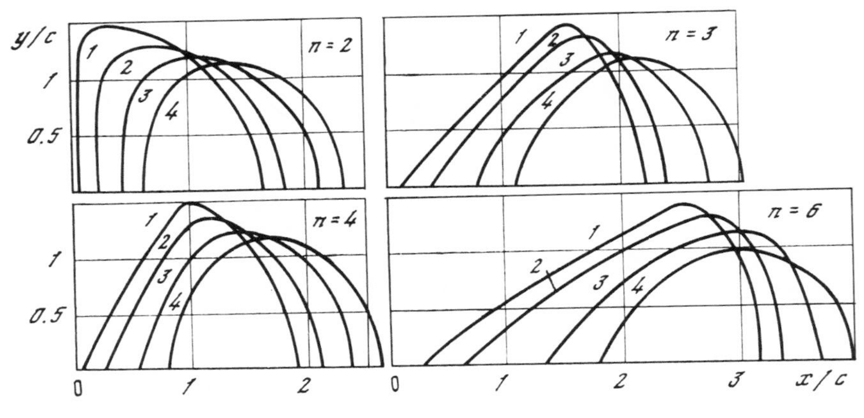

Figure 4.9

1.1, 1.3, and 1.5, with $\tau = 2\pi/n$ and $H = |\zeta_k|$. For $\lambda \gg 1$, and for assigned boundary conditions, we have an obvious asymptotic solution

$$\Lambda(\zeta) = \sum_{k=1}^{n} (\zeta_k - \zeta)^{-1}, \qquad \omega(\zeta) = \sum_{k=1}^{n}{}' (\zeta - \zeta_k)^{-1}, \quad \zeta \in \Gamma_l$$

The prime after the summation symbol indicates that the term with $k = l$ is omitted.

4.6. Optimization of the Shapes of Holes in Plates Subjected to Bending

1. *Equation of plate bending.* First, before we formulate an optimization problem, let us recall the basic equations describing the bending of an infinite plate having a hole. Let Ω denote a doubly-connected region of the xy plane occupied by the material of the plate, where Γ is the boundary of the hole (see Fig. 4.10.) We use Cartesian x-y coordinates and an ns coordinate system re-

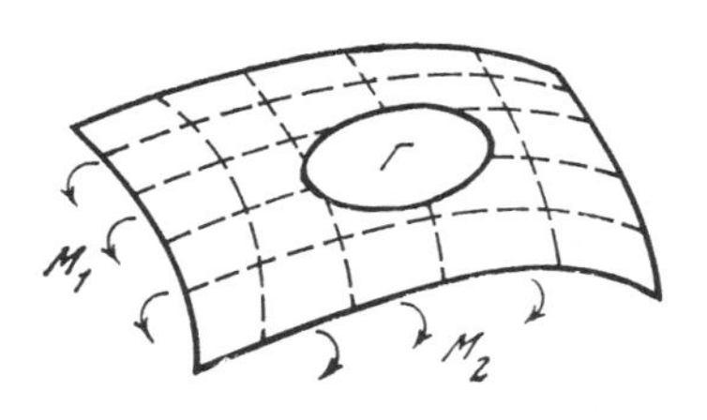

Figure 4.10

lated to the shape of the contour Γ, which measures the distance along the directions normal and tangential to the curve describing the boundary of the hole. We assume that this curve does not contain any corner points. We use the commonly accepted notation for the bending moments and shear forces: M_x, M_y, M_{xy}, Q_x, Q_y, and M_n, M_s, M_{ns}, Q_n, Q_s. The plate is bent by application of the moments $M_x = M_1 \geqslant 0$, $M_y = M_2 \geqslant 0$, and $M_{xy} = 0$, which are applied at infinity. The boundary of the hole is load-free. The displacement function $w = w(x, y)$ satisfies the biharmonic equation $\Delta^2 w = 0$, since we have assumed the absence of transverse loads. Boundary values that reflect our assumption concerning the absence of loads on the boundary of the hole may be stated as

$$M_n = 0, \quad M_{ns} = 0, \quad Q_n = 0, \; (x, y) \in \Gamma \tag{4.93}$$

Bending moments and shear loads are related to the displacement function as follows[119]:

$$\begin{aligned} M_x &= -D\,(w_{xx} + \nu w_{yy}) \\ M_y &= -D\,(w_{yy} + \nu w_{xx}) \\ M_{xy} &= -D\,(1+\nu)\,w_{xy} \\ Q_x &= -D\,(\Delta w)_x \\ Q_y &= -D\,(\Delta w)_y \end{aligned} \tag{4.94}$$

where $D = Eh^3/[12(1-\nu^2)]$ is the cylindrical rigidity of the plate, E is Young's modulus, ν is Poisson's ratio, and h is the thickness of the plate.

In the general case, it is well known[130] that one of the conditions in Eq. (4.93) is superfluous. Here we keep in mind that for a hole of arbitrary shape a solution of the biharmonic equation cannot simultaneously satisfy all three of the above stated boundary conditions. For this reason, in the engineering literature dealing with the theory of thin plates, only approximate boundary conditions are usually given in the form of two equations, $M_n = 0$ and $(Q_n + \partial M_{ns})/\partial s = 0$. Solution of a boundary-value problem with these boundary conditions leads to a definite distortion of a stress field near the boundary of a hole. Further, it will be proved below that for holes with optimum shape it is possible to satisfy all three conditions of Eq. (4.93). Below, we will investigate the optimization problem, using only the conditions $M_n = 0$, and $M_{ns} = 0$. For an optimum shape of the holes, the condition $Q_n = 0$ is automatically satisfied.

If we solve the boundary-value problem with auxiliary boundary conditions assigned at infinity, such as $M_x = M_1$, $M_y = M_2$, and $M_{xy} = 0$, and find the displacement function $w(x, y)$, then all components of the moments and shears can be determined by applying Eq. (4.94). All nonzero components of the stress tensor may then be computed, according to the formulas given below[118,119]

$$\begin{aligned} \sigma_x &= 12\zeta h^{-3} M_x, \quad \sigma_y = 12\zeta h^{-3} M_y, \quad \tau_{xy} = 12\zeta h^{-3} M_{xy} \\ \tau_{xz} &= {}^3/_2\,(h^2 - 4\zeta^2)\,h^{-3} Q_x, \quad \tau_{yz} = {}^3/_2\,(h^2 - 4\zeta^2) h^{-3} Q_y \end{aligned} \tag{4.95}$$

Here the real variable ζ measures distance from the middle surface, in the direction normal to it. It's values are contained in the interval $-h/2 \leqslant \zeta \leqslant h/2$.

We adopt as our function f, which defines the quality criterion of Eq. (4.53), the magnitude of the second invariant of the deviator component for the tensor

$$\begin{aligned} f &= \sigma_x^2 + \sigma_y^2 - \sigma_x\sigma_y + 3\,(\tau_{xy}^2 + \tau_{xz}^2 + \tau_{yz}^2) \\ &= 144\zeta^2 h^{-6}\,[(M_x + M_y)^2 + 3\,(M_{xy}^2 - M_x M_y)] \\ &\quad + \frac{27}{4} h^{-6}\,(h^2 - 4\zeta^2)^2 [Q_x^2 + Q_y^2] \end{aligned} \tag{4.96}$$

where $(x, y) \in (\Omega + \Gamma)$ and $-h/2 \leqslant \zeta \leqslant h/2$. We note that f depends explicitly on ζ, but depends on x and y only implicitly through M_x, M_y, M_{xy}, Q_x, and Q_y.

The optimization problem consists of finding the shape of the contour for which we realize the minimum of the maximum value of f in the region $(\Omega + \Gamma)$ occupied by the plate material

$$J_* = \min_\Gamma J = \min_\Gamma \max_{xy} \max_\zeta f \tag{4.97}$$

This problem differs from the optimization problems considered in Section 4.4, because the function f given in Eq. (4.96) depends not only on the coordinates x and y, but also on the coordinate ζ. In the process of finding the minimum, as given in Eq. (4.97) with respect to Γ, we assume as before that Γ cannot be contracted to a point, i.e., we assume the existence of a hole in our domain.

2. *Optimality criteria.* We represent the function f as a sum of two components

$$\begin{aligned} f &= f_1 + f_2, \qquad f_2 = \frac{27}{4}\,(h^2 - 4\zeta^2)^2\,h^{-6}\,(Q_x^2 + Q_y^2) \\ f_1 &= 144\zeta^2 h^{-6}\,[(M_x + M_y)^2 + 3\,(M_{xy}^2 - M_x M_y)] \end{aligned} \tag{4.98}$$

and consider, an auxiliary problem of minimizing the maximum value of f_1 on the curve Γ

$$J_{1*} = \min_\Gamma \max_{xy} \max_\zeta f_1 \tag{4.99}$$

where $(x, y) \in \Gamma$ and $\zeta \in [-h/2, h/2]$. We observe that the quantity given inside the square brackets in Eq. (4.98) is positive. This property can be deduced as an obvious consequence of positiveness of the function f. Indeed, assuming to the contrary and evaluating f at $\zeta = \pm h/2$, we arrive at a contradiction: $f = f_1 < 0$. Using this property, we conclude that the maximum of Eq. (4.99) occurs when $\zeta = \pm h/2$.

Let us derive a formula for the function f_1, restricted to the boundary curve Γ. To do this, we need some additional arguments. The magnitude of f is invariant as we change one orthogonal coordinate system to another one. Keeping this in mind, plus the fact that $f = f_1 = 36h^{-4} [(M_x + M_y)^2 + 3(M_{xy}^2 - M_x M_y)]$ at $\zeta = \pm h/2$, we arrive at a conclusion that the quantity inside the square brackets in Eq. (4.98) is invariant under orthogonal transformation from the xy coordinate system to the ns coordinates, i.e., $(M_x + M_y)^2 + 3(M_{xy}^2 - M_x M_y) = (M_n + M_s)^2 + 3(M_{ns}^2 - M_n M_s)$. Using the facts that $M_n = 0$ and $M_{ns} = 0$ for all $(x, y) \in \Gamma$, we arrive at the following formula for f_1, which is valid at all points of the boundary Γ:

$$(f_1)_\Gamma = a M_s^2, \qquad a = 144 \zeta^2 h^{-6} \tag{4.100}$$

We introduce an auxiliary function $\chi = M_x + M_y$. Substituting Eq. (4.94) we have $\chi = -D(1 + \nu) \Delta w$. The deflection function w satisfies the biharmonic equation. Therefore, the function χ is harmonic. Since $M_x + M_y = M_n + M_s$, we note that the conditions on Γ and at the point at infinity to derive the following relations for χ

$$\Delta\chi = 0, \qquad (\chi)_\Gamma = M_s, \qquad (\chi)_\infty = M_1 + M_2 \tag{4.101}$$

which are analogous to Eq. (4.59). Considering further Eqs. (4.100) and (4.101), and using again properties [Eq. (4.55)] of bounded harmonic functions, we conclude that the minimum of the maximum value of f_1 is attained on the curve Γ, if and only if

$$(M_s)_\Gamma = M_1 + M_2 \tag{4.102}$$

Equation (4.102) constitutes a necessary and sufficient conditions for optimality in the auxiliary problem described by Eq. (4.99). The formula $J_{1*} = 36h^{-4}(M_1 + M_2)^2$ is a direct consequence of Eqs. (4.100) and (4.102).

3. *Hole shape optimization.* Next, we shall consider the problem of minimizing the maximum value of f_1 in the domain $(\Omega + \Gamma)$, and we shall prove that in this problem the condition of optimality is given by Eq. (4.102). To do this, we study the behavior of the function f_1 in the domain $(\Omega + \Gamma)$ and we prove that for any choice of curves in $(\Omega + \Gamma)$ satisfying Eq. (4.102), the function f_1 attains the minimum of its maximum value on the boundary of a hole.

We invoke a complex representation (as given Ref. 119) of the bending moments M_x, M_y, M_{xy}, and shear loads Q_x, Q_y, in terms of two analytic functions $\Phi(z)$ and $\Psi(z)$

$$\begin{aligned} M_x + M_y &= -2D\,(1 + \nu)[\Phi(z) + \overline{\Phi}(\bar{z})] \\ M_y - M_x + 2iM_{xy} &= 2D\,(1 - \nu)[\bar{z}\Phi'(z) + \Psi(z)] \\ Q_x - iQ_y &= -4D\Phi'(z) \end{aligned} \tag{4.103}$$

where, as usual, $z = x + iy$, $\bar{z} = x - iy$, and $i^2 = -1$. The bar over Φ denotes, as before, complex conjugation.

Let us assume that Eq. (4.102) is satisfied. In that case, the analytic function $\Phi(z)$ is constant in the entire domain of definition $[\Phi'(z) \equiv 0]$ and Eq. (4.103) assumes the form

$$\begin{aligned} M_x + M_y &= M_1 + M_2 \\ M_y - M_x + 2iM_{xy} &= 2D\,(1 - \nu)\,\Psi \end{aligned} \tag{4.104}$$

The second relation given in Eq. (4.104) and its complex conjugate relation $M_y - M_x - 2iM_{xy} = 2D(1 - \nu)\,\bar{\psi}(\bar{z})$ may be multiplied by each other side by side to obtain

$$f_0 \equiv (M_x + M_y)^2 + 4\,(M_{xy}^2 - M_xM_y) = b\Psi\overline{\Psi} \tag{4.105}$$

where $b = 4D^2(1 - \nu)^2$. Let us represent the functions Ψ and $\overline{\Psi}$ in the form $\Psi = \psi_1 + i\psi_2$ and $\overline{\Psi} = \psi_1 - i\psi_2$, where ψ_1 and ψ_2 are real. The quantity f_0 can be transformed into a form that is convenient for estimating its behavior, $f_0 = b(\psi_1^2 + \psi_2^2)$. The functions ψ_1 and ψ_2 satisfy the Cauchy-Riemann equations $\psi_{1_x} = \psi_{2_y}$, $\psi_{1_y} = \psi_{2_x}$, and therefore are harmonic $(\Delta\psi_1 = \Delta\psi_2 = 0)$.

After applying the Laplace operator to f_0, performing some elementary substitutions, and recalling the properties of ψ_1 and ψ_2, we arrive at the inequality $\Delta f_0 = 4b(\nabla\psi_1)^2 \geqslant 0$. Expressing the function f_1 in terms of f_0 and of the constants introduced into this problem, we have $f_1 = 1/4\, a[(M_1 + M_2)^2 + 3f_0]$. Applying Laplace operator to both sides of this equation, we obtain $\Delta f_1 = 3ab(\nabla\psi_1)^2 \geqslant 0$. Consequently, f_1 cannot attain its maximum at an interior point of the domain $(\Omega + \Gamma)$.(74) Substituting the appropriate boundary values and comparing the value of $f_1 = a(M_1 + M_2)^2$ on Γ with the value $a(M_1 + M_2)^2 - 3M_1M_2$ it assumes at infinity, we conclude that f_1 attains its maximum value on Γ. Therefore, if Eq. (4.102) is satisfied, the function f_1 attains its maximum value in the domain $(\Omega + \Gamma)$ at some points of the boundary curve Γ. The maximum value of f_1 on Γ, as we have proved above, attains its minimum if and only if Eq. (4.102) is satisfied. Therefore, in the problem of minimizing the maximum value of f_1 in the region $(\Omega + \Gamma)$, necessary and sufficient conditions for optimality are given by Eq. (4.102).

4. *Equivalence of optimality criteria.* We now return to the original problem of minimizing the maximum value of f. Applying the results obtained in the preceding part, we can prove that the optimality conditions in this case are equivalent to Eq. (4.102). In fact, for holes satisfying Eq. (4.102), we have $Q_x - iQ_y = -4D\Phi'(z) = 0$ and

$$Q_x = Q_y = 0 \tag{4.106}$$

In this case, $f_2 = 0$ for all points in the region $(\Omega + \Gamma)$, and the equality

$$f = f_1, \qquad (x, y) \in \Omega + \Gamma \tag{4.107}$$

is identically satisfied.

For holes of arbitrary shape, it is true that

$$f = f_1 + f_2 \geqslant f_1, \qquad (x, y) \in \Omega + \Gamma \tag{4.108}$$

Comparing Eqs. (4.107) and (4.108) and recalling that the minimum of the greatest value of f_1 in the domain $(\Omega + \Gamma)$ is attained if and only if Eq. (4.102) is satisfied, we assert that this equality constitutes the necessary and sufficient condition for minimization of the maximum value of f.

5. *Multiple holes.* The proof of optimality of holes that satisfy Eq. (4.102) did not depend on connectivity of the region Ω. Therefore, Eq. (4.102) is a necessary and sufficient condition for optimality in the presence of n holes.

These arguments can be generalized to the case in which constant distributed moments are applied to the boundaries of the holes, $M_n = -M_0$ ($M_0 > 0$ is a given constant). It is easy to show that in the presence of distributed moments the optimality condition has the form

$$(M_s)_\Gamma = M_1 + M_2 + M_0 \tag{4.109}$$

Making use of Eq. (4.102) to find the shape of the optimum boundary curve leads to a well-posed boundary-value problem for the biharmonic operator, with boundary conditions $M_n = 0$, $M_{ns} = 0$, and conditions at infinity, while Eq. (4.102) serves the purpose of determining the unknown boundary Γ. With the help of relations $M_x + M_y = -D(1 + \nu)\,\Delta w$ and $M_x + M_y = M_1 + M_2$, this problem can be reduced to the inverse boundary-value problem for Poisson's equation

$$\Delta w = c, \qquad c = -(M_1 + M_2)/D\,(1 + \nu) \tag{4.110}$$

which may be investigated by methods of the theory of functions of complex variables, as given, in Ref. 144 (see Section 4.5). Making use of the results of Ref. 144, we derive the following result: In bending of a plate, optimum holes are ellipses

$$x^2 M_1^{-2} + y^2 M_2^{-2} = \lambda^2 \tag{4.111}$$

where λ^2 denotes an arbitrary positive constant.

Since finding optimum shapes of holes for plates in simultaneous tension and bending leads to an identical mathematical formulation, finding the shapes of systems of holes may be accomplished by the technique already described in Section 4.5.

Because of Eq. (4.106), plates having an optimum hole satisfy the "ignored" boundary condition $Q_n = 0$. It is also a consequence of Eq. (4.106) that all components of the stress tensor and the function f vanish on the middle surface ($\zeta = 0$).

We comment that the mean curvature of the middle surface of a plate with an optimum hole, subjected to bending, is constant on that surface for all $(x, y) \in (\Omega + \Gamma)$

$$\frac{1}{r_x} + \frac{1}{r_y} = \Delta w = c \tag{4.112}$$

where c is a constant that may be determined by Eq. (4.110), and r_x^{-1} and r_y^{-1} appearing in Eq. (4.112) denote the curvatures measured in the x and y directions, respectively.

We remark that, in analogy with our conclusion reached in Section 4.4. concerning pure tension, for plates with optimum holes subjected to bending, the plastic zone in the plate material first appears on Γ (where $\Gamma = \Sigma\Gamma_i$), and it originates simulatneously at every point of the contour Γ.

Problems similar to the ones studied in Section 4.4–4.6 arise in finding optimum shapes of bounded elastic bodies in designing structures having optimum stress distribution in regions with sharply varying geometry and having high stress concentrations (see Refs. 205, 206, 238, and 245).

5

Optimization of Anisotropic Properties of Elastic Bodies

Questions raised in this chapter relate to a new class of problems concerning optimization of the internal composition of elastic bodies. This research was stimulated by the recent interest in utilization of composite materials. In Section 5.1, we formulate some problems in optimization of anisotropic bodies. We pay special attention to finding the optimum orientation of the axes of anisotropy. Analysis of the optimality conditions and of stationary orientation of elastic moduli indicates that the necessary conditions for an extremum do not uniquely determine the best possible orientation for the axes of anisotropy. A basic difficulty in solving this optimization problem arises when we consider different possibilities of static orientation, in different parts of the body and compare the values of the cost functional. For this reason, we first consider some variational problems involving rotation matrices and study some related sufficient conditions for optimality (see Section 5.2). Applying these results in Section 5.3, we derive optimum distributions of moduli in inhomogeneous anisotropic bars having the greatest torsional rigidity. In Section 5.4 we consider questions related to optimizing the anisotropic properties of an elastic medium and their application to the two-dimensional theory of elasticity. In Section 5.5, we present results derived by numerical techniques concerning the optimum orientation of anisotropy axes. In Section 5.6, we make some observations concerning simultaneous optimization of the shape and anisotropic properties of an elastic body. Some results given in this chapter have been published in Refs. 21-24.

5.1. Optimization Problems for Anisotropic Bodies

For a loaded structure, the structural elements are generally in a deformation and stress state that is very complex. The actual stresses and deflections are not the same in different directions or in different parts of the same elastic body. This circumstance is the main reason why anisotropic and nonhomogeneous materials are widely used. Transferring some material from a lowly stressed part of the body, weakening the structure in the direction in which it "does not work," and strengthening certain directions that carry the greatest load, or reinforcing a part that has a high stress concentration, is the general idea of structural reinforcement.

Modern technology makes broad use of structural design with composites. As reinforcing elements (which are, respectively, one-dimensional or two-dimensional), we use high-strength fibers or sheets, such as boron fibers, glass fibers, ceramic fibers, or wooden dowels. The binding material could be a phenol compound, epoxy, some other polymer tarlike substances, or various types of glue. At the present time, extensive research is in progress to find new structural reinforcing materials. The idea of reinforcement opens new vistas in finding materials having desirable mechanical and physical properties. Anisotropic materials are widely used in modern technology. Increasing possibilities of different types of design, using a deliberately designed form of anisotropy, becomes an important practical consideration. We need to decide how to effectively exploit the anisotropic properties of materials in design of elastic structures. Of great interest are the following problems: What is an optimum shape of an elastic anisotropic body (made of a material with known anisotropy)? What is the optimum distribution of the moduli of rigidity in a deformed body? How can we simultaneously optimize the shape and the composition of the material?

In considering problems of optimal design, we first of all need to observe that in formulating the problem of optimum reinforcement of a structure, the choice of a numerical scheme plays a very important part. The following observations may be found in Ref. 41. Two principal numerical techniques are widely applied in such problems. The first type of numerical scheme is founded on a purely phenomenological approach. We apply directly the well-known equations of the theory of anisotropic elasticity. The mechanical constants A_{ijkl} (the elastic moduli) enter into generalized Hooke's law

$$\sigma_{ij} = A_{ijkl}\varepsilon_{kl} \tag{5.1}$$

and other state equations for the structure are determined by performing laboratory experiments on samples of the reinforcing materials. Here, σ_{ij} and ϵ_{ij} in Eq. (5.1) denote the components of the stress and strain tensor, respectively. The quantities A_{ijkl} are elements of the tensor of rank four of the elastic moduli. In the most general case of anisotropy, the number of independent moduli A_{ijkl}

is equal to 21. If the internal structure of an anisotropic body has a symmetry of any kind, this symmetry is reflected in its elastic properties.[156] A symmetry present in the elastic properties of a body is revealed by the fact that at each point inside the body certain directions may be found such that elastic properties are the same in these directions. Then the number of distinct moduli A_{ijkl} is smaller. If a plane of symmetry passes through each point, then the number of elastic constants is reduced to 13. An orthotropic body having the property that two mutually perpendicular planes of symmetry pass through each point is characterized by nine elastic constants. In a transversely isotropic material, where all planes of symmetry are parallel to each other and all such planes of symmetry are also planes of isotropy, the number of independent elastic constants is five. In the limiting case of a fully isotropic material, only two independent elastic constants exist.

We observe that the quantities A_{ijkl} form a tensor of rank four. Therefore, as one rotates the coordinate axes, they transform according to a linear law. The coefficients in this linear expression are products of direction cosines $(n_{i'i}, \ldots, n_{l'l})$ of the angles between the new directions (denoted by primes) and the old directions of the coordinate axes (see Refs. 7, 85, or 87):

$$A_{i'j'k'l'} = A_{ijkl} n_{i'i} n_{j'j} n_{k'k} n_{l'l} \tag{5.2}$$

A second type of computational scheme for evaluating the deformation or fracture of the materials is based on its macrostructural properties. In this type of scheme, the mechanical properties of composites are related to the mechanical properties of the binders and reinforcing materials to the coefficients of the reinforcement, the dimensions of the reinforcing elements, and other macrostructural parameters. The advantage of such schemes consists in the ability to relate deformations to the strength of an elastic body, to predict mechanical properties of composite materials from the mechanical properties of the components (see Ref. 41), to solve problems of optimal design or material selection, etc.

Basic techniques for which equations of reinforced materials were derived were outlined in Refs. 2, 40, 41, 141, and 175–177. Some extremal properties of reinforced materials were derived in these references. Utilizing established mathematical theories of anisotropic media, it is possible to formulate various optimization problems. Problems concerning the optimum shape of an anisotropic body are a natural generalization of corresponding problems for an isotropic body. The essential difficulty in finding the optimum solution is caused by the greater complexity of the state equations for an anisotropic structure. One type of problem that does not fit into the classical model of structural optimization is the problem of optimizing the distribution of elastic moduli in the medium. We shall concentrate on the type of problem in the present chapter. For the sake of convenience, we shall consider a specific problem of minimizing the compliance of a structure for a given weight.

Let the elastic medium consist of identical, infinitely small crystals that have a random orientation relative to each other. Since the crystals are all the same, this means that the axes of symmetry for the elastic properties, relative to some fixed Cartesian coordinates, vary as we move from one point to another inside the medium, but the values of elastic moduli measured along the axes of symmetry remain the same. We describe the orientation of the axes of anisotropy at each point $x = \{x_1, x_2, x_3\}$ in the medium, relative to the fixed $x_1 x_2 x_3$ coordinate system, by the triple of angles $\alpha_1(x)$, $\alpha_2(x)$, $\alpha_3(x)$, which are regarded as components of the control variable vector $\alpha(x)$, i.e., $\alpha(x) = \{\alpha_1(x), \alpha_2(x), \alpha_3(x)\}$. Let $\alpha_j (j = 1, 2, 3)$ denote the angle between the axis x_j of elastic symmetry and the fixed axis x_j of the Cartesian coordinate system. Determining the optimum orientation of the axes of anisotropy, i.e., of the vector function $\alpha(x)$, from the condition of minimal compliance, amounts to finding

$$J_* = \min_\alpha J(\alpha) \tag{5.3}$$

A different type of optimization problem arises if at each point the axes of anisotropy are given, while the magnitudes of elastic moduli along these given axes are regarded as the unknown control variables. In formulating these problems, we must assign bounds on elastic moduli, which are determined by studying the structural properties of composites and mechanical properties of their components. Otherwise, we would permit arbitrarily large values for elastic moduli.

As we optimize anisotropic properties of a material, we may choose as our unknown control variables the distributions of the parameter values of the macrostructure. In fact, if we regard the elastic moduli A_{ijkl} as averages at a point of material properties that depend on the parameters of the macrostructure (i.e., of concentrations of the reinforcing and binding materials, dimensions and positioning of reinforcing elements, etc.), we could regard these parameters as the control quantities. Such formulation of the optimization problem permits us to account for design and technological limitations. Therefore, solutions of these problems are of substantial interest in technological applications. Finding an optimum distribution of the values of elastic moduli permits us to determine the optimum direction of reinforcement and to formulate an opinion with regard to traditionally accepted structural designs. Even when it is difficult to construct a structure having the optimum anisotropy, it is useful to have a theoretical solution to the optimization problem, to clarify limitations and to apply quasi-optimal reinforcement designs.

Before we tackle some specific problems of optimization for structures made of anisotropic materials, we shall follow the ideas of Ref. 23 and consider an auxiliary variational problem, consisting of finding the optimum rotation transforming a given matrix.

5.2. An Extremal Problem for Rotation of a Matrix

1. *Optimality criteria.* Let the state of a system be described by a scalar function φ and a square matrix T. The function $\varphi(x, y)$ and the elements $t_{ij} = t_{ij}(x, y)$ of the matrix T are defined in the region Ω of the x-y plane. The function φ vanishes on the boundary Γ of Ω. The matrix T is orthogonal. We define an integral functional J which depends on φ and T

$$J(\varphi, T) = \iint_{\Omega} [(\nabla\varphi, T^*AT\nabla\varphi) - 2f\varphi]\, dx\, dy \tag{5.4}$$

where $f > 0$ is a scalar function whose values are known in Ω and A is a positive definite symmetric matrix that is given in Ω. The elements a_{ij} $(i, j = 1, 2)$ of A depend on x and y and $a_{11} > 0$ and $a_{11}a_{22} - a_{12}^2 > 0$. The round brackets in Eq. (5.4), under the integral sign, denote a scalar product, the asterisk superscript denotes transpose of a matrix, and ∇ denotes the gradient operation, i.e., $\nabla\phi = \{\varphi_x, \varphi_y\}$. The eigenvalues of the matrix A are denoted by λ_i, with $i = 1, 2$. Since A is positive definite, λ_i are positive real numbers $(\lambda_i > 0, i = 1, 2)$. To be more specific, let us assume that $\lambda_1(x, y) < \lambda_2(x, y)$. The product of matrices that appears in Eq. (5.4) is denoted by $M = T^*AT$. We note that the eigenvalues of the matrix M are the same as those of A, i.e., they are equal to λ_i.

With these assumptions, we consider the minimization of the functional J with respect to φ and T

$$\begin{gathered} J_* = \min_T \min_\varphi J(\varphi, T) \\ (\varphi)_\Gamma = 0, \quad T^*T = E \end{gathered} \tag{5.5}$$

where E denotes the identity matrix. Let us derive some necessary conditions for the variational problem of Eqs. (5.4) and (5.5). We consider the function φ and the matrix T as slightly perturbed values that still satisfy the boundary conditions and the orthogonality condition of Eq. (5.5). We require the variation of

φ to be identically zero on the boundary Γ, while δT should behave in a skew-symmetric manner, i.e.,

$$(\delta\varphi)_{\Gamma} = 0, \qquad (\delta T)^* = -\delta T \tag{5.6}$$

Then the first variation δJ of the functional J may be expressed in the form

$$\delta J = 2\iint\limits_{\Omega} (\delta T \nabla\varphi, M\nabla\varphi)\, dx\, dy + 2\iint\limits_{\Omega} [(\nabla\delta\varphi, M\nabla\varphi) - f\delta\varphi]\, dx\, dy$$

Since δJ must be equal to zero for any scalar function $\delta\varphi$ satisfying the boundary condition of Eq. (5.6) and for any skew-symmetric matrix δT, we obtain the necessary conditions for an extremum in terms of φ and T. A necessary condition for any extremum of J, regarded as a function of φ, is given by the Euler–Lagrange equation

$$\operatorname{div}\,(M\nabla\varphi) = -f \tag{5.7}$$

A necessary condition for the extremum of J, regarded as a function of T, is given by collinearity of $\nabla\varphi$ and $M\nabla\varphi$. Therefore, the vector $\nabla\psi$ is an eigenvector of the matrix M, i.e.,

$$M\nabla\varphi = \lambda_i \nabla\varphi \qquad (i = 1, 2) \tag{5.8}$$

Substituting the relations of Eq. (5.8) into the Euler–Lagrange equation of Eq. (5.7), we have a system of equations that may be used to determine φ, for stationary behavior of J, if T is given. According to Eq. (5.8) we have

$$\begin{gathered} \operatorname{div}\,(\lambda_i \nabla\varphi) = -f \qquad (i = 1, 2) \\ \lambda_i = \frac{1}{2}\left[a_{11} + a_{22} + (-1)^i \sqrt{(a_{11} - a_{22})^2 + 4a_{12}^2}\right] \end{gathered} \tag{5.9}$$

The elements of the orthogonal matrix T are written in the form

$$t_{11} = \cos\alpha, \qquad t_{12} = -t_{21} = \sin\alpha, \qquad t_{22} = \cos\alpha$$

where α is the angle of rotation assigned to the matrix T. Using Eq. (5.8), we obtain an explicit formula relating the angle $\alpha = \alpha(x, y)$ to the function $\varphi = \varphi(x, y)$. To be more specific, let us assume that the vector $\nabla\varphi = \{\varphi_x, \varphi_y\}$ corresponds to the eigenvalue λ_i. Then, corresponding to the eigenvalue λ_j, $j \neq i$, there is an eigenvector $b = \{\varphi_y, -\varphi_x\}$ that is orthogonal to the eigenvector $\nabla\varphi$. We form a scalar product of both sides of Eq. (5.8) with the vector b. Then we have $(b, M\nabla\varphi) = 0$. This relation contains two separate cases:

Case 1

$$\cos 2\alpha = P, \qquad \sin 2\alpha = Q$$

$$P = -\frac{(a_{11} - a_{22})(\varphi_x^2 - \varphi_y^2) + 4a_{12}\varphi_x\varphi_y}{(\nabla\varphi)^2 \sqrt{(a_{11} - a_{22})^2 + 4a_{12}^2}} \tag{5.10}$$

$$Q = -\frac{2(a_{11} - a_{22})\varphi_x\varphi_y - 2a_{12}(\varphi_x^2 - \varphi_y^2)}{(\nabla\varphi)^2 \sqrt{(a_{11} - a_{22})^2 + 4a_{12}^2}}$$

corresponds to the smaller eigenvalue λ_1.

Case 2

$$\cos 2\alpha = -P, \qquad \sin 2\alpha = -Q \tag{5.11}$$

corresponds to the larger eigenvalue λ_2. Therefore, the condition for stationary behavior of J does not offer a technique for uniquely determining the angle α and does not result in formulation of a solvable boundary-value problem for finding the unknown quantities. To derive a single-valued dependence of the function α on the variable φ we need to investigate the sign of the second variation of the cost functional and clarify which of the two cases corresponds to the minimum of J.

2. *Proof of optimality.* We shall prove that the minimum of J is attained if in the entire region Ω the condition of Eq. (5.10), corresponds to the smaller eigenvalue, is satisfied. To carry out this proof, we formulate an expression for the second variation of the functional J, in terms of variations in ψ and T satisfying Eq. (5.5)

$$\begin{aligned}\delta^2 J = \iint_\Omega \{&(\nabla\delta\varphi, M\nabla\delta\varphi) + (\delta T\nabla\varphi, M\delta T\nabla\varphi) \\ &+ (M\nabla\varphi, \delta T\delta T\nabla\varphi) + 2(\delta T\nabla\varphi, M\nabla\delta\varphi) \\ &- 2(\nabla\delta\varphi, \delta T M\nabla\varphi)\}\, dx\, dy\end{aligned} \tag{5.12}$$

Let $\nabla\varphi$ be an eigenvector corresponding to the eigenvalue λ_1. Then the vector $\delta T\nabla\varphi$, orthogonal to it, corresponds to the eigenvalue λ_2. If we represent the matrix δT as $\delta T = \delta\alpha B$ where B is a skew-symmetric matrix with the elements $b_{11} = b_{22} = 0, b_{12} = -b_{21}$, then

$$\delta T \delta T = -(\delta\alpha)^2 E, \qquad \delta T \nabla\varphi = \delta\alpha \{\varphi_y, \ -\varphi_x\} \tag{5.13}$$

Using Eq. (5.13), we transform Eq. (5.12) into the form

$$\delta^2 J = \iint_\Omega (p, Cp)\, dx\, dy, \qquad p = \{\delta\alpha, \delta\varphi_x, \delta\varphi_y\} \tag{5.14}$$

where C is a symmetric matrix with entries given by

$$\begin{aligned}
c_{11} &= (\lambda_2 - \lambda_1)(\nabla\varphi)^2, \qquad c_{12} = c_{21} = (\lambda_2 - \lambda_1)\varphi_y \\
c_{13} &= c_{31} = -(\lambda_2 - \lambda_1)\varphi_x \\
c_{22} &= {}^1\!/_2\,[\lambda_1 + \lambda_2 - (\lambda_2 - \lambda_1)(\varphi_x^2 - \varphi_y^2)(\nabla\varphi)^{-2}] \\
c_{23} &= c_{32} = -(\lambda_2 - \lambda_1)\varphi_x\varphi_y\,(\nabla\varphi)^{-2} \\
c_{33} &= {}^1\!/_2\,[\lambda_1 + \lambda_2 + (\lambda_2 - \lambda_1)(\varphi_x^2 - \varphi_y^2)(\nabla\varphi)^{-2}]
\end{aligned}$$

The integrand in Eq. (5.14) represents a quadratic form in the components of the $p = \{\delta\alpha_1, \delta\varphi_x, \delta\varphi_y\}$. To find the sign of this quadratic form, we compute the principal minors Δ_1, Δ_2, Δ_3 of the matrix C, of order one, two, and three, respectively. We have

$$\begin{aligned}
\Delta_1 &= (\lambda_2 - \lambda_1)(\nabla\varphi)^2, \qquad \Delta_2 = \lambda_1\,(\lambda_2 - \lambda_1)(\nabla\varphi)^2 \\
\Delta_3 &= \lambda_1^2\,(\lambda_2 - \lambda_1)(\nabla\varphi)^2
\end{aligned} \tag{5.15}$$

Our assumption concerning the positive definite property of A and the inequality $\lambda_1(x, y) < \lambda_2(x, y)$ implies that the minors given in Eq. (5.15) are nonnegative ($\Delta_i \geqslant 0$, $i = 1, 2, 3$). The quantities Δ_i are all equal to zero if $\nabla\varphi = 0$. However, vanishing of the gradient of φ in a bounded subdomain $\Omega_0 \subset \Omega$ contradicts Eq. (5.7). Hence, the strong inequalities $\Delta_i > 0$ ($i = 1, 2, 3$) are satisfied almost everywhere in Ω. According to a criterion derived by Sylvester, these inequalities imply that the quadratic form (p, Cp) is positive definite, and therefore that $\delta^2 J > 0$. Thus, if the function $\alpha(x, y)$ depends, in the region Ω, on φ according to Eq. (5.10) (the case of the smaller eigenvalue), then the functional J attains its minimum.

5.3. Optimal Anisotropy for Bars in Torsion

The results derived in Section 5.2 are now applied to find the optimum distribution of the elastic moduli for bars in torsion. We consider the problem of pure torsion of an elastic cylindrical bar. Let the axis of the shaft be parallel to the z axis, in the x-y-z Cartesian coordinate system. The torques applied to the ends of the bar act along that axis. We denote by Ω the x-y plane cross section of the bar and by Γ the boundary of Ω. Material of the shaft is assumed to be linearly anisotropic, having at each point a plane of symmetry perpendicular to the generators (that is to the z axis).

We introduce the stress function $\varphi(x,y)$, which is related to the unit angle of twist θ and to the components of the stress tensor by means of the formulas $\tau_{xz} = \theta\varphi_y$, $\tau_{yz} = -\theta\varphi_x$. The equations of state and the boundary conditions appear in the well-known form (86)

$$
\begin{aligned}
&(m_{11}\varphi_x - m_{12}\varphi_y)_x + (m_{22}\varphi_y - m_{12}\varphi_x)_y = -2 \\
&\varphi = 0, \qquad (x, y) \in \Gamma
\end{aligned}
\tag{5.16}
$$

where m_{11}, m_{12}, and m_{22} are deformation coefficients relative to the x-y-z coordinate system. In addition to the Cartesian coordinates x-y-z, we also introduce ξ-η-ζ coordinates, with the ζ axis parallel to the z axis, where the ξ axis forms the angle $\alpha(x,y)$ with the x axis (see Fig. 5.1). The bar material is characterized by deformation coefficients $a_{11}(x,y)$, $a_{12}(x,y)$, and $a_{22}(x,y)$ measured along the axes ξ, η, and ζ. These coefficients are assumed to be known functions of the variables x and y. The coefficients m_{ij} are related to the coefficients a_{ij}(86)

$$
\begin{aligned}
m_{11} &= a_{11}\cos^2\alpha - a_{12}\sin 2\alpha + a_{22}\sin^2\alpha \\
m_{22} &= a_{11}\sin^2\alpha + a_{12}\sin 2\alpha + a_{22}\cos^2\alpha \\
m_{12} &= {}^1\!/_2\,(a_{11} - a_{22})\sin 2\alpha + a_{12}\cos 2\alpha
\end{aligned}
\tag{5.17}
$$

which may be related as a matrix equation $M = T^*AT$ (for the meaning of this notation, see Section 5.2).

The function $\alpha = \alpha(x,y)$ describing the orientation of the ξ-η-ζ axes will be regarded as the control variable. We consider the following optimization problem. We wish to determine a function $\alpha = \alpha(x,y)$ such that the solution of the boundary-value problem of Eqs. (5.16) and (5.17) assigns a maximum to the

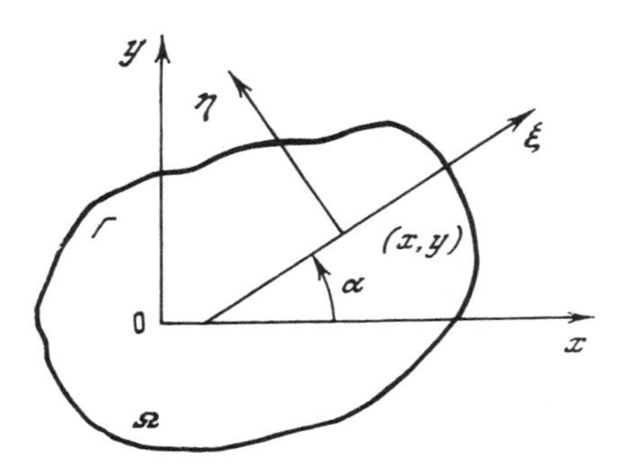

Figure 5.1

functional

$$K(\alpha) = 2 \iint_{\Omega} \varphi(x, y)\, dx\, dy \to \max_{\alpha} \tag{5.18}$$

which represents torsional stiffness of the bar.

We observe that if the elastic material is locally orthotropic and the ξ-η-ζ axes are the axes of orthotropy, then our problem is the one of finding the optimum distribution of the angles of orthotropy throughout the material. The problem of maximizing torsional rigidity of a bar permits us to ignore the differential relation of Eq. (5.16).

We introduce the functional

$$J = \iint_{\Omega} (m_{11}\varphi_x^2 - 2m_{12}\varphi_x\varphi_y + m_{22}\varphi_y^2 - 4\varphi)\, dx\, dy \tag{5.19}$$

and observe that for a given function $\alpha(x, y)$ the solution of the boundary-value problem of Eqs. (5.16) and (5.17) corresponds to the minimum value of the functional J. It is not hard to check that our assertion is true. We write the Euler–Lagrange equation with respect to the variable φ for the functional of Eq. (5.19) and observe that the well-known conditions for the absolute minimum of this functional, namely $m_{11} > 0$, and $m_{11}m_{22} - m_{12}^2 > 0$, are automatically satisfied. Moreover, assuming that the function $\varphi(x, y)$ that minimizes the functional of Eq. (5.19) and satisfies the boundary conditions of Eq. (5.16) also satisfies the equality $J(\varphi) = -K(\varphi)$. This relation can be written in the form $K = -\min_{\varphi} J$. Using this relation and the conditions of Eqs. (5.16) and (5.17), we can reduce our maximization problem to a computation of the minimum of the functional J with respect to the choice of functions φ and α, i.e.,

$$K_* = \max_\alpha K(\alpha) = -\min_\alpha \min_\varphi J(\alpha, \varphi) \tag{5.20}$$

An interior minimum with respect to φ of the functional of Eq. (5.20) is computed for a given value $\alpha = a(x, y)$ and with the assigned boundary condition of Eq. (5.16).

Hence, the finding the optimum orientation of the axes of anisotropy is reduced to the solution of a variational problem that we have already discussed in Section 5.2. Therefore, all results obtained in Section 5.2 may be applied to our study of maximizing the torsional rigidity of a bar. In the general case, finding of φ and α can be reduced to the solving of the boundary-value problem

$$\begin{gathered}(\lambda_1\varphi_x)_x + (\lambda_1\varphi_y)_y = -2, \qquad (\varphi)_\Gamma = 0 \\ \lambda_1 = \frac{1}{2}\left(a_{11} + a_{22} - \sqrt{(a_{11} - a_{22})^2 + 4a_{12}^2}\right)\end{gathered} \tag{5.21}$$

while α is found from Eq. (5.10). The solution of boundary-value problems of the type of Eq. (5.21) may be obtained in various cases concerning the assignment of the coefficient λ_1 (for the torsion of inhomogeneous bars). Such results were given in Refs. 86 and 91. Let the coefficients of deformation, and therefore the eigenvalue λ_1, be independent of position coordinates x and y. The torsion equation becomes the Poisson equation

$$\varphi_{xx} + \varphi_{yy} = -2\lambda_1^{-1} \tag{5.22}$$

describing torsion of an isotropic bar with the shear modulus $G = \lambda_1^{-1}$, subject to the boundary condition $(\varphi)_\Gamma = 0$. Since the theory of torsion of isotropic bars has been well researched and the solution has been found for a majority of practically important shapes, either analytically or numerically (see for example Ref. 6), this representation of our problem permits us to solve directly the optimization problem posed here. Let us study the problem of optimizing a bar made of a locally orthotropic material $a_{11} = 1/G_1$, $a_{22} = 1/G_2$, and $a_{12} = 0$, where $G_1 > G_2$ are shear moduli. In this case the angles of the orthotropic axes are given by

$$\begin{aligned}&\alpha = {}^1\!/_2 \arctan\mu \text{ when } \mu \geq 0,\ \varphi_x\varphi_y \geq 0 \text{ and when } \mu < 0\ \varphi_x\varphi_y < 0 \\ &\varphi_x\varphi_y < 0 \\ &\alpha = {}^1\!/_2 \arctan\mu + \tfrac{1}{2}\pi \text{ when } \mu > 0\ \varphi_x\varphi_y < 0 \text{ and when } \mu < 0 \\ &\varphi_x\varphi_y > 0\end{aligned} \tag{5.23}$$

where μ denotes the quantity $2\varphi_x\varphi_y/(\varphi_x^2 - \varphi_y^2)$.

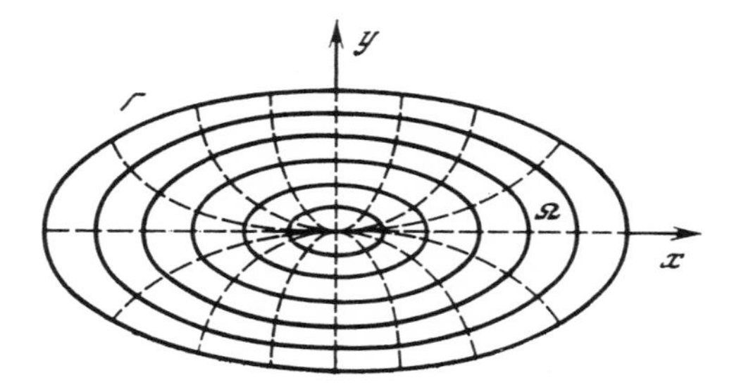

Figure 5.2

It is easy to show[22] that at each point $(x, y) \in \Omega$, the axis η having the largest shear modulus is tangent to a level surface of φ, while the axis ξ corresponding to the smallest shear modulus G_2 is orthogonal to it.

Figures 5.2 and 5.3 illustrate solutions of the optimization problem for bars having an elliptical and a square cross section. The solid and dotted curves represent, respectively, the families of curves of constant stress and lines orthogonal to these families.

Let us evaluate the effectiveness of the optimization process. This means comparing the rigidity K_* of the optimum bar with the rigidity of a homogeneous isotropic bar that has the same cross section Ω and has the torsional rigidity $G_c = (G_1 + G_2)/2$. The gain in rigidity attains by optimizing does not depend on the shape of the cross section and is given by

$$(K_* - K_c)/K_c = (G_1 - G_2)/(G_1 + G_2)$$

It is clear from this formula that the gain varies between 0 and 100%, as the ratio of the moduli G_1/G_2 varies between one and infinity.

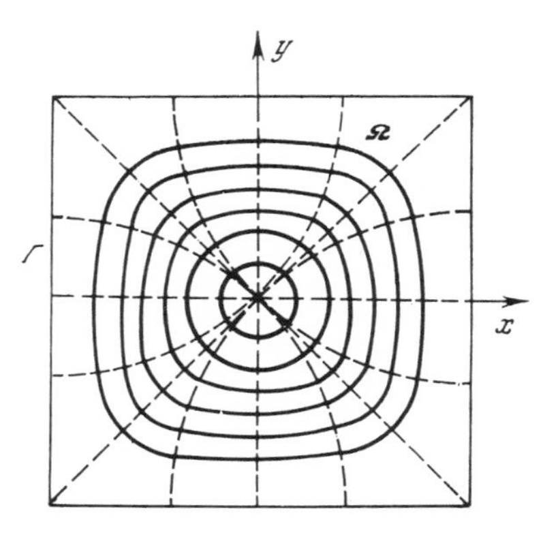

Figure 5.3

5.4. Optimization of Anisotropic Properties of an Elastic Medium in Two-Dimensional Problems of the Theory of Elasticity

We consider the problem of optimizing the anisotropic properties of an elastic body, with two-dimensional deformations and stresses (see Ref. 24).

1. *Formulation.* We adopt a Cartesian x-y coordinate system for our two-dimensional problem in the theory of elasticity, concerning equilibrium of an elastic, anisotropic body loaded by forces (q_x, q_y) on the part Γ_1 of its boundary and rigidly held on the part Γ_2 of the boundary. We assume that the material properties do not change in the direction of the z axis, which is perpendicular to the x-y plane, and therefore that $\epsilon_z = 0$ (a two-dimensional deformation field). We also assume that the elastic medium is locally orthotropic and denote by ξ and η the axes of orthotropy. At the point with coordinates (x, y), the direction of the ξ-η axes of orthotropy, relative to the x-y axes is given by the angle $\alpha = \alpha(x, y)$, where α is the angle between the x and ξ axes. The values of the orthotropic constants $A_{11}^0, A_{12}^0, A_{13}^0, A_{22}^0, A_{23}^0, A_{33}^0, A_{44}^0, A_{55}^0$, and A_{66}^0 (see Ref. 87 for explanations) are known. The equilibrium of this elastic body, for boundary conditions specified above, is described by a variational principle

$$\Pi = \iint_{\Omega} f dx\, dy - \int_{\Gamma_1} (uq_x + vq_y)\, ds \to \min_{u,\, v}$$
$$f = \frac{1}{2}(A_{11}\varepsilon_x^2 + A_{22}\varepsilon_y^2 + A_{66}\gamma_{xy}^2) + A_{12}\varepsilon_x\varepsilon_y + A_{16}\varepsilon_x\gamma_{xy} + A_{26}\varepsilon_y\gamma_{xy} \tag{5.24}$$

where u and v are displacements along the x and y axes, respectively, and $\epsilon_x, \ldots, \epsilon_{xy} = 1/2\, \gamma_{xy}, \sigma_x, \ldots, \tau_{xy}$, are components of the strain and stress tensors. In a fixed x-y coordinate system, the elastic moduli A_{ij} are related to the assigned constant A_{ij}^0 in the $\xi\eta$ system by well-known transformation formulas

$$\begin{aligned} A_{11} &= c_1 \cos^4 \alpha + c_2 \sin^4 \alpha + c_3, \qquad A_{22} = c_1 \sin^4 \alpha \\ &\quad + c_2 \cos^4 \alpha + c_3 \\ A_{12} &= (c_1 + c_2) \sin^2 \alpha \cos^2 \alpha + A_{12}^0 \\ A_{16} &= \sin \alpha \cos \alpha\, (c_2 \sin^2 \alpha - c_1 \cos^2 \alpha) \\ A_{26} &= \sin \alpha \cos \alpha\, (c_2 \cos^2 \alpha - c_1 \sin^2 \alpha) \\ A_{66} &= (c_1 + c_2) \sin^2 \alpha \cos^2 \alpha + A_{66}^0 \end{aligned} \tag{5.25}$$

where we substituted $c_1 = A_{11}^0 - A_{12}^0 - 2A_{66}^0$, $c_2 = A_{22}^0 - A_{12}^0 - 2A_{66}^0$, and $c_3 = A_{12}^0 + 2A_{66}^0$. The quantities A_{ij} are functions of the angle α; i.e., $A_{ij} = A_{ij}(\alpha)$, and the functional Π depends on $\alpha(x, y)$. We note that the minimum with respect to u and v of the functional of Eq. (5.24) is sought in the class of functions $u(x,y)$ and $v(x,y)$ that satisfy the kinematic conditions assigned on Γ_2. The boundary conditions assigned on Γ_1 are the natural conditions for the functional of Eq. (5.24) and there is no need to require, in formulating our problem, that they should be satisfied.

As the quality criterion that we wish to optimize, we adopt the work performed by the external forces acting on the arc Γ_1 (see Section 1.4 of Chapter 1).

$$J(\alpha) = \frac{1}{2} \int_{\Gamma_1} (uq_x + vq_y)\, ds \tag{5.26}$$

We study the problem of minimizing this functional, i.e.,

$$J_* = \min_\alpha J(\alpha) \tag{5.27}$$

as we select the angle formed by the axes of orthotropy with the fixed x-y coordinates, i.e., $\alpha(x,y)$ is the control variable. The functional J, as we have already noted, is the compliance of an elastic body.

2. *Optimality criteria.* As is well-known, a standard technique for obtaining optimality conditions consists of considering first the differential relations (in our cases the equilibrium conditions expressed in terms of u and v), followed by introduction of adjoint variables. However, in the problem being considered, the equations of equilibrium in terms of the displacements are the Euler–Lagrange equations for the functional of Eq. (5.24). Therefore, the problem may be restated in a manner analogous to the preceding paragraph and the differential relations can be eliminated from the problem. To do this we make use of Clapeyron's theorem and complete the following transformation (see Refs. 14 and 22):

$$J_* = \min_\alpha (-\min_{u,\,v} \Pi) = -\max_\alpha \min_{u,\,v} \Pi$$

Thus the problem of Eq. (5.27) is reduced to finding a max-min. For the sake of ease in obtaining the conditions for stationary behavior of Π as we vary α, and to cut short our exposition, we introduce at each point (x,y) a system of principal X-Y axes of deformation and denote the components of the strain tensor in this coordinate system by ϵ_X, ϵ_Y, and ϵ_{XY} (with $\epsilon_{XY} = 0$). Let ψ and χ denote the angles between the X and ξ axes and between the x and X axes, respectively, so that $\psi = \zeta - \chi$ (see Fig. 5.4). The quantities ϵ_x, ϵ_y, ϵ_{xy} and ϵ_X, ϵ_Y,

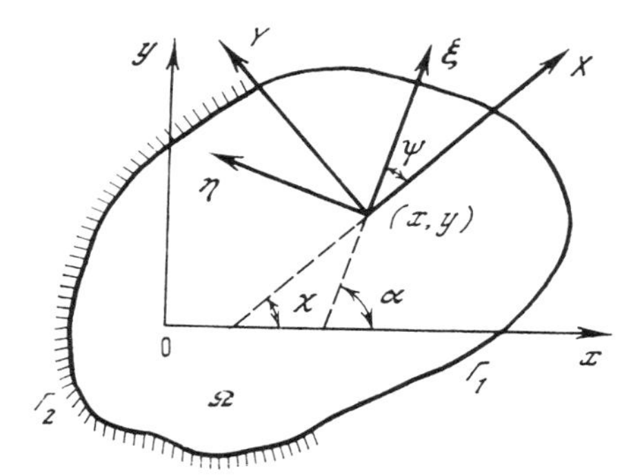

Figure 5.4

ϵ_{XY} are related to each other by the well-known formulas

$$\begin{aligned}\varepsilon_X &= \varepsilon_x \cos^2 \chi + \varepsilon_y \sin^2 \chi + \gamma_{xy} \sin \chi \cos \chi \\ \varepsilon_Y &= \varepsilon_x \sin^2 \chi + \varepsilon_y \cos^2 \chi - \gamma_{xy} \sin \chi \cos \chi \\ \varepsilon_{XY} &= {}^1/_2 (\varepsilon_y - \varepsilon_x) \sin 2\chi + {}^1/_2 \gamma_{xy} \cos 2\chi = 0\end{aligned} \tag{5.28}$$

Equation (5.24) for f may now be rewritten as

$$\begin{aligned}f &= {}^1/_2 A_{11}(\psi)\varepsilon_X^2 + A_{12}(\psi)\varepsilon_X\varepsilon_Y + {}^1/_2 A_{22}(\psi)\varepsilon_Y^2 = N\cos^4\psi \\ &\quad + Q\cos^2\psi + R \\ N &\equiv {}^1/_2 (c_1 + c_2)(\varepsilon_X - \varepsilon_Y)^2, \qquad Q \equiv (\varepsilon_X - \varepsilon_Y)(c_1\varepsilon_Y - c_2\varepsilon_X) \\ R &\equiv {}^1/_2 A_{22}^0 \varepsilon_X^2 + A_{12}^0 \varepsilon_X \varepsilon_Y + {}^1/_2 A_{11}^0 \varepsilon_Y^2\end{aligned} \tag{5.29}$$

where ϵ_X and ϵ_Y are given by Eq. (5.28). Thus, the functional Π is expressed as a function of the angles ψ and χ, by means of Eqs. (5.25), (5.28), and (5.29). These angles satisfy the constraint $\psi + \chi = \alpha$ To derive necessary conditions for an extremum, we require the first variation of the function Π, which depends on the variations $\delta\alpha$, δu, and δv, to be zero. As we vary Π by varying u, v, and α (the procedure for varying functionals depending on vector-valued functions is explained in Refs. 50 and 80), the quantities α, u, and v are regarded as independent of each other. Therefore, the angle χ may also be regarded as independent of α, where χ enters into Eqs. (5.28) and (5.29). The same assumption can be made concerning χ in the third relation of Eq. (5.28); i.e., in this formula χ depends only on the components of the strain tensor $\tan 2\chi = \gamma_{xy}/2(\epsilon_x - \epsilon_y)$. Therefore, in finding the first variation we can substitute $\partial f/\partial\alpha = \partial f/\partial\psi$. As we vary α, the condition of stationary behavior of Π becomes $\sin 2\psi(2N\cos^2\psi + Q) = 0$. This condition contains three different ways of orienting the axes,

$$\begin{aligned} &1)\ \cos\psi = 0, \qquad 2)\ \sin\psi = 0 \\ &3)\ \cos^2\psi = -Q/2N,\ 0 \leqslant -Q/2N \leqslant 1 \end{aligned} \tag{5.30}$$

If we adopt the third way of orienting the axes of orthotropy, we find that this is possible only if the inequality of Eq. (5.30) is satisfied. Let us explain this by representing f by a quadratic formula with three terms $f = Nt^2 + Qt + R$, where the variable t is $t = \cos^2\psi$. Since t varies on the interval $0 \leqslant t \leqslant 1$, the extremum of f, as a function of t may occur either at the boundary points $t = 0$ or $t = 1$, which corresponds to (1) and (2), or it can occur at an interior point. The inequality of Eq. (5.30) simply states that t must be chosen in the interval $[0, 1]$.

Let us formulate the expression for the potential f, corresponding to all three ways of orienting the axes

$$\begin{aligned} (f)_1 &= {}^1/_2 A_{22}^0 \varepsilon_X^2 + A_{12}^0 \varepsilon_X \varepsilon_Y + {}^1/_2 A_{11}^0 \varepsilon_Y^0 = {}^1/_8 A_{22}^0 [\varepsilon_x \\ &\quad + \varepsilon_y + \sqrt{(\varepsilon_x - \varepsilon_y)^2 + \gamma_{xy}^2}]^2 + {}^1/_8 A_{11}^0 [\varepsilon_x + \varepsilon_y \\ &\quad - \sqrt{(\varepsilon_x - \varepsilon_y)^2 + \gamma_{xy}^2}]^2 + A_{12}^0 (\varepsilon_x \varepsilon_y - {}^1/_4 \gamma_{xy}^2) \\ (f)_2 &= {}^1/_2 A_{11}^0 \varepsilon_X^2 + A_{12}^0 \varepsilon_X \varepsilon_Y + {}^1/_2 A_{22}^0 \varepsilon_Y^2 = {}^1/_8 A_{11}^0 [\varepsilon_x + \varepsilon_y \\ &\quad + \sqrt{(\varepsilon_x - \varepsilon_y)^2 + \gamma_{xy}^2}]^2 + {}^1/_8 A_{22}^0 [\varepsilon_x + \varepsilon_y \\ &\quad - \sqrt{(\varepsilon_x - \varepsilon_y)^2 + \gamma_{xy}^2}]^2 + A_{12}^0 (\varepsilon_x \varepsilon_y - {}^1/_4 \gamma_{xy}^2) \\ (f)_3 &= D_1 (\varepsilon_X^2 + \varepsilon_Y^2 + 2D_2 \varepsilon_X \varepsilon_Y) = D_1 [(\varepsilon_x + \varepsilon_y)^2 \\ &\quad + (D_2 - 1)\left(2\varepsilon_x \varepsilon_y - \frac{1}{2}\gamma_{xy}^2\right)] \end{aligned} \tag{5.31}$$

where D_1 and D_2 are constants related to A_{ij}^0

$$D_1 = \frac{A_{11}^0 A_{22}^0 - (A_{12}^0 + 2A_{66}^0)^2}{2(A_{11}^0 + A_{22}^0 - 2A_{12}^0 - 4A_{66}^0)}$$

$$D_2 = \frac{A_{11}^0 A_{22}^0 - (A_{12}^0)^2 - 2A_{66}^0 (A_{11}^0 + A_{22}^0 - 2A_{66}^0)}{A_{11}^0 A_{22}^0 - (A_{12}^0 + 2A_{66}^0)^2}$$

To be more specific, let us assume that $A_{11}^0 > A_{22}^0$.

3. *Infinite plate with a hole.* We consider applications of Eqs. (5.30) and (5.31) to a specific problem of optimizing the anisotropic properties of an elastic infinite plane containing a circular hole. The region Ω is defined by $r \geqslant a$, $0 \leqslant \theta \leqslant 2\pi$, where a is the radius of the hole and r and θ are polar coordinates, with the origin placed at the center of the hole. Constant normal pressure p is applied on the boundary of the hole $\Gamma(r = a)$, i.e.,

$$\sigma_r = p, \qquad \tau_{r\theta} = 0 \tag{5.32}$$

If axial symmetry exists, then the angles of inclination of the axes of orthotropy α and the radial displacements u are independent of θ, i.e., $\alpha = \alpha(r)$, $u = u(r)$. In this case, shear stresses, shear strains and displacements in the circumferential direction are all equal to zero ($\tau_{r\theta} = 0$, $\gamma_{r\theta} = 0$, $v = 0$). The principal axes of the strain tensor have, at each point (r, θ) in the region Ω, radial and circumferential components. The minimum functional of Eq. (5.26), which is computed by integrating along the boundary curve Γ, turns out to be proportional to the radial displacement $u(a)$ of the boundary points and its value $J = 2\pi rpu(a)$ can be regarded as a measure of rigidity.

Attempting to solve the problem of optimizing rigidity, we consider first the case in which the same method of orienting the axes of orthotropy can be realized in the entire domain Ω. Let $\cos\psi = 0$ everywhere in Ω. This means that the axis of orthotropy, with the maximum modulus A^0_{11}, is oriented in the circumferential direction, while the axis with the smallest modulus A^0_{22} points in the radial direction. The formula for f, the equation of equilibrium, and the equation for radial displacements have the form

$$\begin{gathered} f = \frac{1}{2} A^0_{22}\varepsilon_r^2 + A^0_{12}\varepsilon_r\varepsilon_\theta + \frac{1}{2} A^0_{11}\varepsilon_\theta^2, \qquad u_{rr} + \frac{u_r}{r} - \varkappa^2 \frac{u}{r^2} = 0 \\ u = C_1 r^{\varkappa} + C_2 r^{-\varkappa}, \qquad \varkappa = \sqrt{A^0_{11}/A^0_{22}} > 1 \end{gathered} \tag{5.33}$$

Determining the unknown constants C_1 and C_2 from the first boundary condition of Eq. (5.32) and the condition $\sigma_r = 0$ at the point at infinity, we have

$$u = \frac{pa^{\varkappa+1}}{\gamma r^{\varkappa}}, \qquad J = \frac{2\pi a^2 p^2}{\gamma}, \qquad \gamma = \sqrt{A^0_{11}A^0_{22}} - A^0_{12} \tag{5.34}$$

In Fig. 5.5(a), the solid and dotted lines indicate, respectively, the directions with the largest and smallest modulus.

If $\sin\psi = 0$ for $r \geqslant a$, then the axis of orthotropy with the largest modulus

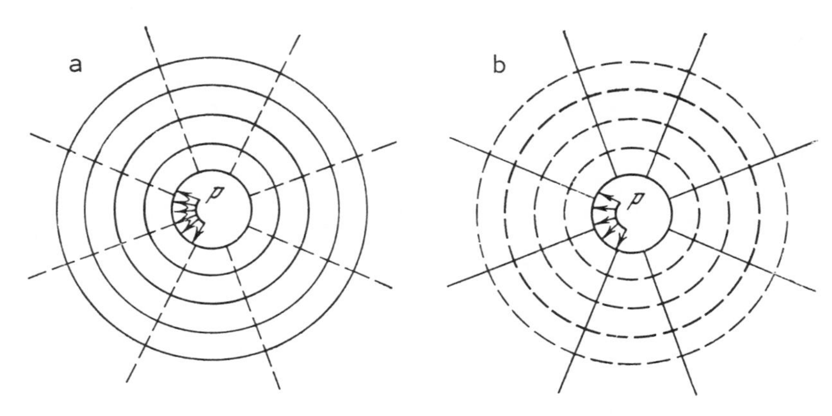

Figure 5.5

A_{11}^0 has a radial direction. In this case, the directions with greatest and smallest modulus are shown by the solid and dotted lines, respectively, in Fig. 5.5(*b*).

We display the formula for *f*, the equation of equilibrium, and the formula for the radial displacements for the case

$$f=\frac{1}{2}A_{11}^0\varepsilon_r^2+A_{12}\varepsilon_r\varepsilon_\theta+\frac{1}{2}A_{22}\varepsilon_\theta^2,\qquad u_{rr}+\frac{u_r}{r}-k^2\frac{u}{r^2}=0$$
$$u=C_1r^k+C_2r^{-k},\qquad k=1/\varkappa=\sqrt{A_{22}^0/A_{11}^0}<1 \tag{5.35}$$

Substituting the boundary conditions $(\sigma_r)_{r=a}=p$ and $(\sigma_r)_{r=\infty}=0$, and manipulating Eq. (5.35), we derive

$$u=pa^{k+1}/\gamma r^k,\qquad J=2\pi a^2p^2/\gamma \tag{5.36}$$

Comparing Eqs. (5.34) and (5.36), we observe that either method (1) or (2) of orienting the axes of orthotropy leads to exactly the same value of the functional *J*.

If the third way of orienting the axes of anisotropy is considered, with $\cos^2\psi=-Q/(2N)$ in Ω, we arrive at the following formulas for *f* and *u* and the following form of the equation of equilibrium:

$$f=D_1(\varepsilon_r^2+\varepsilon_\theta^2+2D_2\varepsilon_r\varepsilon_\theta),\qquad u_{rr}+u_r/r-u/r^2=0$$
$$u=C_1r+C_2r^{-1} \tag{5.37}$$

Computing the values of constants C_1 and C_2, we uniquely determine the deflection function, the magnitude of $\cos^2\psi$, and the value of the minimized functional J,

$$u = pa^2/2A_{66}^0 r, \qquad J = \pi a^2 p^2/A_{66}^0, \qquad \cos^2\psi = 1/2 \tag{5.38}$$

In this case, the axes of orthotropy are directed along the tangents of the solid lines drawn in Fig. 5.6. By comparing values of the functional J given in Eqs. (5.34), (5.36), and (5.38), we see that if

$$\sqrt{A_{11}^0 A_{22}^0} - A_{12}^0 > 2A_{66}^0 \tag{5.39}$$

then methods (1) and (2) of orienting the axes of anisotropy correspond to the greatest rigidity. If, in Eq. (5.39), the inequality is reversed, then the smallest value of the functional J is realized by method (3) or orienting the axes.

In the discussion given above, we considered only purely static solutions obtained from the hypothesis that each of the above stated methods of orienting the axes of anisotropy may be realized in the entire domain Ω. However, it is possible that a solution that turns out to be optimum is obtained by "patching together" different regions with different methods of distributing the elastic moduli. Therefore, we should consider the possibility of coexistence in Ω of methods (1)-(3), assuming that Ω may be decomposed into annular regions bounded by circles ($r = r_i$), such that in each region we realize different methods of orienting the axes of anisotropy. The solution of this problem leads to postulation with an unknown boundary, where on $r = r_i$ the stresses σ_r are continuous, but discontinuities occur in the derivatives of displacements u_r. Applying the Erdman-Weierstrass corner condition at the points $r = r_i$ and carrying out some elementary, but extremely tedious computations, it is possible to show that if the elastic moduli A_{11}^0, A_{22}^0, and A_{66}^0 satisfy the inequality of Eq. (5.39), then

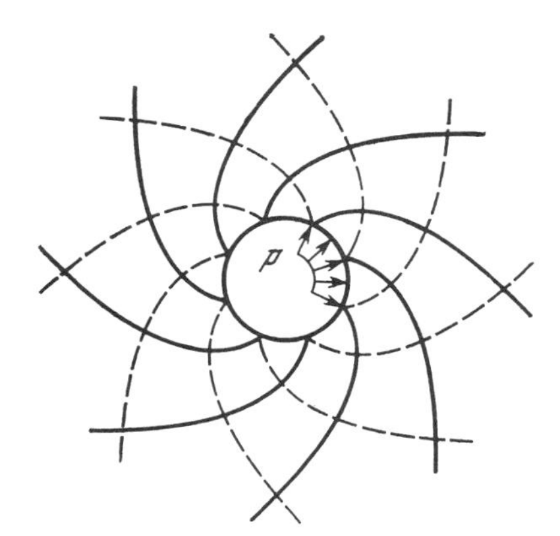

Figure 5.6

"patching" together regions with a different kind of orientation of the axes of orthotropy does not contribute to the value of the cost functional when it is smaller than the functional J given by Eqs. (5.34) and (5.36). If the inequality of Eq. (5.39) is not satisfied, then combining such regions does not permit us to attain a value of the cost functional that is smaller than the value of J given by Eq. (5.39). We observe that "stitching" together an arbitrary number of regions of types (1) and (2) (i.e., the regions in which orientation of the first and second kind has been realized), we attain a value of the cost functional that is equal to the value of J given by Eqs. (5.34) and (5.36), i.e., the value remains the same as for the cases in which one of the kinds of orientation has been realized in the entire domain Ω. Hence, if the inequality of Eq. (5.39) is satisfied, methods (1) and (2) are optimum, while in the contrary case method (3) is optimum.

In this problem, orientations (1) and (2) assigned to the entire region Ω—or an arbitrary combination of regions in which axes of orthotropy are oriented according to types (1) and (2)—result in an identical value of the cost functional. However, this property is true only for unbounded regions Ω, and is not true for bounded annular regions. In fact, let us consider an annular region $\Omega\ \{a \leqslant r \leqslant b, 0 \leqslant \theta \leqslant 2\pi\}$ with conditions of Eq. (5.32) assigned on the inner boundary, while the outer boundary is stress-free. We obtain the following result: For any a and b, the orientation of axes of type (1) results in a smaller value of the functional J.

5.5. Computation of Optimum Anisotropic Properties for Elastic Bodies

In this section, we present some research results derived by the author, A. V. Albul and D. M. Epurash, which concern numerical solution of specific problems of optimizing anisotropic properties of two-dimensional elastic bodies. A general formulation of such problems was presented in the preceding paragraph, where we have already shown that maximizing the compliance (i.e., the rigidity) can be reduced to finding the maximum value of the potential energy for the system, i.e.,

$$J_* = -\max_\alpha \min_{u,v} \Pi \tag{5.40}$$

To numerically solve the problem of Eq. (5.40), we apply the technique of successive optimization, which was explained in Section 1.10 of Chapter 1. Using this technique, the variation of the control variable $\delta\alpha$ is taken to be equal to

$\delta\alpha = (\partial f/\partial\alpha)\, t$, to assure the positive sign of the variation of the functional Π

$$\delta_\alpha \Pi = t \iint_\Omega (\partial f/\partial\alpha)^2 \, dx \, dy \geqslant 0 \tag{5.41}$$

where $t > 0$ is the step size in the direction of the gradient and

$$\begin{aligned}
\partial f/\partial\alpha &= {}^1/_2 \, (\partial A_{11}/\partial\alpha)\, \varepsilon_x^2 + (\partial A_{12}/\partial\alpha)\, \varepsilon_x \varepsilon_y + {}^1/_2 \, (\partial A_{22}/\partial\alpha)\, \varepsilon_y^2 \\
&\quad + (\partial A_{16}/\partial\alpha)\, \varepsilon_x \gamma_{xy} + (\partial A_{26}/\partial\alpha)\, \varepsilon_y \gamma_{xy} + {}^1/_2 \, (\partial A_{66}/\partial\alpha)\, \gamma_{xy}^2 \\
\partial A_{11}/\partial\alpha &= 4 \sin\alpha \cos\alpha \, (c_2 \sin^2\alpha - c_1 \cos^2\alpha) \\
\partial A_{22}/\partial\alpha &= 4 \sin\alpha \cos\alpha \, [c_1 \sin^2\alpha - c_2 \cos^2\alpha] \\
\partial A_{12}/\partial\alpha &= \partial A_{66}/\partial\alpha = 2\,(c_1 + c_2)(\sin\alpha \cos^3\alpha - \cos\alpha \sin^3\alpha) \\
\partial A_{16}/\partial\alpha &= -\, c_2 \sin^4\alpha - c_1 \cos^4\alpha + 3\,(c_1 + c_2) \sin^2\alpha \cos^2\alpha \\
\partial A_{26}/\partial\alpha &= c_2 \cos^4\alpha + c_1 \sin^4\alpha - 3\,(c_1 + c_2) \sin^2\alpha \cos^2\alpha
\end{aligned} \tag{5.42}$$

Computations are conducted for rectangular regions ($0 \leqslant x \leqslant a$, $-b/2 \leqslant y \leqslant b/2$), for various loading conditions. We map our region into a square Ω of unit size and seek the solution for different values of the parameter $\lambda = b/a$. In all cases, we assume $E_1 = 17.5$, $E_2 = 13.1$, $E_3 = 5.3$, $G_{12} = 2.82$ (kg/cm$^2 \cdot 10^4$), $\nu_{12} = 0.1$, $\nu_{23} = 0.17$, and $\nu_{31} = 0.229$ (glasslike texture). To compute the constant A_{ij}^0, we make use of the formulas

$$\begin{aligned}
A_{11}^0 &= e_{22}(e_{11}e_{22} - e_{12}^2), \qquad A_{22}^0 = e_{11}(e_{11}e_{22} - e_{12}^2) \\
A_{12}^0 &= -e_{12}(e_{11}e_{22} - e_{12}^2), \qquad A_{66} = G_{12} \\
e_{11} &= \frac{E_3 - \nu_{31}^2 E_1}{E_1 E_3}, \qquad e_{12} = \frac{\nu_{12} E_2 + E_1 \nu_{23} \nu_{31}}{E_1 E_2} \qquad e_{22} = \frac{E_2 - \nu_{23}^2 E_3}{E_2^2}
\end{aligned}$$

Here E_1, E_2, and E_3 are Young's moduli, G_{12} is the shear modulus, and ν_{12}, ν_{23}, and ν_{31} are Poisson's ratios. Numerically computed optimum distributions for the angle $\alpha(x, y)$ of inclination of the axes of orthotropy are illustrated in Figs. 5.7–5.11. Directions tangent to the solid lines shown in these figures represent the directions with maximum elastic modulus.

The distributions of the angle α illustrated in Figs. 5.7 and 5.8 correspond to rigid fastening along the edge $x = 0$, $-{}^1/_2 \leqslant y \leqslant {}^1/_2$, and tension applied by forces $q_x = 1$, $q_y = 0$ on the boundary $x = 1$, $-{}^1/_2 \leqslant y \leqslant {}^1/_2$. Along the boundaries

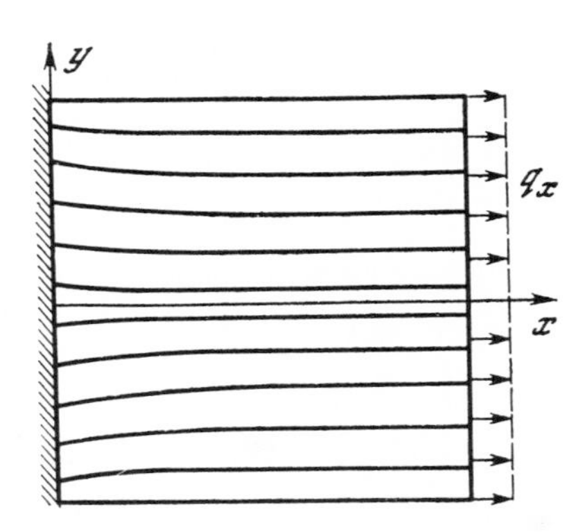

Figure 5.7

$y = \pm^1/_2$, $0 \leqslant x \leqslant 1$, no forces are applied; i.e., $q_x = q_y = 0$. For the distributions shown in Figs. 5.7 and 5.8, the parameter λ was equal to $\lambda = 2$ and $\lambda = 1$, respectively. As can be seen, in a considerable portion of the domain the lines with the greatest elastic modulus are oriented in a direction parallel to the action of the exterior loads and are, in the entire region occupied by the material, symmetric with respect to the x axis. These lines (see Figs. 5.7 and 5.8) begin to curve only in the vicinity of the clamped edge. As we decrease λ, the optimum distribution of the angle $\alpha(x,y)$ approaches the function $\alpha(x,y) \equiv 0$, corresponding to a homogeneous orthotropic body with the axis of its maximum modulus parallel to the x axis. The relative gain obtained by optimization, as compared to a homogeneous orthotropic body ($\alpha \equiv 0$), decreases with decreasing value of λ. For values of λ equal to 2, 1, $^1/_2$, and $^1/_6$, it is equal to 0.23, 0.11, 0.05, and 0.012%, respectively.

Figures 5.9-5.11 illustrated the function $\alpha(x,y)$ in the case when the

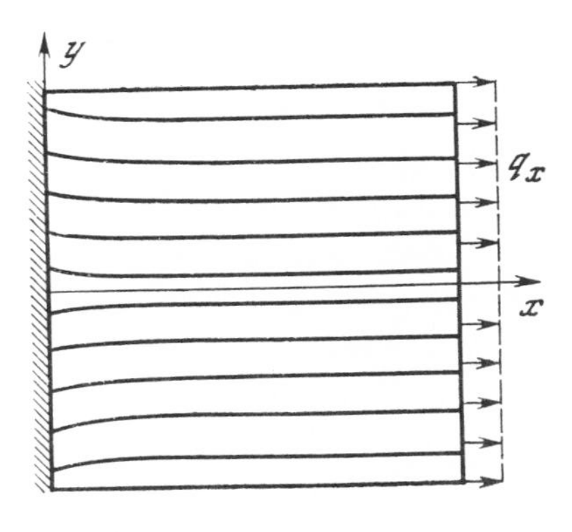

Figure 5.8

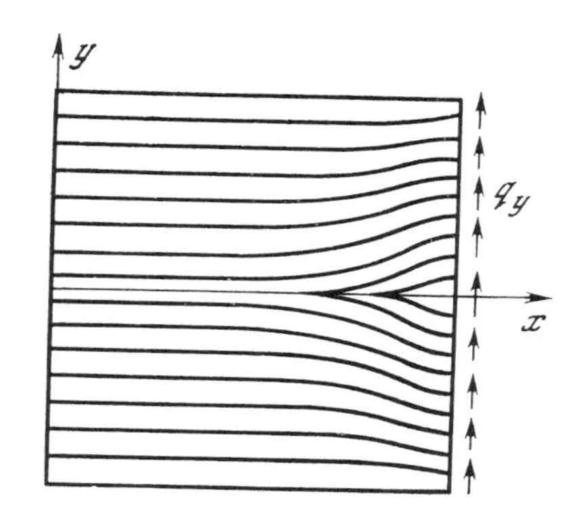

Figure 5.9

load $q_y = 0.001$, $q_x = 0$ is applied to the edge $x = 1$, $-1/2 \leqslant y \leqslant 1/2$. The remaining boundary conditions are the same as in the preceding examples. The values of the parameter λ for Figs. 5.9 to 5.11 are, respectively, $\lambda = 1.0$, 0.5, and 0.1. Orientation of the elastic moduli turns out to be symmetric about the axis $y = 0$.

These computations indicate that as λ is decreased, the solid lines in our figures straighten out and only in the vicinity of the edge $x = 1$ are they noticeably different from horizontal lines. Let us compare, for a fixed value of λ, an optimum body with a homogeneous orthotropic body having the axis of orthotropy with the maximum modulus parallel to the x axis. For values of $\lambda = 1.0$, 0.5, and 0.1, the relative gain obtained by optimization amounted to about 32, 10, and 0.6%, respectively. This comparison shows that for a case $\lambda = 0.1$ the optimum body differs insignificantly from a homogeneous orthotropic body with uniform horizontal direction for the axis of the greatest modulus.

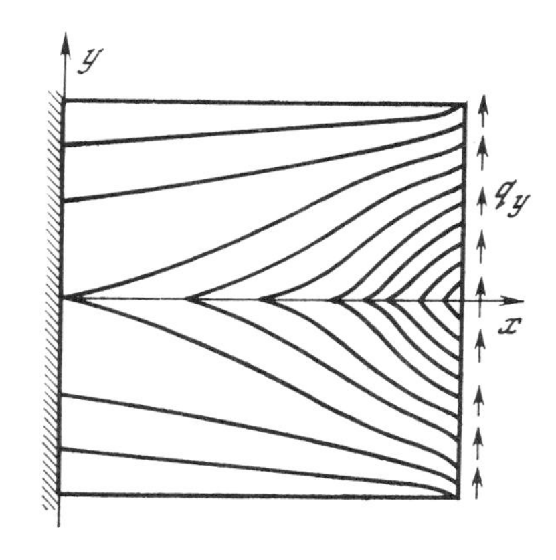

Figure 5.10

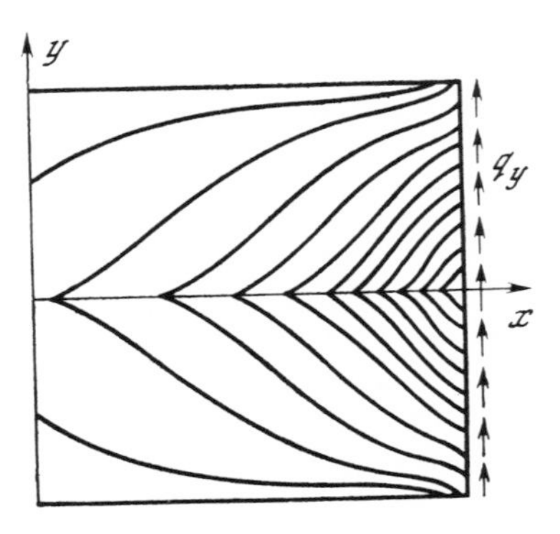

Figure 5.11

5.6. Some Comments Concerning the Shapes of Anisotropic Bodies and Problems of Simultaneous Optimization of the Shape and of the Orientation of Axes of Anisotropy

In the preceding paragraphs, we considered questions of optimum distribution of the anisotropic properties throughout the volume of an elastic body. We found that the function giving the angle of inclination of the axes of anisotropy was not constant. We comment, in this respect, that in practice much simpler problems of optimization for homogeneous, linearly anisotropic materials are of interest. Such problems reduce to optimization with respect to some parameters. To be more specific, let us look at the following problem: From a homogeneous, linearly anisotropic sheet we wish to cut out a rectangular plate of a given size having the best rigidity properties. Clearly, the problem consists of finding the angle of rotation of one of the axes of anisotropy of the sheet relative to the system of coordinate axes assigned to the plate.

In practical applications, questions concerning optimization of anisotropic properties arise not only in this pure form. We usually consider problems of finding the best shape of structural parts, using materials with given anisotropic properties, so we need to optimize simultaneously the shape and the orientation of the axes of anisotropy for the elastic structural element. Let us explain how to formulate this type of problem for the example of maximizing the rigidity of a bar subject to torsion.

The problem of maximizing a shaft rigidity by choosing an optimum shape of the cross section was formulated in Section 1.3 of Chapter 1. An optimality condition for the boundary curve was formulated in Section 1.8 of Chapter 1 [Eq. (1.122)]. Using this condition, we arrive at a solvable boundary-value problem for which the unknowns are the torsion function $\varphi(x, y)$ and the shape of the cross section

$$a_{11}\varphi_{xx} - 2a_{12}\varphi_{xy} + a_{22}\varphi_{yy} = -2, \qquad (x, y) \in \Omega, \qquad (\varphi)_\Gamma = 0$$
$$(a_{11}\varphi_x^2 + a_{22}\varphi_y^2 - 2a_{12}\varphi_x\varphi_y)_\Gamma = \text{const}, \qquad \text{mes}\,\Omega = S \tag{5.43}$$

By solving Eq. (5.43), we find the function φ, the shape of the boundary curve Γ, and therefore we are able to compute the torsional rigidity K_*

$$\varphi = \frac{1}{2(ab - c^2)}\left[\frac{S}{\pi}\sqrt{a_{11}a_{22} - a_{12}^2} - a_{22}x^2 - 2a_{12}xy - a_{11}y^2\right]$$
$$\Gamma: a_{22}x^2 + 2a_{12}xy + a_{11}y^2 = \frac{S}{\pi}\sqrt{a_{11}a_{22} - a_{12}^2} \tag{5.44}$$
$$K_* = \frac{S^2}{2\pi\sqrt{a_{11}a_{22} - a_{12}^2}}$$

To evaluate the gain obtained by optimization, we compare the derived value of K_* with the torsional rigidity $K_0 = S^2/\pi(a_{11} + a_{22})$ of a round shaft, having the same cross-sectional area.

$$\frac{K_* - K_0}{K_0} = \frac{1}{\beta} - 1, \qquad \beta = \frac{2\sqrt{a_{11}a_{22} - a_{12}^2}}{a_{11} + a_{22}} \tag{5.45}$$

Accounting for the fact that the deformation coefficients a_{11}, a_{22}, and a_{12} of an anisotropic material satisfy the conditions[86] $a_{11} > 0$ and $a_{11}a_{22} - a_{12}^2 > 0$, it is easy to show that $0 \leqslant \beta \leqslant 1$.

For an orthotropic material $a_{12} = 0$, $a_{11} = 1/G_1$, and $a_{22} = 1/G_2$, where G_1 and G_2 are the shear moduli along the x and y axes, respectively. We substitute these values of the coefficients into Eqs. (5.44) and (5.45) and arrive at formulas for an orthotropic material

$$\varphi = \frac{1}{2}\left(\frac{S}{\pi}\sqrt{G_1G_2} - G_1x^2 - G_2y^2\right)$$
$$\Gamma: G_1x^2 + G_2y^2 = \frac{S}{\pi}\sqrt{G_1G_2}$$
$$K_* = \frac{S^2\sqrt{G_1G_2}}{2\pi}, \qquad K_0 = \frac{S^2G_1G_2}{\pi(G_1 + G_2)} \tag{5.46}$$
$$\frac{K_* - K_0}{K_0} = \frac{G_1 + G_2}{2\sqrt{G_1G_2}} - 1$$

Looking at the last relation in Eq. (5.46), it is easy to see that the gain obtained by optimizing increases as either $G_1/G_2 \to 0$ or as $G_1/G_2 \to \infty$, i.e., the gain increases as we increase the degree of anisotropy. Minimum gain (equal to zero) is obtained for $G_1 = G_2 = G$, i.e., for an isotropic material. In this case, $K_* = K_0 = GS^2/2\pi$ and the optimum cross section is a circular disk.

We now study the problem of simultaneous optimization of the shape of the region Ω and of the angles of inclination of the axes of anisotropy $\alpha = \alpha(x, y)$, making use of the condition $K \to \max_{\Gamma_0} \max_\alpha$, where Γ_0 is the unknown part of the boundary of the region Ω. We utilize the relations derived in Section 1.3 and employ the following formulation of the simultaneous optimization problem:

$$K_* = -\min_{\Gamma_0} \min_\alpha \min_\varphi J \tag{5.47}$$

where J is defined by Eq. (5.19). We assume that the deformation coefficients a_{ij} do not depend on x and y. We seek the minimum with respect to Γ_0 and we impose the isoperimetric condition that the cross-sectional area of the bar remains constant. The minimum with respect to φ is computed in the class of functions satisfying the condition $(\varphi)_\Gamma = 0$. To find the unknown "control" quantities α and Γ_0, we derive a system of two necessary conditions for optimality, consisting of Eq. (5.10) and the relation $(m_{11}\varphi_x^2 - 2m_{12}\varphi_x\varphi_y + m_{22}\varphi_y^2)_{\Gamma_0} = \lambda^2$ (where λ^2 is a constant). Analysis of the optimality conditions, and of the basic relations of this problem, permits us to transform the condition used for determining the contour Γ_0 into the form

$$(\nabla\varphi)^2_{\Gamma_0} = \lambda^2 \tag{5.48}$$

Thus, finding $\varphi(x, y)$, $\alpha(x, y)$, and Γ_0 for the optimum bar is reduced to the solving the boundary-value problem

$$\Delta\varphi = -2\lambda_1^{-1}, \qquad (\varphi)_\Gamma = 0$$

with the auxiliary condition of Eq. (5.48). Following this we determine the angles $\alpha(x, y)$ by recalling Eq. (5.10), or for an orthotropic material by using Eq. (5.23). If the cross section is simply connected and Γ_0 denotes the entire boundary of $\Omega(\Gamma_0 = \Gamma)$, then an optimum orthotropic bar has a circular cross section, while the axes of orthotropy point in the radial and circumferential directions.

In the general case for which a part of the boundary of Ω is assigned, the problem of simultaneous optimization may be displayed directly, if one finds the solution of an auxiliary problem that consists of finding the shape of the cross section for a homogeneous isotropic bar (with a shear modulus G) having the greatest torsional rigidity. The optimum shape of the region Ω remains the same in our problem as it was in the auxiliary problem. In the expressions for the stress function $\varphi(x, y)$ for an orthotropic medium, we need to replace G by G_1. Equation (5.23) determines the expression for optimum distribution of the angle $\alpha(x, y)$ of orthotropy in terms of the stress function.

6

Optimal Design for Problems in Hydroelasticity

In this chapter we discuss optimal design of elastic plates interacting with an ideal fluid. In Section 6.1 we introduce the basic relations describing small vibrations of elastic plates submerged in an ideal fluid and study some essential properties of the corresponding boundary-value problem. Section 6.2 contains the formulation and the study of problems of frequency optimization for plates vibrating in a fluid. In Section 6.3 we derive an analytic solution of the exterior hydrodynamic problem with two-dimensional plane motions of the plate and of the fluid. In Section 6.4 we derive the optimum shapes of long rectangular plates simply supported along the longer edges. In Section 6.4 we consider a specific case of a trilayered plate whose optimum shape is determined by analytic techniques. In Section 6.5 we study a two-dimensional parallel flow problem concerning divergence (i.e., the loss of stability caused by the action of hydrodynamic forces) of an elastic plate. We seek an optimum distribution of thickness for plates having a maximal rate of divergence. To determine nontrivial equilibrium shapes for rectangular plates that are clamped along one edge, we present in Section 6.6 a description of a solenoidal flow in the presence of an infinite cavity, and we study some properties of the corresponding boundary-value problem. We consider an optimization problem of finding the thickness function for plates whose nontrivial equilibrium shape coincides with the largest value of the velocity of the impacting fluid. Results presented in this chapter were partially published in Refs. 31-34.

6.1. State Equations for Plates That Vibrate in an Ideal Fluid

1. *Small vibrations of a plate in an ideal fluid.* We consider the problem of small vibrations of a plate submerged in an ideal fluid. We assume that the plate

is simply supported along the smooth boundary curve Γ, which lies in the plane $z = 0$ of the Cartesian x–y–z coordinate system. It forms the boundary of a simply connected region Ω. We denote, respectively, by $h = h(x, y)$, $w = w(x, y, t)$, t, and $q = q(x, y, t)$, the thickness, deflection, time, and force of interaction between the plate and the fluid. The equation of small vibrations and the boundary conditions are

$$\rho_p h w_{tt} + aL(h)w = q \tag{6.1}$$

$$(w)_\Gamma = 0, \qquad \left(h^3 \left[w_{xx} + w_{yy} - \frac{1-\nu}{R}\frac{\partial w}{\partial n}\right]\right)_\Gamma = 0 \tag{6.2}$$

where $a = E/12(1 - \nu^2)$, E is Young's modulus, ν is Poisson's ratio, ρ_p is specific density of the plate material, n is the unit normal to the contour Γ (in the x–y plane), and R is the radius of curvature Γ.

We make the assumption that the fluid is at rest at infinity. To describe the fluid flow we introduce a velocity potential $\varphi = \varphi(x, y, z, t)$, assuming the absence of curl. The potential φ satisfies the Laplace equation and the following boundary conditions: Velocity vanishes at infinity and linearized boundary conditions are assigned to the upper [denoted by (+)] and the lower [denoted by (–)] boundary for the cut $z = 0$, $(x, y) \in \Omega$

$$\Delta\varphi = 0 \tag{6.3}$$

$$(\varphi_z)_\Omega^\pm = w_t \tag{6.4}$$

$$(\nabla\varphi)_\infty = 0 \tag{6.5}$$

The boundary condition of Eq. (6.4) translates to the x–y plane the statement that the fluid flow does not cross the surface of the plate. We shall use assumptions of "smallness" of displacements w and thickness h and continuity of the motion of the liquid and the plate. As before, Δ and ∇ denote, respectively, the Laplace operator and the gradient.

We denote by $p = p(x, y, z, t)$ the pressure and write an expression for the interaction force between the fluid and the plate, $q = p^- - p^+$. An integral of the Cauchy–Lagrange type relates the pressure function $p(x, y, z, t)$ to the velocity potential $\varphi(x, y, z, t)$ (see Refs. 59 and 81):

$$p = p_\infty - \rho_f\left[\varphi_t + \frac{1}{2}(\nabla\varphi)^2\right] \tag{6.6}$$

where ρ_f and p_∞ denote the fluid density and the pressure at infinity, respectively. In the Cauchy–Lagrange integral we may ignore the term $(\nabla\varphi)^2$, which is small of second order of smallness, so we have $p = p_\infty - \rho_f\varphi_t$. Using this relation and $q = p^- - p^+$, we arrive at the following formula for the reaction force exerted by the fluid:

$$q = \rho_f(\varphi_t^+ - \varphi_t^-) \tag{6.7}$$

The well-posed boundary-value problem of Eqs. (6.1)-(6.5), and (6.7) uniquely determines the functions $w(x, y, z, t)$ and $\varphi(x, y, z, t)$. We make use of Eqs. (6.1)-(6.5), and (6.7), and try to find a solution of the form

$$\begin{aligned} &x' = x/l, \quad y' = y/l, \quad z' = z/l, \quad u' = u/l, \quad h' = l^2h/V \\ &\omega' = [12\rho_p l^8(1-\nu^2)/EV^2]^{1/2}\,\omega, \qquad \rho = \rho_f l^3/\rho_p V \end{aligned} \tag{6.8}$$

where ω denotes the frequency of vibration and $i^2 = -1$. For the sake of convenience, we transform our problem to a system of dimensionless variables

$$\begin{aligned} w &= u(x, y)\exp i\omega t \\ \varphi &= i\omega\Phi(x, y, z)\exp i\omega t \end{aligned} \tag{6.9}$$

where V stands for the total volume of the plate and l is a characteristic linear dimension for the region Ω. As we change to the dimensionless variables of Eq. (6.9), the region Ω is transformed into a region Ω' and the curve Γ into a curve Γ'. We omit the primes in our future discussions. Our transformation results in the following relations, which may be used to determine the amplitude functions $u(x, y)$ and $\Phi(x, y, z)$:

$$L(h)u - \omega^2[hu - \rho(\Phi^+ - \Phi^-)] = 0 \tag{6.10}$$

$$(u)_\Gamma = 0, \qquad h^3\left(u_{xx} + u_{yy} - \frac{1-\nu}{R}\frac{\partial u}{\partial n}\right)_\Gamma = 0 \tag{6.11}$$

$$\Delta\Phi = 0, \qquad (\Phi_z)_\Omega^\pm = u, \qquad (\nabla u)_\infty = 0 \tag{6.12}$$

The boundary-value problem of Eqs. (6.10)-(6.12) is an eigenvalue problem with ω^2 playing the part of an eigenvalue. Finding the spectrum for a given function $u(x, y)$ is greatly facilitated if we succeed in solving Eq. (6.12) for an arbitrary function $u(x, y)$. In that case, we eliminate the potential difference $\Phi^+ - \Phi^-$ from Eq. (6.10) and obtain an eigenvalue boundary-value problem with only a single equation. We analyze the boundary-value problem Eq. (6.12) and represent the solution of this problem in the form

$$\Phi(x, y, z) = \iint_\Omega K'(\xi, \eta, x, y, z)\,u(\xi, \eta)\,d\xi\,d\eta$$

where K' is Green's function. The function K' does not depend on either h or u and is uniquely determined by the shape of the region Ω. We denote by $K(\xi, \eta, x, y) = -2K'(\xi, \eta, x, y, 0)$, use our formula for Φ, and obtain a formula for the discontinuity in the velocity potential

$$\Phi^+ - \Phi^- = 2\Phi^+ = -\iint_\Omega K(\xi, \eta, x, y)\, u(\xi, \eta)\, d\xi\, d\eta \qquad (6.13)$$

Substituting Eq. (6.13) into Eq. (6.10), we arrive at the following integrodifferential equation:

$$L(h)u - \omega^2 \Big[hu + \rho \iint_\Omega K(\xi, \eta, x, y)\, u(\xi, \eta)\, d\xi\, d\eta\Big] = 0 \qquad (6.14)$$

We shall write Eq. (6.14) in a symbolic form

$$L(h)u - \omega^2 N(h)u = 0 \qquad (6.15)$$

denoting by $N(h)\,u$ the expression inside the square brackets in Eq. (6.14).

2. *The eigenvalue problem.* We study the boundary-value–eigenvalue problem for the linear integrodifferential equation of Eq. (6.15) with boundary values of Eq. (6.11). We shall prove that the operators $L(h)$ and $N(h)$ are self-adjoint and positive.

To check the self-adjoint property, it suffices to verify the equality

$$\iint_\Omega u_1 L(h)\, u_2 dx\, dy = \iint_\Omega u_2 L(h)\, u_1 dx\, dy$$

and a similar equality for the operator $N(h)$, where u_1 and u_2 are arbitrary functions that satisfy Eq. (6.11). The self-adjoint property of the operator $L(h)$ is well known (see, for example, Ref. 99). Let us prove that the operator $N(h)$ is also self-adjoint. We introduce auxiliary harmonic functions Φ^1 and Φ^2 that solve the boundary-value problem of Eq. (6.12), with u equal to u_1 and u_2, respectively. Because of Eq. (6.12), we have

$$\begin{aligned}\iint_\Omega u_1 N(h)\, u_2 dx\, dy &= \iint_\Omega [hu_1u_2 - \rho\Phi_z^1((\Phi^2)^+ - (\Phi^2)^-)]\, dx\, dy \\ &= \iint_\Omega hu_1u_2 dx\, dy - \rho \iint_\Omega (\Phi^2\Phi_z^1)^+\, dx\, dy + \iint_\Omega (\Phi^2\Phi_z^1)^-\, dx\, dy \\ &= \iint_\Omega hu_1u_2 dx\, dy - \rho \iint_\Sigma \Phi^2 \frac{\partial \Phi^1}{\partial n_p}\, d\sigma\end{aligned} \qquad (6.16)$$

where n_p is the outward unit normal to the plate surface and n is the normal to Γ. The double integral

$$-\rho \iint_{\Sigma} \Phi^2 \frac{\partial \Phi^1}{\partial n_p} d\sigma$$

is taken over the entire surface of the cut $[z=0, (x,y) \in \Omega]$, i.e., over both boundaries. Noting the harmonic property of the functions Φ^1 and Φ^2, we transform this integral to

$$-\rho \iint_{\Sigma} \Phi^2 \frac{\partial \Phi^1}{\partial n_p} d\sigma = -\rho \iint_{\Sigma} \Phi^1 \frac{\partial \Phi^2}{\partial n_p} d\sigma$$

Transforming the right-hand side of this equality, according to Eq. (6.16), and reversing our sequence of arguments, we obtain

$$\iint_{\Omega} h u_1 u_2 dx\, dy - \rho \iint_{\Sigma} \Phi^1 \frac{\partial \Phi^2}{\partial n_p} d\sigma = \iint_{\Omega} u_2 N(h)\, u_1 dx\, dy$$

Thus, the operator $N(h)$ is proved to be self-adjoint.

Now, let us prove that the operators $L(h)$ and $N(h)$ are positive, i.e.,

$$\iint_{\Omega} uL(h)\, u\, dx\, dy > 0, \qquad \iint_{\Omega} uN(h)\, u\, dx\, dy > 0$$

for any nontrivial function $u (u \neq 0)$ satisfying the boundary conditions of Eq. (6.11). This property of the operator L is also well-known (see Ref. 99). Let us prove the positive property of the operator $N(h)$. Let Φ denote a harmonic function that solves the problem of Eq. (6.12). We complete a transformation identical with the one that was used to prove the self-adjointness of N, giving

$$\iint_{\Omega} uN(h)\, u dx\, dy = \iint_{\Omega} [hu^2 - \rho(\Phi^+ - \Phi^-)\, \Phi_z]\, dx\, dy$$
$$= \iint_{\Omega} hu^2 dx\, dy - \rho \iint_{\Sigma} \Phi \frac{\partial \Phi}{\partial n_p} d\sigma$$

Applying Green's formula to the surface integral and making use of Eq. (6.12), we have

$$-\rho \iint_{\Sigma} \Phi \frac{\partial \Phi}{\partial n_p} d\sigma = \rho \iiint (\nabla \Phi)^2 \, dx \, dy \, dz$$

These relations imply the positive property of the operator N. Because the operators in the problem given by Eqs. (6.11) and (6.15) are self-adjoint and positive, the eigenvalues are real and positive, and we can apply Rayleigh's variational principle to find the smallest eigenvalue. The square of the natural frequency can be determined from

$$\iint_{\Omega} u \, [L(h) u - \omega^2 N(h) u] \, dx \, dy = 0$$

We introduce the following notation:

$$\begin{aligned} J_1(h, u) &= \iint_{\Omega} h^3 [(u_{xx} + u_{yy})^2 - 2(1-\nu)(u_{xx}u_{yy} - u_{xy}^2)] \, dx \, dy \\ J_2(h, u) &= \iint_{\Omega} hu^2 dx \, dy \\ J_3(u) &= \iint_{\Omega} u(x, y) \left(\iint_{\Omega} K(\xi, \eta, x, y) \, u d\xi \, d\eta \right) dx \, dy \end{aligned} \tag{6.17}$$

to derive

$$\omega^2 = J(h, u) = J_1(h, u)/[J_2(h, u) + \rho J_3(u)] \tag{6.18}$$

It follows from our discussion that for any sufficiently smooth function $u \neq 0$ that satisfies the boundary conditions of Eq. (6.11), the quadratic functionals J_1 and $(J_2 + \rho J_3)$ are strictly positive and consequently the Rayleigh quotient J is positive. For a given function $h = h(x, y)$, according to Rayleigh's variational principle the smallest eigenvalue of the problem of Eqs. (6.11) and (6.15) is equal to

$$\omega_0^2(h) = \min_u J(h, u) \tag{6.19}$$

The minimum of J with respect to u, as given in Eq. (6.19), is computed in the class of smooth functions $u(x,y)$ that satisfy the first boundary condition of Eq. (6.11). The second boundary condition given in Eq. (6.11) is a "natural" condition for the functional of Eq. (6.18), so there is no need to satisfy it *a priori*. If the minimum of the functional J is found in the class of functions satisfying the first boundary condition of Eq. (6.11), then the minimizing function automatically satisfies the second boundary condition in Eq. (6.11). This function $u(x,y)$, which makes the functional J attain its minimum, is an eigenfunction corresponding to the smallest eigenvalue.

3. *Monotone property of* ω_0^2. The fundamental frequency (or rather its square) ω_0^2 depends on the parameter ρ. Using the symmetric and positive properties of the operators of our problem, we shall prove that for a given function $h = h(x,y)$, ω_0^2 is a monotone decreasing function of the parameter ρ. To prove it we consider infinitesimally close values, ρ and $(\rho + d\rho)$, of the parameter and the corresponding values of ω_0^2, u and $(\omega_0^2 + d\omega_0^2, u + du)$. The fundamental frequencies and displacements (ω_0^2, u), $(\omega_0^2 + d\omega_0^2,\ u + du)$ solve Eqs. (6.18) and (6.19) for the values of the parameter ρ and $(\rho + d\rho)$, respectively, with $h = h(x,y)$ regarded as given. Equation (6.18) must be satisfied if we substitute into it the quantities $(\rho, \omega_0^2,\ u)$ and $(\rho + d\rho,\ \omega_0^2 + d\omega_0^2,\ u + du)$. Let us write Eq. (6.18) twice with these quantities and subtract one result from the other. After some elementary manipulations, disregarding infinitesimals of higher order, we obtain

$$d\omega_0^2\,[J_2(h,u) + \rho J_3(u)] + d\rho\,\omega_0^2 J_3(u) + 2\iint_{\Omega} [L(h)\,u - \omega_0^2 N(h)\,u]\,du\,dx\,dy = 0$$

Since the quantities u, h, and ω_0^2 are related by virtue of Eq. (6.15), we derive the following formula for the derivative $d\omega_0^2/d\rho$:

$$d\omega_0^2/d\rho = -\omega_0^2 J_3(u)/[J_2(h,u) + \rho J_3(u)] \tag{6.20}$$

We have already discussed the fact that the functionals in the numerator and denominator of the right-hand side of Eq. (6.20) are positive. Therefore, $d\omega_0^2/d\rho < 0$, which proves that as the parameter ρ increases $[\rho = \rho_f l^3/(\rho_p V) > 0]$, the fundamental frequency decreases.

6.2. Optimizing the Frequency of Vibrations

1. *Maximizing the smallest eigenvalue.* The Rayleigh quotient of Eq. (6.18) is a functional whose values depend on $h(x,y)$ and $u(x,y)$. Consequently, the

value of the fundamental frequency ω_0^2 determined by Eq. (6.19) depends on $h(x, y)$. We shall consider the following optimization problem in the space of continuous functions. We wish to find a function $h(x, y)$ which maximizes the smallest eigenvalue, i.e.,

$$\omega_{0*}^2 = \max_h \omega_0^2(h) = \max_h \min_u J(h, u) \tag{6.21}$$

and satisfies the constant volume condition

$$\iint_\Omega h\,dx\,dy = 1 \tag{6.22}$$

where we have made the volume equal to unity. This can be explained by our formulation of Section 6.1, in which we introduced dimensionless variables in Eq. (6.9), making the volume dimensionless (in variables having dimensions, the volume of the plate was equal to V). Minimizing Eq. (6.21) with respect to u is carried out in a space of sufficiently smooth functions satisfying the first condition of Eq. (6.11).

The problem formulated in Eqs. (6.22), (6.23), (6.11), and (6.18) is isoperimetric with a single parameter $\rho = \rho_f l^3/(\rho_p V)$. For $\rho = 0$, it reduces to an optimization of the fundamental frequency of a freely vibrating plate.

Let us derive some necessary conditions for optimality in the problem of Eqs. (6.22), (6.23), (6.11), and (6.18). We state a formula for the first variation, with h satisfying the condition of Eq. (6.22). Setting $\delta J = 0$, we arrive at the following necessary condition for the existence of an extremum:

$$3h^2 [(u_{xx} + u_{yy})^2 - 2(1 - \nu)(u_{xx}u_{yy} - u_{xy}^2)] - \omega_0^2 u^2 = c^2 \tag{6.23}$$

where $c^2 > 0$ is a constant that is to be determined as part of the solution.

2. *Boundary properties of optimum design.* We study the behavior of the functions h, u, and Φ in the vicinity of the boundary curve Γ for an optimum plate. We make use of our assumptions concerning the smoothness of the boundary, which permits us to introduce a unique orthogonal ξ-η-ζ coordinate system at an arbitrary point O_Γ of the contour Γ, such that the coordinate η is tangent to the curve Γ at the point O_Γ, ξ points in the interior direction, and ζ is perpendicular to the x-y plane (see Fig. 6.1). In the vicinity of the point O_Γ, i.e., for small values of ξ, the optimality condition of Eq. (6.23) and the second boundary condition of Eq. (6.11) have asymptotic representations

$$3h^2 u_{\xi\xi}^2 - \omega_0^2 u^2 = c^2, \qquad h^3 u_{\xi\xi} = 0 \tag{6.24}$$

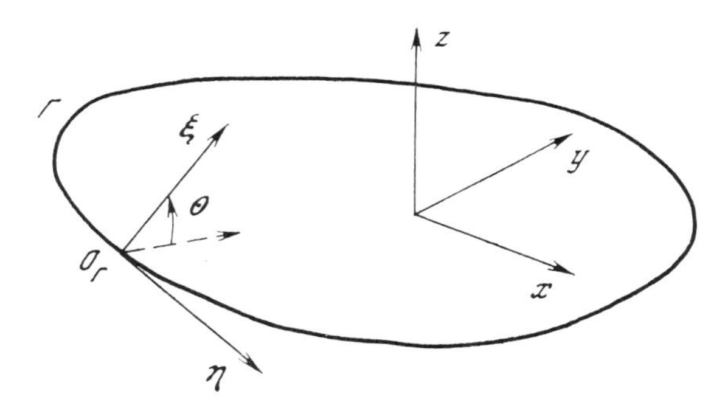

Figure 6.1

From Eq. (6.24) and the first boundary condition in Eq. (6.11), we derive the following boundary conditions for the deflection and thickness function at $\xi = 0$:

$$u_{\xi\xi}^{-2} = 0, \qquad h = 0 \tag{6.25}$$

The second condition in Eq. (6.25) is typical of optimization problems in beam and plate theory.

Using the first condition of Eq. (6.24), Eq. (6.13), and the boundary conditions for the function u, we rewrite Eq. (6.10) for small values of ξ as

$$(u_{\xi\xi}^{-2})_{\xi\xi} - \omega_0^2\left(\frac{c}{\sqrt{3}}\, u u_{\xi\xi}^{-1} + 2\rho\Phi^+\right) = 0 \tag{6.26}$$

We take note of the first equality given in Eq. (6.25) and observe that u vanishes on the boundary curve Γ to write the asymptotic representation

$$u = a_1\xi + a_2\xi^2 + \cdots + \xi^\mu(b_0 + b_1\xi + \cdots) \tag{6.27}$$

The parameter μ has the following constraints

$$0 < \mu < 2 \tag{6.28}$$

The inequality $0 < \mu$ is a consequence of the function u vanishing on Γ, while the upper bound $\mu < 2$ follows from the first condition in Eq. (6.25).

We study the asymptotic behavior of the function Φ near the boundary of the plate. We introduce a cylindrical polar r-θ-η coordinate system. The axis r

lies in the ξ-ζ plane, while θ denotes the angle between the r and ξ axes. The origins of both coordinate systems coincide.

It is easy to confirm that for small values of r, the following asymptotic representation is valid for Eq. (6.12):

$$\frac{1}{r}\left[(r\Phi_r)_r + \left(\frac{1}{r}\Phi_\theta\right)_\theta\right] = 0$$

We look for a solution of this equation in the form

$$\Phi = \psi(\theta) r^m + o(r^m)$$

where m is a parameter and ψ is a function of the angle θ. We substitute this expression into the potential equation and derive the following equation for the function ψ:

$$\psi_{\theta\theta} + m^2 \psi = 0$$

A general solution of this equation is $\psi = A \sin m\theta + B \cos m\theta$. The constants A, B, and m are determined from the relations

$$\begin{aligned} \psi(\pi) &= 0 \\ \xi^{m-1}(\psi_\theta)_{\theta=0} &= a_1\xi + a_2\xi^2 + \cdots + \xi^\mu(b_0 + b_1\xi + \cdots) \end{aligned} \tag{6.29}$$

which can be derived, respectively, from the boundary conditions $\Phi = 0$ for $\xi < 0, \zeta = 0$ and $\Phi_\zeta = r^{-1}\Phi_0$, for $\xi > 0, \zeta = 0$.

We could carry out the usual arguments of asymptotic analysis to determine the parameters μ and m, consisting in successive examinations of different cases corresponding to different hypothesis regarding the magnitudes of the parameters. Without carrying out in detailed analysis each case, we shall state the final conclusion. It is possible to determine uniquely m and μ from Eqs. (6.26)-(6.29). Their values are $m = 2$, and $\mu = 3/2$.

Substituting the asymptotic expression for the deflection function into the first Eq. (6.24), we obtain the asymptotic formula

$$h = \frac{c}{\sqrt{3}}\sqrt{\xi} + \cdots \tag{6.30}$$

Substituting $\mu = 3/2$ into the asymptotic expressions of Eqs. (6.30) and (6.27), we see that these formulas have a general form that is the same as the formulas for optimum plates vibrating in a vacuum. We observe that these asymptotic formulas may be applied to numerical solutions of optimization problems, to predict the behavior of the relevant functions in cells adjoining the contour Γ.

3. *Dependence of optimum design on* ρ. The optimum shape of a plate and the magnitude of the first eigenvalue (which is the functional we wish to maximize) depend on the values of the parameter ρ. We now study this dependence $\omega_{0*}^2 = \omega_{0*}^2(\rho)$. Let us investigate the effect of close values of the parameter ρ and $(\rho + d\rho)$ on the optimum solutions (ω_{0*}^2, u, h) and $(\omega_{0*}^2 + d\omega_{0*}^2, u + du, h + dh)$. These solutions satisfy Eq. (6.18). We substitute $(\omega_{0*}^2 + d\omega_{0*}^2, u + du, h + dh)$ into Eq. (6.18) and retain in this relation only the terms that are linear in $d\omega_{0*}$, du, and dh. After some elementary manipulations, we have

$$d\omega_{0*}^2\,[J_2(h, u) + \rho J_3(u)] + d\rho\,\omega_{0*}^2 J_3(u) + 2\iint_\Omega [L(h)\,u - \omega_{0*}^2 N(h)\,u]\,du\,dx\,dy + \iint_\Omega \{3h^2[(u_{xx} + u_{yy})^2 - 2(1-\nu)(u_{xx}u_{yy} - u_{xy}^2)] - \omega_{0*}^2 u^2\}\,dh\,dx\,dy = 0$$

We can transform this formula, using Eq. (6.15), the optimality condition of Eq. (6.23), and the constraints on the variation dh, which are a consequence of the isoperimetric equation of Eq. (6.22). Also noting that $J_1(h,u)$, $J_2(h,u)$ and $J_3(h,u)$, are positive functionals, we have

$$d\omega_{0*}^2/d\rho = -\omega_{0*}^2 J_3(u)/[J_2(h, u) + \rho J_3(u)] \leqslant 0 \tag{6.31}$$

This inequality shows that the maximum of the optimized functional (i.e., of the fundamental frequency) is a monotone decreasing function of the parameter ρ.

We observe that if we assign the value of the parameter $\rho = \rho_1$ and find a solution of the optimization problem of Eqs. (6.21), (6.22), (6.11), and (6.18), i.e., if we find the value of the functional $(\omega_{0*}^2)_1$ and the functions u_1 and h_1, then we can find an approximate value of the maximum for the fundamental frequency that corresponds to a value $\rho = \rho_2$ of the parameter, if it is close to ρ_1. We use the formula

$$(\omega_{0*}^2)_2 = (\omega_{0*}^2)_1\left[1 - \frac{(\rho_2 - \rho_1)\,J_3(u_1)}{J_2(h_1, u_1) + \rho_1 J_3(u_1)}\right]$$

which follows directly from Eq. (6.31).

6.3. Determining the Reaction of a Fluid When the Flow Field and the Motion of the Plate are Two-Dimensional and the Flow is Solenoidal

We shall restrict our study to vibrations of long rectangular plates that are cantilevered from an edge parallel to the y axis. We assume that the thickness function is independent of the y coordinate and the fluid flow lines lie in planes parallel to the x–z plane. We study solenoidal flows and compute the hydrodynamic reactions, utilizing techniques of functions of complex variables.[81]

The problem of determining the kernel K in a two-dimensional solenoidal fluid flow reduces to solving a boundary-value problem for a two-dimensional Laplace equation, with boundary conditions assigned on the boundary of the section $-1 \leqslant x \leqslant 1, z = 0$. These conditions contain an arbitrary function $u(x)$

$$\Phi_{xx} + \Phi_{zz} = 0 \tag{6.32}$$

$$(\Phi_z)\pm = u, \qquad -1 \leqslant x \leqslant 1, \qquad z = 0 \qquad (\nabla\Phi)_\infty = 0 \tag{6.33}$$

We introduce an auxiliary function

$$W = \Psi + i\Phi$$

of the complex variable $\zeta = x + iz$, where $i^2 = -1$. W is analytic in this complex plane, with a slit at $-1 \leqslant x \leqslant 1, z = 0$. The Cauchy–Riemann equations and the boundary conditions of Eq. (6.33) imply that $\Psi_x = \Phi_z = u$. Therefore

$$\Psi = \int_{-1}^{x} u(\xi)\, d\xi + C = U(x) + C \tag{6.34}$$

where C is a constant of integration. Finding the potential Φ reduces to the computation of the imaginary part of the analytic function W, whose real part on $[-1, +1]$ is

$$\operatorname{Re} W = \Psi = U(x) + C$$

We make use of results given in Ref. 153 and give the solution of this problem as

$$W = \frac{1}{2\pi i}\left(\frac{\zeta - 1}{\zeta + 1}\right)^{1/2} \int_{-1}^{1} \left(\frac{t+1}{t-1}\right)^{1/2} \frac{U(t) + C}{t - \zeta}\, dt \tag{6.35}$$

$$\frac{1}{2\pi i}\int_{-1}^{1}\frac{U(t)+C}{\sqrt{t^2-1}}\,dt=0 \tag{6.36}$$

Equation (6.36) can be used to determine the constant C. It also represents a regularity condition for the function W at the point $\zeta = 1$. Observing that

$$-\frac{1}{\pi i}\int_{-1}^{1}\frac{dt}{\sqrt{t^2-1}}=1$$

we obtain from Eq. (6.36)

$$C=\frac{1}{\pi i}\int_{-1}^{1}\frac{U(\xi)\,d\xi}{\sqrt{\xi^2-1}} \tag{6.37}$$

Using Eq. (6.37) for C and the formula

$$\frac{1}{2\pi i}\int_{-1}^{1}\left(\frac{\xi+1}{\xi-1}\right)^{1/2}\frac{d\xi}{\xi-\zeta}=\frac{1}{2}\left(\frac{\zeta+1}{\zeta-1}\right)^{1/2}-\frac{1}{2}$$

we complete substitutions into Eq. (6.35) for W, to obtain

$$\begin{aligned} W&=\frac{1}{2\pi i}\left(\frac{\zeta-1}{\zeta+1}\right)^{1/2}\int_{-1}^{1}\left(\frac{\xi+1}{\xi-1}\right)^{1/2}\frac{U(\xi)\,d\xi}{\xi-1}+\frac{C}{2}\left[1-\left(\frac{\zeta-1}{\zeta+1}\right)^{1/2}\right]\\ &=\frac{\sqrt{\zeta^2-1}}{2\pi i}\int_{-1}^{1}\frac{U(\xi)\,d\xi}{(\xi-\zeta)\sqrt{\xi^2-1}} \end{aligned} \tag{6.38}$$

We also compute the quantity Φ^+ by proceeding to the limit in Eq. (6.38) for $\zeta = x + iz \to x + i0$ $(z \to +0)$ and separating the imaginary part

$$\Phi^+=\lim_{z\to+0}\,[\operatorname{Im} W(x+iz)]=\text{v. p.}-\frac{\sqrt{1-x^2}}{2\pi}\int_{-1}^{1}\frac{U(\xi)\,d\xi}{(\xi-x)\sqrt{1-\xi^2}} \tag{6.39}$$

Integration in Eq. (6.39) is understood in the sense of Cauchy's principal value (p.v.). The difference in the value of the potential at the upper and lower side of the cut is given by

$$\Phi^+ - \Phi^- = 2\Phi^+ = \text{v. p.} - \frac{\sqrt{1-x^2}}{\pi} \int_{-1}^{1} \frac{U(\xi)\, d\xi}{(\xi - x)\sqrt{1-\xi^2}} \tag{6.40}$$

Let us transform the integral of Eq. (6.40). By definition of the principal value, we have

$$2\Phi^+ = \text{v. p.} \frac{1}{\pi} \int_{-1}^{1} \left(\frac{1-x^2}{1-t^2}\right)^{1/2} \frac{U(x)\, dx}{t-x}$$

$$= \lim_{\varepsilon \to 0} \frac{1}{\pi} \left[\int_{-1}^{x-\varepsilon} \left(\frac{1-x^2}{1-t^2}\right)^{1/2} \frac{U(t)\, dt}{t-x} + \int_{x+\varepsilon}^{1} \left(\frac{1-x^2}{1-t^2}\right)^{1/2} \frac{U(t)\, dt}{t-x} \right]$$

Integrating by parts and substituting Eq. (6.34) for $U(t)$, we have

$$2\Phi^+ = \lim_{\varepsilon \to 0} \left[T(x-\varepsilon, x) \int_{-1}^{x-\varepsilon} u(t)\, dt - T(x+\varepsilon, x) \int_{-1}^{x+\varepsilon} u(t)\, dt \right.$$
$$\left. - \int_{-1}^{x-\varepsilon} T(t, x)\, u(t)\, dt - \int_{x+\varepsilon}^{1} T(t, x)\, u(t)\, dt \right] \tag{6.41}$$

where we use the notation

$$T(t, x) = \frac{1}{\pi} \ln \left| \frac{1+\chi}{1-\chi} \right|, \qquad \chi = \left[\frac{(1-x)(1+t)}{(1-t)(1+x)} \right]^{1/2} \tag{6.42}$$

We observe that all terms on the right-hand side of Eq. (6.41) (i.e., expressions inside the square brackets) are finite; therefore the integration by parts is legitimate. As $\epsilon \to 0$, the sum of the first two terms in Eq. (6.41) approaches zero, while the last two integrals converge. Therefore, the required functional dependence of the jump in the potential value of the plate deflection function is of the form

$$2\Phi^+ = - \int_{-1}^{1} K(t, x)\, u(t)\, dt, \qquad K(t, x) = T(t, x) \tag{6.43}$$

6.4. Finding the Optimum Shape of a Vibrating Plate

We consider a long rectangular plate that is simply supported along the longer edges, which are parallel to the y axis (see Fig. 6.2). We assume that the plate thickness does not change in the y direction, i.e., $h = h(x)$, and that our discussion is restricted only to the x–z plane, since the derivatives of all relevant functions with respect to y are equal to zero. The intersection of the plate with the x–z plane is denoted by S. We regard the area of S, rather than the plate volume V as given. A transition to dimensionless variables is accomplished by the use of Eq. (6.9), with a formal replacement of V in Eq. (6.40) by lS. In our new variables, the supported edges of the plate in the x–z plane correspond to the points $(x = -1, z = 0)$, and $(x = 1, z = 0)$.

In this case, Eq. (6.15), the boundary conditions of Eq. (6.11), the formulas for J_1, J_2, and J_3, and the isoperimetric condition of Eq. (6.22) are rewritten in the form

$$(h^3 u_{xx})_{xx} - \omega^2 \left[hu + \rho \int_{-1}^{1} K(t, x)\, u(t)\, dt \right] = 0 \tag{6.44}$$

$$u(-1) = u(1) = 0, \qquad (h^3 u_{xx})_{x=-1} = (h^3 u_{xx})_{x=1} = 0 \tag{6.45}$$

$$\int_{-1}^{1} h\, dx = 1 \tag{6.46}$$

$$J_1 = \int_{-1}^{1} h^3 u_{xx}^2\, dx, \quad J_2 = \int_{-1}^{1} hu^2\, dx, \quad J_3 = \int_{-1}^{1} K(t, x)\, u(t)\, u(x)\, dt\, dx \tag{6.47}$$

The kernel K is determined by Eqs. (6.42) and (6.43).

We derive separately the solutions for homogeneous and for trilayer plate optimization. In the case of a trilayer plate, h^3 in the first term of Eq. (6.44) and in the expression of Eq. (6.47) for J_1 is replaced by h.

We carried out some numerical computations for the optimum thickness

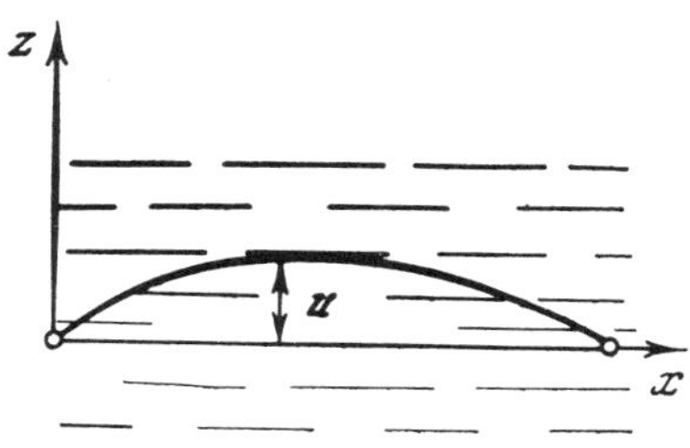

Figure 6.2

distribution in homogeneous plates using a successive optimization algorithm described in Section 1.10. We give some basic relations that were used in these computations, with some explanation. The formulas for variation of the payoff functional and for the integral on the left-hand side of Eq. (6.46) assume the form

$$\delta\omega_0^2 = (\psi, \delta h) = \int_{-1}^{1} \psi \delta h \, dx, \qquad \psi = \frac{3h^2 u_{xx}^2 - \omega_0^2 u^2}{J_2(h, u) - \rho J_3(u)}$$
$$(\psi_1, \delta h) = \int_{-1}^{1} \delta h \, dx = 0, \qquad \psi_1 = 1 \tag{6.48}$$

To find a new thickness function, we apply Eq. (1.145), changing ψ to $-\psi$, since we are solving a maximization, rather than a minimization problem. After this substitution and the substitution for formulas of ψ and ψ_1, we arrive at the following computational scheme:

$$h^{k+1} = h^k + \delta h,$$
$$\delta h = \tau \left\{ \psi - \frac{(\psi, \psi_1)}{(\psi_1, \psi_1)} \psi_1 \right\} = \tau \left\{ \psi - \frac{1}{2} \int_{-1}^{1} \psi \, dx \right\} \tag{6.49}$$

Choosing a sufficiently small positive number τ and computing the initial distribution of thickness according to Eqs. (6.48) and (6.49), we may increase the value of the functional, while retaining the isoperimetric condition.

To find the plate deflection, for the current choice of thickness function, we use the technique of local variations. We also make use of symmetry of our problem with respect to the point $x = 0$ and carry out our computations only in the interval $-1 \leqslant x \leqslant 0$, which is decomposed into 30 subintervals. The trapezoidal rule was used in computing the integrals J_1, J_2, and J_3.

Computations were carried out for values of $\rho = 0$ and $0.1 \cdot 2^n$, $n = 1, 2, \ldots, 7$. For each value of ρ, we choose as the initial thickness function h, the constant function $h \equiv 1/2$. The first stage of calculation consists of finding the deflection function and the square of the fundamental frequency ω_0^2, corresponding to the constant thickness.

Figure 6.3 illustrates the dependence of the square of the frequency ω_0^2 on the value of ρ, for the optimum plate (curve 1) and for a plate of constant thickness (curve 2).

Figure 6.4 illustrates the optimum thickness function for $\rho = 1.6$. The function $h(x)$ attains its maximum at $x = 0$ and tends to zero as x approaches

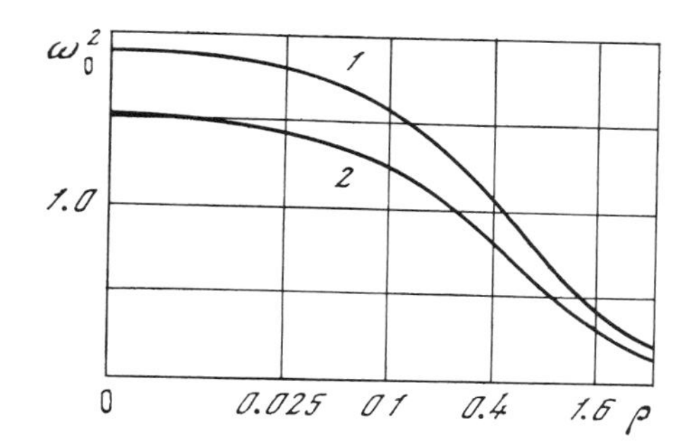

Figure 6.3

± 1, i.e., as we approach the simply supported edges of the plate. The dotted line indicates the initial approximation for the thickness function. We note that for $\rho = 0$ the gain in value of ω_0^2 is 23.6%, while for $\rho = 1.6$ it is 38%.

Now we determine the optimum distribution of thickness for the load-bearing layers of a trilayer plate. Let h_d denote the distance between the load-bearing layers ($h_d = \text{const}$). We introduce a dimensionless quantity ω' (where as before we shall omit the primes in future formulas)

$$\omega' = \left[\frac{4\rho_f l^6 (1 - \nu^2)}{S E h_d^2}\right]^{1/2} \omega \tag{6.50}$$

Let us recall the basic relationships of this problem: the vibration equation, the boundary conditions, and the optimality condition,

$$(h u_{xx})_{xx} = \omega^2 \left[\frac{hu}{\rho} + \int_{-1}^{1} K(t, x)\, u(t)\, dt\right] \tag{6.51}$$

$$u(-1) = u(1) = 0 \tag{6.52}$$

$$(h u_{xx})_{x=-1} = (h u_{xx})_{x=+1} = 0 \tag{6.53}$$

$$(u_{xx})^2 - \frac{\omega^2}{\rho} u^2 = c^2 \quad (c^2 \neq 0) \tag{6.54}$$

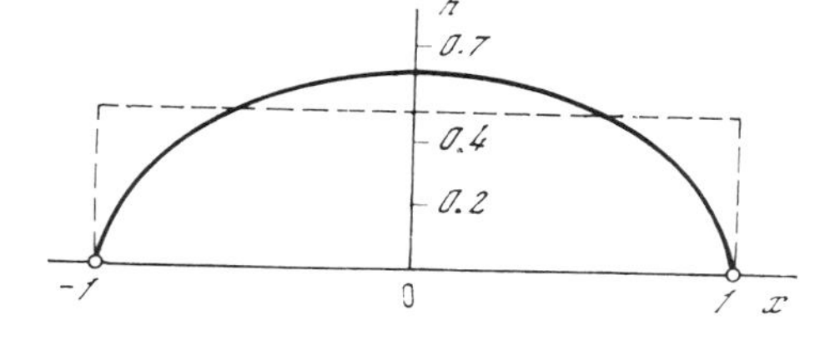

Figure 6.4

We consider the ideal case $\rho = \rho_f l^2/(\rho_d S) \to \infty$ corresponding to either a very thin plate or to a very dense liquid. Equation (6.51) and the optimality condition of Eq. (6.54) are written in the form

$$(hu_{xx})_{xx} = \omega^2 \int_{-1}^{1} K(t, x)\, u(t)\, dt \tag{6.55}$$

$$(u_{xx})^2 = c^2 \tag{6.56}$$

Since our problem is symmetric with respect to the point $x = 0$, we can study it on the interval $[-1, 0]$, assigning to the point $x = 0$ the following boundary conditions

$$(u_x)_{x=0} = 0 \tag{6.57}$$

$$(h_x)_{x=0} = 0 \tag{6.58}$$

We observe a remarkable property of this problem, namely that the optimality condition of Eq. (6.56) does not contain explicit dependence on h. This permits us to employ the following technique: Solving the boundary-value problem of Eq. (6.56) and the boundary conditions of Eqs. (6.52) and (6.57), we determine (with an accuracy up to a factor of c^2) the function $u(x)$, corresponding to the deflection of an optimum plate. Substituting this function $u(x)$ into Eq. (6.55) and into the boundary conditions of Eqs. (6.53) and (6.58) and solving this system for h, we find the optimum distribution of thickness. The function h so constructed depends on the parameter ω^2, which occurs in Eq. (6.55). The quantity ω^2 is determined by substituting our expression for h into the isoperimetric condition

$$\int_{-1}^{0} h\, dx = \frac{1}{2} \tag{6.59}$$

and performing some elementary computations. Thus we complete the solution of our problem.

Equations (6.56) and the boundary conditions of Eqs. (6.57) and (6.52) (the first equality) permit us to find the deflection function u that corresponds to the optimum distribution of thickness

$$u = c\,(x^2 - 1)/2$$

Substituting this function u into the vibration equation of Eq. (6.55) and into Eq. (6.53), which states the absence of a bending moment at the point $x=-1$, and also noting the condition of Eq. (6.58), we arrive at the following boundary-value problem:

$$h_{xx} = \frac{\omega_{0*}^2}{2} \int_{-1}^{1} K(t, x)(t^2 - 1)\, dt$$

$$h(-1) = 0, \quad (h_x)_{x=0} = 0$$

A solution of this boundary-value problem has the general form

$$h = \frac{\omega_{0*}^2}{2} H(x), \quad H(x) \equiv \int_{-1}^{x} \int_{0}^{\eta} \int_{-1}^{1} K(t, \xi)(t^2 - 1)\, dt\, d\xi\, d\eta \qquad (6.60)$$

We make use of Eq. (6.60) for the optimum thickness function $h(x)$ and of the isoperimetric condition of Eq. (6.59) to derive the following formula for the fundamental frequency:

$$\omega_{0*}^2 = \left[\int_{-1}^{0} H(x)\, dx \right]^{-1} \qquad (6.61)$$

The corresponding value of ω_{0*}^2, computed by means of Eqs. (6.60) and (6.61) is equal to $\omega_{0*}^2 = 1.121$. A graph illustrating the optimum distribution of thickness $h(x)$ is given in Fig. 6.5.

To find the optimum solutions, we have used in each case the optimality condition of Eq. (6.56). We shall prove that in the case considered here, Eq. (6.56) is not only necessary but is also a sufficient condition for optimality.[223]

Let h^*, u^* satisfy Eqs. (6.55) and (6.56), the isoperimetric condition of Eq. (6.59), and the boundary conditions. Let us also consider functions h and u that satisfy all of these conditions, with the exception of the conditions of

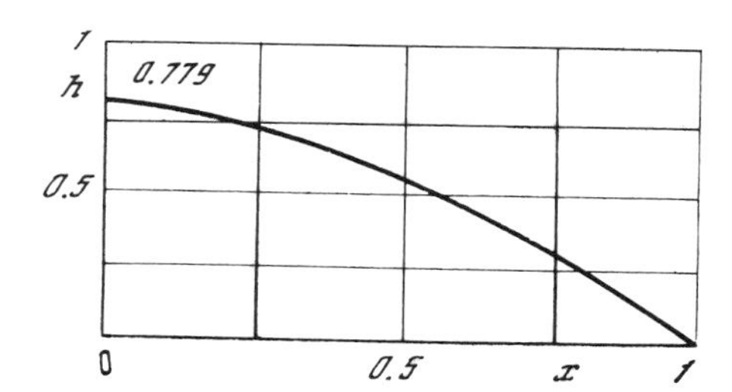

Figure 6.5

optimality of Eq. (6.56). The set of all such functions h and u contains h^* and u^* as a subset. Let us prove that the quantity

$$\Delta\omega_0^2 = \omega_0^2(h^*) - \omega_0^2(h) \equiv \frac{J_1(h^*, u^*)}{J_3(u^*)} - \frac{J_1(h, u)}{J_3(u)}$$

is positive. To prove this we make use of the isoperimetric condition of Eq. (6.59) and of the properties of h^* and u^*, forming the following estimates:

$$\Delta\omega_0^2 = \frac{J_1(h^*, u^*)}{J_3(u^*)} - \frac{J_1(h, u)}{J_3(u)} \geqslant \frac{J_1(h^*, u^*)}{J_3(u^*)} - \frac{J_1(h, u^*)}{J_3(u^*)} = \frac{c^2}{J_3(u^*)} \int_{-1}^{1} (h^* - h)\, dx = 0$$

Thus

$$\omega_{0*}^2 = \omega_0^2(h^*) \geqslant \omega_0^2(h)$$

and, consequently, we have shown that the fundamental frequency ω_0 attains a maximum if the condition of Eq. (6.56) is satisfied.

6.5. Maximizing the Divergence Velocity of a Plate Subjected to the Flow of an Ideal Fluid

The treatment of optimization of a plate subjected to a continuous flow of an ideal fluid presented here is based on the research of A. A. Mironov. We consider an ideal fluid flowing over the surface of a plate, with a zero angle of attack. It is well known that if the surface of the plate curves, then the pressure on the convex surface is smaller than on the concave side. Therefore, the reaction force is proportional to the kinetic energy density ρv_∞^2 (where ρ is the fluid density and v_∞ the speed). For sufficiently large values of the speed v_∞, this force may overcome the resisting elastic force and a loss of stability may result.

Let us formulate this problem mathematically. We shall limit our study to a solenoidal flow in two dimensions (a long rectangular plate). We introduce a Cartesian x–z coordinate system and direct the x axis parallel to the flow of the fluid. We denote by w, φ, and Φ, respectively, plate deflection, the velocity potential, and a function related to φ by the formula $\varphi = xv_\infty + \Phi$.

The total reaction force q exerted by the fluid is equal to the difference of pressures acting on the upper and lower faces of the plate, $q = p^- - p^+$. The pressure p depends on the velocity potential φ. In terms of Bernoulli's law, we have $p = p_\infty - \frac{1}{2}\rho(\nabla\varphi)^2$. The function $\varphi = \varphi(x, y)$ can be found by solving Neumann's boundary-value problem for the region exterior to the plate. We shall consider, instead of φ, the auxiliary potential function Φ related to φ by the formula $\varphi = xv_\infty + \Phi$. Regarding φ and w as small, we linearize the hydrodynamic problem and introduce a boundary condition that the fluid does not cross the surface of the plate, relating it to the boundary of the cut $z = 0$, $0 \leqslant x \leqslant l$, where l is the length of the plate. We introduce the dimensionless variables $x' = x/l$, $z' = z/l$, $w' = w/l$, $\Phi' = \Phi(lv_\infty)$, and $h' = h/S$. The linearized equations of hydroelasticity can be written as

$$(h^\alpha w_{xx})_{xx} = \lambda\,(\Phi_x^+ - \Phi_x^-) \tag{6.62}$$

$$w\,(0) = w\,(1) = 0, \qquad (h^\alpha w_{xx})_{x=0} = (h^\alpha w_{xx})_{x=1} = 0 \tag{6.63}$$

$$\Delta\Phi \equiv \Phi_{xx} + \Phi_{zz} = 0 \tag{6.64}$$

$$\Phi_z^\pm = w_x \quad (z = 0, \quad 0 \leqslant x \leqslant 1), \quad (\nabla\Phi)_\infty = 0 \tag{6.65}$$

where S is the area of perpendicular cross section of the plate and λ in Eq. (6.62) denotes

$$\lambda = \rho v_\infty^2 l^{\alpha+3}/A_\alpha S^\alpha$$

The hydrodynamic problem of Eqs. (6.64) and (6.65), concerning the potential Φ, and the bending problem of Eqs. (6.62) and (6.63) are coupled to each other, since the boundary conditions of Eq. (6.65) for Φ contain a derivative of the deflection function, while the right-hand side of the elastic bending equation for the plate contains derivatives of the potential function Φ. The problem of Eqs. (6.62) to (6.65) is a homogeneous eigenvalue problem, the parameter λ playing the part of the eigenvalue.

The quantity Φ_x^+ and Φ_x^- that occur in the right-hand side of the plate bending equation of Eq. (6.62) can be determined by applying the results of Section 6.3. Specifically, we observe that Φ satisfies the boundary-value problem

$$\begin{gathered} \Delta\Phi_x = 0 \\ (\Phi_x)_z^\pm = w_{xx} \quad (z = 0,\ 0 \leqslant x \leqslant 1), \quad (\Phi_x)_\infty = 0 \end{gathered} \tag{6.66}$$

whose solution may be found in Section 6.3. The pressure jump on the plate surfaces is given by

$$\Phi_x^+ - \Phi_x^- = \int_0^1 K(t, x)\, w_{tt}(t)\, dt$$
$$K(t, x) = \frac{2}{\pi} \ln \left| \frac{1+r}{1-r} \right|, \quad r = \left[\frac{t(1-x)}{x(1-t)} \right]^{1/2} \tag{6.67}$$

Thus, the problem of determining the plate deflection is reduced to the solution of the integrodifferential equation

$$(h^\alpha w_{xx})_{xx} = \lambda \int_0^1 K(t, x)\, w_{tt}(t)\, dt \tag{6.68}$$

with boundary conditions of Eq. (6.63).

The integral operator on the right-hand side of Eq. (6.68) is positive and self-adjoint. Proof of these properties can be carried out in the same manner as we proved positiveness and self-adjointness of the operator N in Section 6.1. It is therefore omitted. Positiveness and self-adjointness of the operator in the left-hand side of Eq. (6.68) is well-known. Consequently, the problem of Eqs. (6.63) and (6.68) has only positive eigenvalues.

The first eigenvalue for a plate of constant thickness is computed numerically, using the technique of successive approximations. It is equal to $\lambda_0 = 39.9693$. The critical divergence speed is given by the formula $v_\infty = [\lambda_0 A_\alpha S^\alpha/(\rho_f l^{\alpha+3})]^{1/2}$.

Let us now look at the problem of maximizing the flutter speed by optimizing the plate distribution of thickness, i.e.,

$$\lambda_* = \max_h \lambda_0 \tag{6.69}$$

$$\int_0^1 h\, dx = 1 \tag{6.70}$$

$$(h^\alpha u)_{xx} = \lambda_0 \int_0^1 K(t, x)\, u(t)\, dt \tag{6.71}$$

$$(h^\alpha u)_{x=0} = (h^\alpha u)_{x=1} = 0 \tag{6.72}$$

where $u = w_{xx}$. We derive necessary conditions for optimality in the problem given by Eqs. (6.69)-(6.72). To do this, we rewrite Eq. (6.69) in the variational form

$$(h^{\alpha}\delta u)_{xx} - \lambda_0 \int_0^1 K(t, x)\,\delta u(t)\,dt = \delta\lambda \int_0^1 K(t, x)\,u(t)\,dt - \alpha(h^{\alpha-1}u\delta h)_{xx} \tag{6.73}$$

Since λ_0 is an eigenvalue of Eq. (6.68), it is necessary for solvability of this problem that the right-hand side of Eq. (6.73) must be orthogonal to the eigenfunction of Eq. (6.68), i.e.,

$$\delta\lambda \int_0^1\int_0^1 K(t, x)\,u(t)\,u(x)\,dt\,dx - \alpha \int_0^1 h^{\alpha-1}u^2\delta h\,dx = 0$$

Hence

$$\delta\lambda = \alpha \frac{\int_0^1 h^{\alpha-1}u^2\delta h\,dx}{\int_0^1\int_0^1 K(t, x)\,u(t)\,u(x)\,dt\,dx}$$

Since the function h satisfies the isoperimetric condition of Eq. (6.70), the integral of the variation δh must be equal to zero. Using this fact and assuming $\delta\lambda = 0$, we derive a necessary condition for optimality as

$$h^{\alpha-1}u^2 = c^2 \tag{6.74}$$

Here c is a Lagrange multiplier that can be determined from the condition of Eq. (6.70).

The optimum distribution of thickness for a composite plate ($\alpha = 3$) can be found numerically by using the expression for the variation of the pay-off functional $\delta\lambda$ and the algorithm for successive optimization described in Section 1.10 (see also Section 6.4). The optimum distribution of thickness found as a result of this computation is shown in Fig. 6.6. The corresponding value of λ_* is $\lambda_* = 47.37$, while the gain in comparison with a plate of constant thickness is 28.15%.

For a trilayer plate ($\alpha = 1$), the optimality condition $u^2 = c^2$ determines the deflection function w. This permits us to affect some simplifications and to offer an analytic solution of the optimization problem. Using the optimality conditions, we can represent the equalibrium equation for the plate, Eq. (6.71),

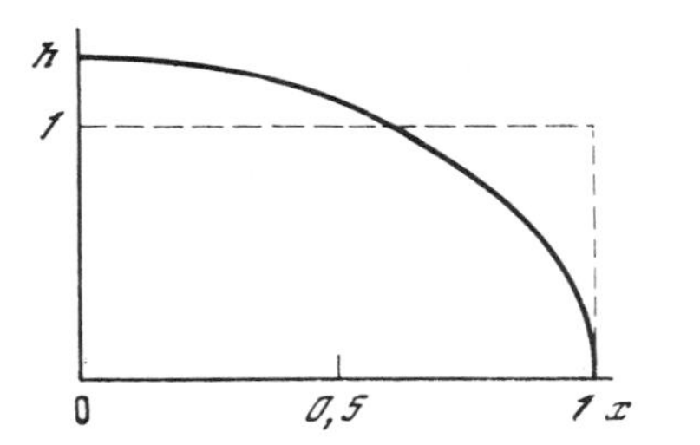

Figure 6.6

and the boundary conditions of Eq. (6.72) in the form

$$h_{xx} = \lambda_* \int_0^1 K(t, x)\, dt, \quad h(0) = h(1) = 0$$

Making use of the isoperimetric condition, it is easy to obtain

$$\begin{gathered} h = \lambda_* H(x), \qquad H(x) \equiv \int_0^1 \int_0^1 T(x, \xi) K(t, \xi)\, dt\, d\xi \\ T(x, \xi) = \begin{cases} x - 1, & \xi < x, \\ 1 - \xi, & \xi > x, \end{cases} \qquad \lambda_* = \left(\int_0^1 H(x)\, dx \right)^{-1} \end{gathered} \tag{6.75}$$

In a manner similar to Section 6.4, it is possible to show that a constant value of u^2 is not only a necessary but also a sufficient condition for the existence of a global optimum.

6.6. A Scheme in Solenoidal Flow for Investigating Equilibrium Shapes of Elastic Plates and a Problem of Optimization

In the preceding discussion, we solved problems of maximizing critical velocities or frequencies of vibration, by making the assumption that the fluid flow over the surface of the plate is continuous. In this section, we follow Ref. 34. As we determine the hydrodynamic forces acting on the plate, we consider a scheme for a solenoidal flow with an infinite cavity (the Kirchhoff scheme).

1. *An ideal fluid flow problem.* The fluid flows over the surface of an elastic plate OA (see Fig. 6.7). In an undeformed state, the plate lies in the plane perpendicular to the z axis and the leading edge of the plate (i.e., the point A'

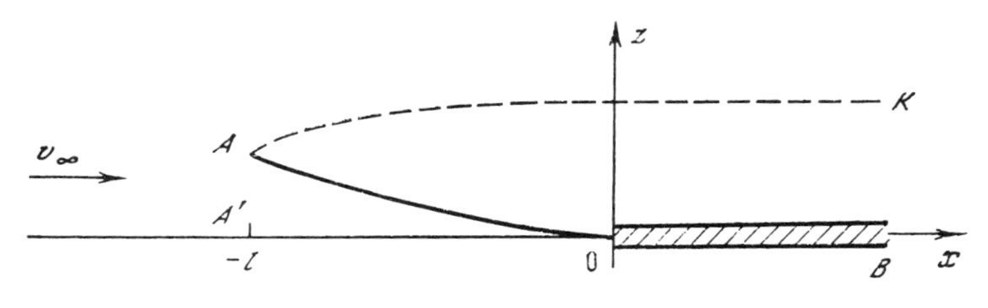

Figure 6.7

in Fig. 6.7) is free. The trailing edge ($x = 0, z = 0$) of this plate is rigidly fastened to a rigid infinite plate OB, which lies on the semi-axis $x > 0, z = 0$.

To investigate nontrivial equilibrium positions of the plate, compared to the initial undisturbed position ($z = 0$), we consider a possible equilibrium position of a bent plate with the shape indicated by OA. The deflection function $w(x)$ ($w \ll l$) satisfies the plate equation of equilibrium and the boundary conditions for a clamped edge, $w = w_x = 0$ at $x = 0$, and conditions indicating the absence of shear forces or moments $h^\alpha w_{xx} = h^\alpha (w_{xx}) = 0$ at the free end $x = -l$.

To determine the reaction force exerted by an ideal fluid flowing over the curve, we consider the hydrodynamic equations. We assume that the flow lines, for fluid flowing over the bent plate, have a discontinuity and that an infinite cavity $BOAK$ is created. The velocity vector is parallel to the x axis at infinity and its magnitude is equal to v_∞.

The fluid flow is curl-free and has the velocity potential $\varphi(x, z)$ ($v = \nabla\varphi$), which satisfies Laplace equation $\Delta\varphi = 0$. As in Section 6.5, we represent the potential φ as a sum of two terms $\varphi = xv_\infty + \Phi$, where Φ is a harmonic function that tends to zero at infinity, i.e.,

$$\Delta\Phi = 0, \qquad (\Phi)_\infty = 0 \tag{6.76}$$

The function Φ satisfies boundary conditions on the plate surface and on the free surface of the fluid (i.e., on the boundary of the cavity). Assuming that plate deflections and the distance of the free surface from the z axis are "small", we linearize the problem and transfer the boundary conditions to the x axis ignoring terms of order $O(h_{\max})$ and $O(w_{\max})$, where $h_{\max} = \max_x h$, and $w_{\max} = \max_x w$.

We make a cut of zero width along the semi-infinite ray $x \geqslant -l$ of the x axis and we denote the values of functions on the upper and lower sides of this cut by superscripts (+) and (−), respectively. Following linearization and assignment of suitable conditions to Γ, expressing the fact that the fluid does not cross the surface of the plate, we have

$$(\Phi_z)^- = v_\infty w_x \quad (-l \leqslant x \leqslant 0), \qquad (\Phi_z)^- = 0 \quad (x \geqslant 0) \tag{6.77}$$

We use the assumption that the representative thickness of the plate is smaller than the typical deflection.

Similarly, we write the condition transferred to the free boundary on Γ^+

$$(\Phi_z)^+ = v_\infty f_x \tag{6.78}$$

The kinematic condition $(\nabla\varphi)^2 = \text{const}$, which follows from Bernoulli's law, expresses the fact that pressure on the boundary of the cavity is constant. It assumes the form

$$(\Phi_x)^+ = 0, \qquad x \geqslant -l \tag{6.79}$$

If the plate deflection function is known, $w = w(x)$, the boundary-value problem of Eq. (6.76), (6.77) and (6.78) is solvable. Solving it, we can determine the function $\Phi(x, z)$. After finding $\Phi(x, z)$, we can determine the shape of the cavity, using quadratures [this is a consequence of Eq. (6.78)] to obtain.

$$f = w(-l) + \frac{1}{v_\infty} \int_{-l}^{x} (\Phi_z(t, 0))^+ \, dt \tag{6.80}$$

The distribution of reaction forces between the fluid and the plate can be found from Bernoulli's law. After linearization and some elementary manipulations, we obtain

$$q = p^- - p^+ = -\rho_f v_\infty (\Phi_x)^-$$

Furthermore we change to dimensionless variables, as introduced in Section 6.5, and rewrite the basic equations of hydroelasticity in the form

$$(h^\alpha w_{xx})_{xx} = -\lambda (\Phi_x)^-, \qquad -1 \leqslant x \leqslant 0 \tag{6.81}$$

$$w(0) = w_x(0) = 0, \qquad (h^\alpha w_{xx})_{x=-1} = [(h^\alpha w_{xx})_x]_{x=-1} = 0 \tag{6.82}$$

$$\Delta\Phi = 0, \qquad (\Phi)_\infty = 0 \tag{6.83}$$

$$(\Phi_x)^+ = 0, \qquad -1 \leqslant x \tag{6.84}$$

$$(\Phi_z)^- = w_x, \quad -1 \leqslant x \leqslant 0; \quad (\Phi_z)^- = 0, \quad x \geqslant 0 \tag{6.85}$$

where λ denotes the following combination of dimensionless parameters of our problem: $\lambda = \rho_f v_\infty^2 S^{-\alpha} A_\alpha^{-1} l^{\alpha+3}$ and S is the area of the cross section of the elastic plate. Thus we have arrived at a solvable boundary-value problem for the functions w and Φ.

The hydrodynamic problem of Eqs. (6.83)-(6.85) concerning the distribution of the potential Φ and the problem of determining the deflection, Eqs. (6.81) and (6.82) are coupled. Also the derivative of the plate deflection function enters into the boundary conditions of Eq. (6.85) for the function Φ, while the right-hand side of the plate deflection equation of Eq. (6.81) contains a derivative of the potential Φ. The problem of Eqs. (6.81)-(6.85) is homogeneous and therefore admits the trivial solution $w = \Phi \equiv 0$. We look for nontrivial solutions of this system of equations and transform them into an eigenvalue problem, where λ assumes the role of an eigenvalue. We look for the first (that is smallest) eigenvalue λ_0, which corresponds to the critical values of the parameters ρ_f, v_∞, l, S, and A_α.

Existence of nontrivial solutions of Eqs. (6.81)-(6.85), for $\lambda = \lambda_0$, may be regarded as instability of the undeformed state of the plate.

2. *Study of the boundary-value problem.* The potential problem of Eqs. (6.83)-(6.85) is linear in w and does not depend on h. Therefore, the expression on the right-hand side of Eq. (6.81) may be written in the form

$$(\Phi_x)^- = Mw$$

where M is a linear operator. We shall prove that this operator is self-adjoint and positive. Let $w^1(x)$ and $w^2(x)$ be quite arbitrary smooth functions defined on the interval $[-1, 0]$ and satisfying appropriate boundary conditions, and let Φ^1 and Φ^2 be solutions of the boundary-value problem of Eqs. (6.83)-(6.85), corresponding, respectively, to $w = w^1$ and $w = w^2$. We define the functions w^1 and w^2 on an interval $[0, \delta]$, where δ is a positive number, by setting $w^1(x) = w^2(x) \equiv 0$ on this interval.

We have

$$\int_{-1}^{0} w^1 M w^2 \, dx = \int_{-1}^{\delta} w^1 M w^2 \, dx = -\int_{-1}^{\delta} w^1 (\Phi_x^2)^- \, dx$$

The last integral is obtained by integration by parts. We note that $(\Phi^1)^+ = (\Phi^2)^- = 0$, for $x = -1$, and insert the boundary conditions for Φ^1 and Φ^2 in the integration by parts formula. We obtain

$$\int_{-1}^{0} w^1 M w^2\, dx = \int_{-1}^{\delta} w_x^1 (\Phi^2)^- \, dx = \int_{-1}^{\delta} (\Phi_z^1)^- (\Phi^2)^- \, dx$$
$$= \int_{-1}^{\delta} [(\Phi_z^1)(\Phi^2)^- - (\Phi_z^1)(\Phi^2)^+]\, dx$$

We denote by Γ_δ^+ and Γ_δ^- the upper and lower boundary of the cut with $-1 \leqslant x \leqslant \delta$, and by Σ_δ a neighborhood of radius δ, whose center lies at the point $(-1, 0)$ and by n an interior unit normal to the boundary $\Gamma_\delta = \Gamma_\delta^+ + \Gamma_\delta^- + \Sigma_\delta$. We have then

$$\int_{-1}^{0} w^1 M w^2\, dx = -\int_{\Gamma_\delta} \frac{\partial \Phi^1}{\partial n} \Phi^2 \, d\sigma + \int_{\Sigma_\delta} \frac{\partial \Phi^1}{\partial n} \Phi^2 \, d\sigma \tag{6.86}$$

Applying Green's formula to the first integral on the right-hand side of Eq. (6.86), we obtain

$$\int_{-1}^{0} w^1 M w^2\, dx = -\int_{\Gamma_\delta} \frac{\partial \Phi^2}{\partial n} \Phi^1 \, d\sigma + \int_{\Sigma_\delta} \frac{\partial \Phi^1}{\partial n} \Phi^2 \, d\sigma$$

Taking care of the boundary conditions for the potentials Φ^1 and Φ^2, we can write this equality in the form

$$\int_{-1}^{0} w^1 M w^2\, dx = \int_{-1}^{\delta} w_x^1 (\Phi^2)^- \, dx + \int_{\Sigma_\delta} \left(\frac{\partial \Phi^1}{\partial n} \Phi^2 - \frac{\partial \Phi^2}{\partial n} \Phi^1 \right) d\sigma$$

Again we integrate by parts and proceed to the limit as $\delta \to \infty$, obtaining the relation

$$\int_{-1}^{0} w^1 M w^2\, dx = \int_{-1}^{0} w^2 M w^1\, dx$$

Therefore, M is self-adjoint. Similarly, we can prove the positive property of M. Let us set $w^1 = w^2 = w$ and $\Phi^1 = \Phi^2 = \Phi$ in Eq. (6.86). Applying Green's lemma, we obtain

$$\int_{-1}^{0} wMw\,dx = \int_{V_\delta} (\nabla\Phi)^2\,d\tau + \int_{\Sigma_\delta} \frac{\partial\Phi}{\partial n}\,\Phi\,d\sigma$$

The region V_δ is bounded by the surface Γ_δ. Letting δ approach infinity in this formula, we arrive at the inequality

$$\int_{-1}^{0} wMw\,dx = \int_{V_0} (\nabla\Phi)^2\,d\tau > 0$$

where V_0 denotes the region complementary to the semi-infinite ray $-1 < x$, $z = 0$.

Thus we have shown that the operator M is self-adjoint and positive. The self-adjoint and positive properties of the operator in the left-hand side of Eq. (6.81) is well known. Therefore, the eigenvalues of Eqs. (6.81)-(6.85) are all real and positive.

3. *Reaction force exerted on the plate by the fluid.* Bending of the plate by the fluid flow is given by the right-hand side of Eq. (6.81). To compute the derivative of the potential $(\Phi_x)^-$, we need to examine the exterior hydrodynamic problem of Eqs. (6.83)-(6.85). Let us introduce an auxiliary function $W = \Phi + i\Psi$, which depends on the variable $\zeta = x + iz$ $(i^2 = -1)$. We assume that W is analytic in the x-z plane, with a semi-infinite cut $-1 \leqslant x, z = 0$. The derivative of W is given by $W' = \Phi_x + i\Psi_x$. From the Cauchy-Riemann equations and the boundary conditions, we derive

$$(\Psi_x)^- = -(\Phi_y)^- = -\chi(x),$$

$$\chi = \begin{cases} w_x & -1 \leqslant x \leqslant 0 \\ 0 & x \geqslant 0 \end{cases}$$

These relations and the boundary condition of Eq. (6.84) imply that for $x \geqslant -1$

$$\operatorname{Re}(W')^+ = 0, \qquad \operatorname{Im}(W')^- = -i\chi \tag{6.87}$$

To find the derivative W' of the analytic function W, we need to solve the mixed boundary-value problem of Eq. (6.87). A solution of this problem was found by D. I. Sherman. It is of the form

$$W' = -\frac{1}{2\pi i\,(\zeta+1)^{1/4}} \int_{-1}^{0} \frac{(t+1)^{1/4}\, w_t}{t-\zeta}\, dt + \frac{1}{2\pi i\,(\zeta+1)^{3/4}} \int_{-1}^{0} \frac{(t+1)^{3/4}\, w_t}{t-\zeta}\, dt \tag{6.88}$$

Proceeding to the limit $\zeta = x + iz \to x - i0$ (with $0 > z$) in the expression given in the right-hand side of Eq. (6.88) and making use of the Sokhotski–Plimelj formula, we have

$$(W')^{-} = -\,iw_x + \frac{1}{2\pi\,(1+x)^{1/4}} \int_{-1}^{0} \frac{(1+t)^{1/4}\, w_t}{t-x}\, dt + \frac{1}{2\pi\,(1+x)^{3/4}} \int_{-1}^{0} \frac{(1+t)^{3/4}\, w_t}{t-x}\, dt$$

It follows that the unknown quantity $(\Phi_x)^{-}$ is equal to

$$(\Phi_x)^{-} = \frac{1}{2\pi\,(1+x)^{1/4}} \int_{-1}^{0} \frac{(1+t)^{1/4}\, w_t}{t-x}\, dt + \frac{1}{2\pi\,(1+x)^{3/4}} \int_{-1}^{0} \frac{(1+t)^{3/4}\, w_t}{t-x}\, dt$$

where the integrals are regarded in the sense of Cauchy's principal value. Thus, the unknown function $(\Phi_x)^{-}$ is given by

$$(\Phi_x)^{-} = -\int_{-1}^{0} K(t,x)\, w_t\, dt$$
$$K(t,x) \equiv \frac{1}{2\pi\,(t-x)} \left[\left(\frac{1+t}{1+x}\right)^{1/4} + \left(\frac{1+t}{1+x}\right)^{3/4} \right] \tag{6.89}$$

4. *Integrodifferential formulation.* The expression we obtained for the reaction force exerted by the fluid may now be substituted into the plate bending equation of Eq. (6.81). We thus obtain a homogeneous integrodifferential

equation describing the displacement of the plate

$$(h^{\alpha}w_{xx})_{xx} = \lambda \int_{-1}^{0} K\,(t,\,x)\,w_t\,dt \qquad (6.90)$$

The solution of the boundary-value problem of Eqs. (6.90) and (6.82), for a plate with constant thickness ($h = 1$), was found numerically using a technique described in Ref. 67. The eigenvalue found as a result of these computations is equal to $\lambda = 5.132$. The corresponding deflection function is shown in Fig. 6.8 by a broken line.

As an example we consider a steel plate, 1 m wide and 1 cm thick. The critical speed for forward motion of this plate in water is

$$v_{\infty} = \sqrt{\lambda S^3/12\rho_f l^6} \approx 10\ \text{м/с}$$

Next, we discuss a plate with variable thickness and find a distribution of thickness for which the first eigenvalue is maximized. We recall the self-adjoint and positive property of the operators in the boundary-value problem. We may thus use Rayleigh's variational principle to find the first eigenvalue λ_0

$$\lambda_0 = \min_w J\,(h,\,w), \qquad J = J_1\,(h,\,w)/J_2\,(w)$$

$$J_1 = \int_{-1}^{0} h^{\alpha}w_{xx}^2\,dx, \qquad J_2\,(w) = \int_{-1}^{0}\int_{-1}^{0} K\,(t,\,x)\,w\,(x)\,w_t\,(t)\,dt\,dx$$

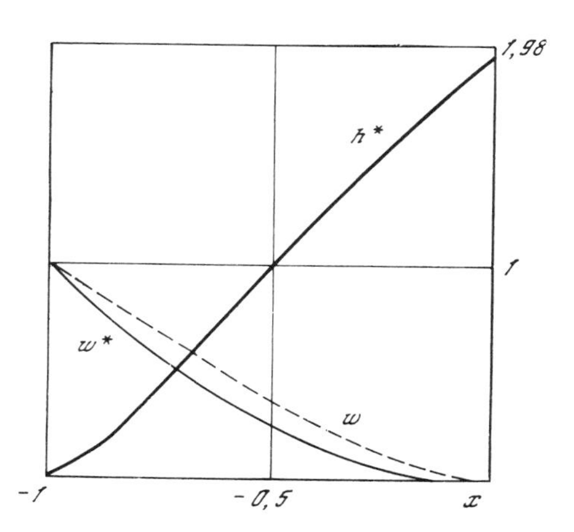

Figure 6.8

The minimum is attained in the class of functions satisfying the boundary conditions of Eq. (6.82) for $x = 0$. The other two boundary conditions given in Eq. (6.82) are natural for the functional J and are therefore automatically satisfied whenever J assumes an extremal value.

Let us examine the optimization problem in the space of continuous functions satisfying the isoperimetric condition

$$\int_{-1}^{0} h(x)\,dx = 1 \tag{6.91}$$

We seek a function $h(x)$ that maximizes the first eigenvalue λ_0, i.e.,

$$\lambda_* = \max_h \min_w J(h, w) \tag{6.92}$$

A necessary condition for optimality is

$$h^{\alpha-1} w_{xx}^2 = c^2 \tag{6.93}$$

where c is a constant Lagrange multiplier arising from the isoperimetric condition of Eq. (6.93).

We shall restrict our discussion to the case $\alpha = 1$ (the trilayer plate with reinforcing layers having variable thickness). From Eq. (6.93), with the boundary condition of Eq. (6.82) for $x = 0$, we find the displacement function $w^*(x)$ for the optimum plate (shown in Fig. 6.8 by the solid line) $w^* = cx^2/2$. Substituting the expression for displacement $w^*(x)$ of the optimum plate into Eq. (6.90) and into the boundary condition of Eq. (6.85), we arrive at the following boundary-value problem for ordinary differential equation of second order:

$$h_{xx}^* = \lambda_* \int_{-1}^{0} K(t, x)\, t\, dt, \qquad h^*(-1) = h_x^*(-1) = 0$$

The optimum thickness function is derived by integrating this equation. It has the form

$$h^* = \lambda_* H(x), \qquad H(x) \equiv \int_{x}^{0}\int_{-1}^{0} (x - \eta) K(t, \eta)\, dt\, d\eta \tag{6.94}$$

The eigenvalue λ_* is determined from the isoperimetric condition of Eq. (6.91)

$$\lambda_* = \left(\int_{-1}^{0} H(x)\, dx \right)^{-1} = 7.567 \tag{6.95}$$

The gain due to optimization, compared with the plate of constraint thickness, amounts to 47.4%. The optimum distribution of thickness $h^*(x)$ is illustrated in Fig. 6.8. In a manner analogous to that presented in Section 6.4, we can prove that, in this case ($\alpha = 1$), the condition of Eq. (6.93) is a sufficient condition for the existence of a global optimum, while Eqs. (6.94) and (6.95) give the solution of the optimization problem.

7

Optimal Design under Conditions of Incomplete Information Concerning External Actions and Problems of Multipurpose Optimization

In this chapter we investigate problems of optimal design of structures under conditions of incomplete information regarding the external actions. In formulating and solving problems, we shall apply the minimax ("guaranteed") approach of game theory. The application of the minimax approach reflects realistic situations that arise in design situations and permit us to consider simultaneously optimization problems under conditions of incomplete information and problems of multipurpose optimization. We introduce techniques for finding shapes of structures that are optimum for certain sets of external loads. The presentation follows research results of Refs. 9, 10, 12, 16, 17, 56, 124, 161, 164, and 227.

7.1. Formulation of Optimization Problems under Conditions of Incomplete Information

The majority of problems in the theory of optimal structural design, in particular problems discussed in the preceding chapters, are considered within the framework of the deterministic approach, i.e., we assume that the loads applied are known and that we have complete knowledge regarding structural materials and boundary conditions. To solve such problems, we apply techniques of the calculus of variations and of optimal control theory. Problems of optimal design with incomplete information are entirely different in their formulation and solution techniques. We shall study here various approaches to optimization problems with incomplete information, and their specific properties. To be specific, we shall consider the problem of optimizing the shape of an elastic body, having a minimum weight and satisfying certain constraints with regard to strength and rigidity.

Formulating and solving optimization problems using a deterministic approach results in an optimum shape that as a rule has the property that, for even very small changes in the external conditions (e.g., a small shift of the point of application of the external force), the design fails to satisfy strength and geometric constraints. In many cases, either we do not have complete information with regard to the applied forces or different forces may act on the structure. Therefore, in parallel with the deterministic formulation we should be interested in a more general formulation of problems of optimization of structures in which optimization is carried out for a whole class of admissible loads.

One of the possible approaches to formulation and solution of such problems (that is problems "with incomplete information") is application of the minimax approach. In application of the minimax (also called the "guaranteed") approach, we have to specify the set containing all admissible loads. The structure of a given shape is optimum if for any other structure having a smaller weight it is possible to find a force in the class of admissible loads for which some strength or geometric constraint is violated. As we solve problems using such criteria, one of two possibilities emerges. There may exist a "worst load", for which the structure has a minimum weight, this weight being computed specifically for that load, while the constraints on strength and rigidity are satisfied for this and all other admissible loads. Otherwise, the structure of a given shape is optimum for the class of admissible loads, i.e., it is the solution of the original optimization problem, but there does not exist a "worst load" and the optimum design for the class of admissible loads is not optimum for any specific load in the admissible class. In this chapter we offer examples of both types. We observe that the minimax approach may also be applied to problems with incomplete information concerning the boundary conditions or properties of structural material. We also remark that the minimax approach is not the only available one that may be applied to problems with incomplete information. It is also possible to apply a probabilistic approach, in which the applied loads are regarded as random variables with a given probability distribution and we minimize the mathematical expectation of the weight of the structure.

Let the equations of equilibrium for an elastic body and boundary conditions be of the form

$$\begin{aligned} L(h)\,u &= q, \quad x \in \Omega \\ N(h)\,u &= 0, \quad x \in \Gamma \end{aligned} \tag{7.1}$$

where the functions u, h, and q are, respectively, the state variable, the control variable, and the externally applied force, and $L(h)$ and $N(h)$ are differential operators whose coefficients depend on h. The external forces q may also enter into the boundary conditions.

The type of load applied to the body is not known *a priori.* We specify only the set R_q, which contains all admissible loads. We denote this by writing

$$q \in R_q \tag{7.2}$$

In solving the problem of optimizing the shape of the body, we only admit admissible forces defined by Eq. (7.2). If, for example, we optimize a plate of variable thickness and the external loads are loads applied from only one side, whose total magnitude does not exceed P, then the set R_q in Eq. (7.2) is given by

$$R_q\{q \geqslant 0, \qquad \int q(x)\,dx \leqslant P, \qquad x \in \Omega\}$$

For given q and h, we assume that the boundary-value problem of Eq. (7.1), consisting of finding the function u, has a unique solution. The optimization problem consisting of finding a function $h(x)$ that minimizes the functional $J(h)$ (which is the weight of the body) and satisfies, for an arbitrary q in Eq. (7.2), the strength and geometric constraints

$$\psi(x, u, h) \leqslant 0 \tag{7.3}$$

where ψ is a known vector-valued function. The constraints of Eq. (7.3) are represented by a system of scalar inequalities.

Because of indeterminacy existing in the system, the formulated problem belongs to the game theory (we play a game with nature). To solve this problem, we can use the minimax (or so-called "guaranteed" solution) method.

Let us explain the main features of this approach (as presented in the Ref. 9). We obtain a solution of the boundary-value problem of Eq. (7.1) of the form $u = u(x, h, q)$. The dependence of u on h and q is, in general, a functional dependence. Now, substituting this expression for u into the left-hand side of the inequality of Eq. (7.3), we arrive at a system of functional inequalities $\Psi(x, h, q) \leqslant 0$, where $\Psi(x, h, q) \equiv \psi[x, u(x, h, q), h]$. Denoting by Ψ_j the components of the vector Ψ, we find the maxima of $\Psi(x, h, q)$ for $q \in R_q$. This maximization process is carried out with the values of h and $x \in \Omega$ regarded as constants. Let us assume that the maximum of the jth component Ψ_j of the vector ψ is attained for $q_j = q_j^*$, i.e.,

$$\Psi_j (x, h, q_j^*) = \max_{q \in R_q} \Psi_j (x, h, q)$$

We also use the notation

$$\Psi_j^* (x, h) \equiv \Psi_j (x, h, q_j^* (x, h)) \tag{7.4}$$

If the maximum of Ψ with respect to q is attained for several different functions q chosen from Eq. (7.2) then any of these functions can be taken as q^*. Utilizing the notation of Eq. (7.4), we arrive at a system of inequalities

$$\Psi_j^* (x, h) \leqslant 0 \tag{7.5}$$

Thus, the original problem has been transformed into a variational problem of minimizing the functional $J(h)$ with respect to h, with inequality constraints of Eq. (7.5). To solve it we can make use of calculus of variations techniques. We comment that the minimax approach can also be applied to problems with incomplete information concerning the boundary conditions or concerning properties of the structural material.

7.2. Design of Beams Having the Smallest Weight for Certain Classes of Loads and with Constraints on Strength

The equilibrium state of a simply supported beam of length l, situated in the x–z plane, with its axis lying on the x axis and loaded by exterior distributed load $q(x)$ parallel to the z axis, is given by the following system of equations and boundary conditions (see Fig. 7.1):

$$\begin{aligned} M_x &= Q, \qquad Q_x = -q \\ M (0) &= M (l) = 0 \end{aligned} \tag{7.6}$$

where $M = M(x)$ and $Q = Q(x)$ denote, respectively, bending moment and shear load acting on the cross section of the beam, perpendicular to the x axis, and $M_x = dM/dx$, $Q_x = dQ/dx$. The beam has a rectangular cross-sectional

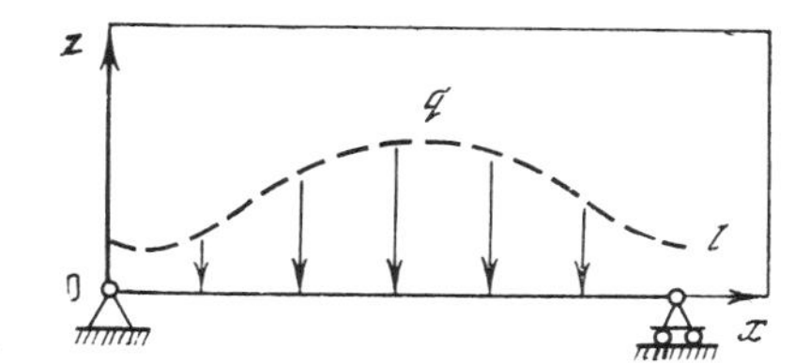

Figure 7.1

area of constant width b and variable height $h = h(x)$. The function $h = h(x)$ determining the shape of the beam is the unknown function.

We assume that the load applied to the beam is positive (the direction of load action coincides with the positive direction of the z axis) and the resultant force does not exceed a given magnitude P, i.e.,

$$q(x) \geqslant 0, \qquad \int_0^l q(x)\,dx \leqslant P \tag{7.7}$$

For an arbitrary load satisfying Eq. (7.7), the normal and tangential stresses σ_x and τ_{xz} must satisfy the strength conditions

$$\psi_1 \equiv |\sigma_x| - \sigma_0 \leqslant 0, \quad \psi_2 \equiv |\tau_{xz}| - \tau_0 \leqslant 0$$
$$\sigma_x = \frac{M\zeta}{I}, \quad \tau_{xz} = \frac{1}{b}\left(\frac{Md}{I}\right)_x', \quad I = \frac{bh^3}{12}, \quad d = \frac{b}{2}\left(\frac{h^2}{4} - \zeta^2\right) \tag{7.8}$$

where σ_0 and τ_0 are given constants. The coordinate ζ is measured from the centroid of the beam cross-sectional area and varies between the limits $-h/2 \leqslant \zeta \leqslant h/2$.

The problem of optimizing the beam shape consists of finding a function $h = h(x)$ that for an arbitrary load $q = q(x)$ satisfying Eq. (7.7), satisfies the constraints of Eq. (7.8) and minimizes the integral

$$J = \gamma b \int_0^l h(x)\,dx \tag{7.9}$$

i.e., the weight of the beam, where γ denotes the specific weight of the material. In this exposition we shall follow Ref. 9.

Before we attempt to solve this problem we need to explain some properties of the functions $M(x)$ and $Q(x)$ that we shall need in our study. Integrating

Eq. (7.6), with the indicated boundary conditions, we derive the following formulas for M and Q:

$$M(x)=\int_0^l \omega_1(x,t)\,q(t)\,dt, \qquad Q(x)=\int_0^l \omega_2(x,t)\,q(t)\,dt$$
$$\omega_1 = t(l-x)/l,\ 0\leqslant t\leqslant x;\ \omega_1 = x(l-t)/l,\ x\leqslant t\leqslant l \tag{7.10}$$
$$\omega_2 = -t/l,\ 0\leqslant t\leqslant x;\ \omega_2 = 1-t/l,\ x<t\leqslant l$$

Since the functions $\omega_1(x,t)$ and $q(x)$ are positive, it follows that $M(x)\geqslant 0$. We fix a point $x\in[l/2,l]$ and consider the set of values that can be attained by $M(x)$ and $Q(x)$ for all possible admissible loads $q=q(t)$ satisfying Eq. (7.6). We denote by $\max_q M(x)$ and $\max_q |Q(x)|$, respectively, the maximum values of the bending moment and of the shear force and we claim that, for $l/2\leqslant x\leqslant l$,

$$\max_q M(x) = Px(1-x/l), \quad \max_q |Q(x)| = Px/l \tag{7.11}$$

To prove this assertion, we use Eq. (7.10) and make the following estimates:

$$M(x)\leqslant \max_t \omega_1(x,t)\int_0^l q(t)\,dt = Px\left(1-\frac{x}{l}\right)$$

The maximum with respect to t is computed for $0\leqslant t\leqslant l$. We also observe that for the admissible load $q(t)=P\delta(t-x)$, the value of the bending moment is $M(x)=Px(l-x)/l$, where δ denotes the Dirac delta function. Combining this with the estimates given above, we see that the assertion of Eq. (7.11) is true for $M(x)$.

To prove the validity of Eq. (7.11) for Q, we carry out analogous estimates,

$$|Q(x)|\leqslant \operatorname{vrai\,max}_t|\omega_2(x,t)|\left|\int_0^l q(t)\,dt\right| = \frac{xP}{l}$$

where $\operatorname{vrai\,max}_t|\omega_2|$ denotes the essential maximum with respect to t $(0\leqslant t\leqslant l)$ of the piecewise continuous function $\omega_2(x,t)$, which has a discontinuity at $t=x$. We substitute the admissible load $q(t)=P\delta(t-x^1)$ with $l/2\leqslant x^1\leqslant x$ into Eq. (7.10) as the shear load Q and compute the corresponding integral. We obtain $|Q(x)|=Px^1/l$. In the limit as $x^1\to x-0$ we have $\lim|Q(x)|=Px/l$. Taking into account the inequality $|Q(x)|\leqslant Px/l$, we obtain Eq. (7.11) for Q.

Since all conditions for this problem are symmetric with respect to the point $x = l/2$, we can restrict discussion of our problem to the interval $l/2 \leqslant x \leqslant l$. We utilize the derived properties of the functions $M(x)$ and $Q(x)$ to find explicit formulas for Ψ_1^* and Ψ_2^* (also see Section 7.1). First, let us find formulas for Ψ_1 and Ψ_2. Making use of Eqs. (7.8) and (7.10), we have

$$\Psi_1 = \frac{12\zeta}{bh^3} \int_0^l \omega_1(x, t)\, q(t)\, dt - \sigma_0$$

$$\Psi_2 = \frac{6}{b} \left| \left(\frac{1}{4h} - \frac{\zeta^2}{h^3} \right) \int_0^l \omega_2(x, t)\, q(t)\, dt + \left(\frac{3\zeta^2}{h^4} - \frac{1}{4h^2} \right) h_x \int_0^l \omega_1(x, t)\, q(t)\, dt \right| - \tau_0$$

We compute the maximum of the function Ψ_1, with respect to q by using Eq. (7.7) and with respect to ζ on the interval $-h/2 \leqslant \zeta \leqslant h/2$, by using the estimate of Eq. (7.11) for max M

$$\Psi_1^* = \frac{6Px}{bh^2}\left(1 - \frac{x}{l}\right) - \sigma_0 \tag{7.12}$$

We shall also determine the maximum of the function Ψ_2 with respect to q and ζ. We note that the expression inside the absolute value signs in the formula for Ψ_2 is a linear function of ζ^2. Consequently, the maximum of Ψ_2, regarded as a function of ζ^2, is attained on the interval $-h/2 \leqslant \zeta \leqslant h/2$, either if $\zeta^2 = h^2/4$ or else if $\zeta^2 = 0$. Following this observation, we have

$$\Psi_2^* = \max(\chi_1, \chi_2) - \tau_0, \qquad \chi_2 = \frac{3Px}{bh^2}\left(1 - \frac{x}{l}\right)|h_x|$$

$$\chi_1 = \max\left(\frac{3P}{2bh}\left| \frac{x}{l} + \frac{x}{h}\left(1 - \frac{x}{l}\right) h_x \right| \times \frac{3P}{2bh}\left| \frac{x}{l} - 1 + \frac{x}{h}\left(1 - \frac{x}{l}\right) h_x \right| \right) \tag{7.13}$$

Similar, but more detailed, arguments were offered in Ref. 9.

We apply the expressions for functions Ψ_1^* and Ψ_2^* derived above and formulate conditions that the function $h = h(x)$ must satisfy, if the inequalities of Eq. (7.5) are to be satisfied. Substituting Eq. (7.12) for Ψ_1^* into the first inequality of Eq. (7.5), we obtain the condition

$$h(x) \geqslant \Phi_1(x) \equiv \left[\frac{6Px}{b\sigma_0}\left(1 - \frac{x}{l}\right)\right]^{1/2} \tag{7.14}$$

Substituting the formula for Ψ_2^* given in Eq. (7.13) into the second inequality in Eq. (7.5) we have

$$h_x \leqslant \frac{1}{x(l-x)} \min\left(\lambda h^2, 2\lambda h^2 - xh, 2\lambda h^2 + (l-x)h\right) \tag{7.15}$$

$$h_x \geqslant \frac{1}{x(l-x)} \max\left(-\lambda h^2, -2\lambda h^2 - xh, -2\lambda h^2 + (l-x)h\right) \tag{7.16}$$

where $\lambda = bl\tau_0/(3P)$. The inequalities of Eqs. (7.15) and (7.16) may be simplified if we observe that the third expression inside the round brackets in Eq. (7.15) is always larger than the second and that the third expression inside the brackets of Eq. (7.16) is larger than the second. Therefore, the inequalities of Eqs. (7.15) and (7.16) can be written in the following manner:

$$h_x \leqslant \frac{1}{x(l-x)} \min\left(\lambda h^2, 2\lambda h^2 - xh\right) \tag{7.17}$$

$$h_x \geqslant \frac{1}{x(l-x)} \max\left(-\lambda h^2, -2\lambda h^2 + (l-x)h\right) \tag{7.18}$$

Now, let us decompose the domain $\Lambda(l/2 \leqslant x \leqslant l, h \geqslant 0)$, in which we seek the solution of our optimization problem, into three sub-domains

$$\Lambda_1\,(l/2 \leqslant x \leqslant l,\ h \geqslant x/\lambda), \qquad \Lambda_2\,(l/2 \leqslant x \leqslant l$$
$$(l-x)/\lambda \leqslant h \leqslant x/\lambda), \quad \Lambda_3\,(l/2 \leqslant x \leqslant l,\ 0 \leqslant h \leqslant (l-x)/\lambda)$$

In each of these subregions, the inequalities of Eqs. (7.17) and (7.18) assume the form

$$-\frac{\lambda h^2}{x(l-x)} \leqslant h_x \leqslant \frac{\lambda h^2}{x(l-x)}, \qquad (x, h) \in \Lambda_1 \tag{7.19}$$

$$-\frac{\lambda h^2}{x(l-x)} \leqslant h_x \leqslant \frac{2\lambda h^2 - xh}{x(l-x)}, \qquad (x, h) \in \Lambda_2 \tag{7.20}$$

$$-\frac{2\lambda h^2-(l-x)h}{x(l-x)}\leqslant h_x\leqslant\frac{2\lambda h^2-xh}{x(l-x)},\qquad (x,h)\in\Lambda_3 \tag{7.21}$$

For any $(x, h)\in\Lambda_1$, the inequalities given by Eq. (7.19) cannot be unsolvable. For inequalities of Eq. (7.20) to be solvable in the region Λ_2 and for inequalities of Eq. (7.21) to be solvable in the region Λ_3, the following conditions must be satisfied:

$$h\geqslant x/3\lambda,\qquad h\geqslant l/4\lambda \tag{7.22}$$

Thus, the original problem of optimizing the shape of a rectangular beam with variable thickness has been reduced to finding a continuous function $h(x)$, satisfying the constraints of Eqs. (7.14) and (7.22) and the differential inequalities of Eqs. (7.19)-(7.21), such that $h(\alpha)$ minimizes the integral of Eq. (7.9), in which the interval of integration has been reduced to $[l/2, l]$ from $[0, l]$, because of symmetry.

The functions $h(x)$ satisfying the inequalities of Eqs. (7.14) and (7.19)-(7.21) are called *admissible*. Some properties of admissible functions shall be derived and applied to future studies. Let us consider a function $h = h(x)$ that passes through the point $(x^0, h^0)\ \Lambda_2+\Lambda_3$, where $h^0 = h(x^0)$. It is a consequence of Eqs. (7.20) and (7.21) that $h(x)\leqslant v(x)$ for all $x > x^0$, where $v(x)$ stands for a solution of the differential equation

$$v_x=(2\lambda v^2-xv)/x(l-x)$$

satisfying the initial condition $v(x^0) = h^0$. Integrating this differential equation, making use of the initial condition, we obtain

$$\begin{gathered} v(x)=\frac{l-x}{2\lambda\ln(c/x)}\\ c=x^0\exp\left[\frac{l-x^0}{2\lambda h^0}\right] \end{gathered} \tag{7.23}$$

Let us consider the behavior of integral curves $v(x)$ depending on the position of the initial point (x^0, h^0). If the values (x^0, h^0) are such that $c > l$, then it is easy to see from Eq. (7.23) that $v(x)\to 0$ as $x\to l$ (see Fig. 7.2). For initial conditions (x^0, h^0) such that $c < l$, the integral curves approach infinity as x approaches c. However, if the magnitudes of x^0 and h^0 satisfy the relation $x^0\exp[(l-x^0)/(2\lambda h^0)] = l$, then as $x\to l$ both the numerator and denominator

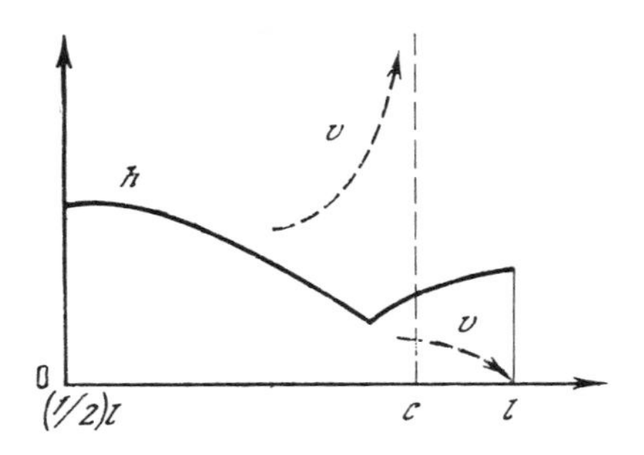

Figure 7.2

in Eq. (7.23) for v approach zero. Applying L'Hospital's rule, we have in the limit $v(l) = l/(2\lambda)$. For $c = l$, we denote the function $v(x)$ by $\Phi_2(x)$.

Using these properties of the function $v(x)$, we can show that the solution of our optimal problem assumes the form (see Fig. 7.2)

$$h = \begin{cases} \Phi_1(x), & l/2 \leqslant x \leqslant x^* \\ \Phi_2(x), & x^* \leqslant x \leqslant l \end{cases}$$
$$\Phi_1(x) \equiv \left[\frac{6Px}{\sigma_0 b}\left(1 - \frac{x}{l}\right)\right]^{1/2}, \qquad \Phi_2(x) \equiv \frac{l - x}{2\lambda \ln(l/x)} \tag{7.24}$$

or formulating it differently $h = \max[\Phi_1(x), \Phi_2(x)]$.

The value x^* may be determined as follows: If the parameters of our problem are such that $\Phi_2(x) \geqslant \Phi_1(x)$ everywhere on the interval $l/2 \leqslant x \leqslant l$, then $h = \Phi_2(x)$ and x^* in Eq. (7.24) may be taken as $x^* = l/2$. It is easy to check that this is true for

$$\varkappa \equiv 6P\sigma_0/lb\tau_0^2 \geqslant 16 \ln^2 2$$

In a different situation, when $\kappa < 16 \ln^2 2$ there may exist a subinterval of $l/2 \leqslant x \leqslant l$ in which $\Phi_2(x) < \Phi_1(x)$ and another subinterval in which $\Phi_2(x) > \Phi_1(x)$. In this case, the optimum design consists of two parts, determined by our formula, and the point x^* may be found from the equality $\Phi_1(x^*) = \Phi_2(x^*)$, which may be restated as

$$[x^*/\varkappa\,(l - x^*)]^{1/2} \ln(x^*/l) = 0.25 \tag{7.25}$$

The proof of optimality for a solution of Eqs. (7.24) and (7.25) consists of showing that the function $h(x)$ is admissible, i.e., satisfies the condition of Eqs. (7.14) and (7.19)-(7.22) and that no other admissible function $h(x)$ exists for which the functional J has a smaller value than for h given by Eq. (7.24).

Let us first consider the case $\kappa < 16 \ln^2 2$ and determine the region in which x can vary so that the inequalities of Eqs. (7.19)-(7.21) are true if we substitute $h = \Phi_1(x)$. We carry out some elementary substitutions and conclude that the inequality of Eq. (7.19) is satisfied if $l/2 \leqslant x \leqslant \beta_1 \equiv l\{1 + [4(4+\kappa)]^{1/2}\}/2$, the inequality of Eq. (7.20) is satisfied if $l/2 \leqslant x \leqslant \beta_2 \equiv l[4 + (16 - \kappa)]^{1/2}/8$, and the inequality of Eq. (7.21) is satisfied if $l/2 \leqslant x \leqslant \beta = \min(\beta_1, \beta_2)$. It is easy to check that $x/(2\lambda) < \Phi_2(x) < x/\lambda$, if $l/2 \leqslant x < l$, and that $\Phi_2(x) = x/(2\lambda)$, if $x = l$. In particular, it follows that $x^1 = 16l/(\kappa + 16)$ is a root of the equation $x^1/(2\lambda) = \Phi_1(x^1)$ that satisfies the inequality $x^1 \geqslant x^*$. Therefore, to prove the inequality $\beta \geqslant x^*$, it suffices to confirm the inequality $\beta \geqslant x^1$ for $0 \leqslant \kappa \leqslant 16 \ln^2 2$, which can be done by elementary manipulations. The curve $\Phi_2(x)$ defined in Eq. (7.24) lies either in the region Λ_2 or in $\Lambda_2 + \Lambda_3$ $[\Phi_2(x) \leqslant x/\lambda)]$, depending on the value of the parameter κ. The inequalities of Eqs. (7.20) and (7.21) are satisfied, since for $h = \Phi_2(x)$ the right-hand side inequalities in Eqs. (7.20) and (7.21) become exact equalities. This follows from our construction of the function $\Phi_2(x)$. Consequently, the function $h(x)$ given by the inequality is admissible if $0 \leqslant \kappa \leqslant 16 \ln^2 2$. For $\kappa \geqslant 16 \ln^2 2$, the function $\Phi_2(x)$ $(l/2 \leqslant x \leqslant l)$ is also admissible, because for $h = \Phi_2(x)$ the conditions of Eqs. (7.20) and (7.21) are satisfied.

We shall now prove that the admissible function $h(x)$ given by Eq. (7.24) is optimum. It suffices to show that the graph of any other admissible function $h(x)$ lies below the curves $\Phi_1(x)$ and $\Phi_2(x)$. For $l/2 \leqslant x \leqslant x^*$, the admissible functions $h(x)$ lie below the graph of $\Phi_1(x)$, as can be seen from Eq. (7.14), i.e., $h(x) \geqslant \Phi_1(x)$. We shall prove that for $x^* \leqslant x \leqslant l$ the inequality $h(x) \geqslant \Phi_2(x)$ is true. Let us assume the contrary; i.e., we assume that at some point $x = x'$ $(x^* \leqslant x' \leqslant l)$ an admissible function $h(x)$ satisfies the inequality $h(x') < \Phi_2(x')$. However, as we have already indicated for any admissible function $h(x)$ passing through the point $[x', h(x')]$, we have the limit $h(x) \to 0$ as $x \to l$. Hence the trajectory $h(x)$ originating at the point $[x', h(x')]$ must enter the forbidden region defined by the inequality of Eq. (7.22) and violates the hypothesis that the function $h(x)$ is admissible. This contradiction proves the inequality $h(x) \geqslant \Phi_2(x)$ for $x^* \leqslant x \leqslant l$. Consequently, Eq. (7.24) provides a solution to our optimization problem.

For the optimum solution, we estimate the magnitude of the stresses σ_x and τ_{xz} that result from application of a concentrated load P. Let us denote by σ_x the maximum value of the normal stress acting on the perpendicular cross section (i.e., along the ζ axis), while τ_{xz} is the maximum value of the tangential stresses. If the concentrated load P is applied to the optimum beam at the point $\xi(l/2 \leqslant \xi \leqslant x^*)$, the magnitudes of $\sigma_x(x)$ and $\tau_{xz}(x)$ satisfy the inequalities $\sigma_x(x) \leqslant \sigma_0$, $\tau_{xz}(x) \leqslant \tau_0(l/2 \leqslant x \leqslant l)$, as a consequence of Eqs. (7.8), (7.10), and (7.24). The equality $\sigma_x = \sigma_0$ is satisfied when $x = \xi$. If $x^* \leqslant \xi \leqslant l$, then

$\sigma_x \leqslant \sigma_0$ and $\tau_{xz} \leqslant \tau_0$. In this case, the upper bound value of $\tau_{xz} = \tau_0$ for the tangential stress is attained at $x = \xi$ and the equality $\sigma_x = \sigma_0$ takes place when $x = \xi = x^*$.

Thus, when a concentrated load is applied to the optimum beam, at an arbitrary point x in the interval $[l/2, l]$, the upper bound stress condition is attained only at that point. Consequently, when we design a beam for a fixed total load, we have additional ways of optimization. This has been confirmed in cases for which such optimum shapes were found (we shall not present the details here). Consequently, for problems considered here (in the general sense indicated above), there is no "worst" load and the optimum beam design for the class of admissible loads is not optimum for any specific load distribution that belongs to that class.

7.3. Optimization of Rigidity for Beams

In Section 7.2 we solved the problem of optimizing the shape of a beam having least weight (or volume), with constraints on its strength. It is interesting to consider an analogous problem, with constraints on the maximum deflection of the beam. This problem is the dual of the problem of minimizing the maximum deflection, with a given weight (or total volume) of the beam. A solution of each of these problems may be obtained from the other by means of a simple computation. In the following discussion we present the research results published in Ref. 12. We consider the problem of minimizing the maximum deflection, assuming as before that the applied load belongs to a class defined by the inequalities of Eq. (7.7).

The deflection function $w(x)$ of an elastic beam supported at the points $x = 0$ and $x = l$ obeys the following differential equation and boundary conditions:

$$\begin{gathered}(EIw_{xx})_{xx} = q \\ w\,(0) = (Iw_{xx})_{x=0} = w\,(l) = (Iw_{xx})_{x=l} = 0\end{gathered} \tag{7.26}$$

The beam length l and volume V are given. The function $S(x)$ (area of the cross section) is related to l and V by the equality

$$\int_0^l S\,(x)\,dx = V \tag{7.27}$$

In the problem treated here, the magnitude of the cross-sectional area $S(x)$ is the unknown function, and the constraint of Eq. (7.27) is regarded as an isoperimetric condition assigned to $S(x)$. The function $S(x)$ is related to $I(x)$ by the equality $EI(x) = C_\alpha S^\alpha(x)$, where α can assume either one of the values $\alpha = 1, 2, 3$ (see Section 1.3).

Let us consider $S(x)$ as given and determine the number J that represents the absolute value of the greatest deflection attained, for any admissible load q (satisfying Eq. (7.7), i.e.,

$$J = \max_q \max_x w$$

The maximum with respect to x can be computed for all x in the interval $0 \leqslant x \leqslant l$, while the maximum with respect to q is taken over all possible admissible loads, i.e., loads satisfying Eq. (7.7). The number J is a functional defined over a class of functions satisfying Eq. (7.27).

Let us now formulate the optimization problem. Among all admissible functions $S(x)$ defining the shape of the beam and satisfying Eq. (7.27), we wish to find the one that minimizes J, i.e.,

$$J_* = \min_S J = \min_S \max_q \max_x w \tag{7.28}$$

In other words, we need to find the optimum shape of a beam having the smallest maximum deflection. Below we determine the optimum shape, using the minimax approach.

Let us integrate both sides of Eq. (7.26), with the following formula for w:

$$\begin{gathered}
w = \int_0^l \Phi(\xi, x)\, q(\xi)\, d\xi \\
\Phi(\xi, x) = \chi(\xi, x), \quad 0 < \xi \leqslant x \\
\Phi(\xi, x) = \chi(x, \xi), \quad x \leqslant \xi < l \\
\chi(\xi, x) = \left(1 - \frac{x}{l}\right)\left(1 - \frac{\xi}{l}\right) \int_0^\xi \frac{t^2\, dt}{EI(t)} + \xi\left(1 - \frac{x}{l}\right) \int_\xi^x \frac{(1 - t/l)\, dt}{EI(t)} \\
+ \xi x \int_x^l \frac{(1 - t/l)^2}{EI(t)}\, dt
\end{gathered} \tag{7.29}$$

Using Eq. (7.29) and the constraints of Eq. (7.8) it is easy to show that $w \leqslant P \max_\xi \Phi(\xi, x) = P\Phi(c, x)$ for any admissible $q(\xi)$ satisfying Eq. (7.7) and that

$$w = P\Phi\,(c,\, x)$$

when $q(\xi) = P\delta(\xi - c)$. Here, c denotes the coordinate of the point where $\Phi(\xi, x)$ attains its maximum with respect to ξ, with $0 < c < l$. It follows from our estimates that the maximum deflection at the point x is attained when a concentrated load is applied at the point $\xi = c$, chosen by applying the condition that Φ attains it maximum with respect to ξ at c. Because of this property, in solving our problem, we only need to consider the effects of concentrated loads on the deflection of the beam.

The scheme for solving the optimization problem consists of the following steps. First let us determine the function $S_0(x)$ that minimizes the deflection w at the point $x = l/2$ when a concentrated load is applied at this point. The magnitude of this deflection is denoted by w_0. We shall prove later in this section that for $S = S_0(x)$, when the point load P is applied at any point ξ in the interval $0 < \xi < l$, the magnitude of any deflection satisfies the inequality $w \leqslant w_0$.

Because of this property of the function $S_0(x)$, it is easy to conclude that $S_0(x)$ solves our original problem and that $w_0 = J_*$. To complete this argument let us assume that there exists another design $S_1(x)$ for which $J < w_0$. Then the deflection w (at the point $x = l/2$ with a load P applied at $\xi = l/2$) corresponding to the design $S_1(x)$, satisfies the inequality $w \leqslant J \leqslant w_0$. But this is a contradiction, because $S_0(x)$ was the design that minimized the deflection w at the point $x = l/2$. Consequently, $S_0(x)$ solves our original problem of Eqs. (7.26)-(7.28), and (7.7). Finding the function $S_0(x)$ is reduced to the solution of an isoperimetric variational problem of minimizing the deflection $w(l/2)$ for all admissible choices of $S(x)$, under the condition of Eq. (7.27). Applying the Lagrange multiplier technique and performing some elementary substitutions, we derive

$$S_0(x) = \frac{(\alpha+3)\,V}{(\alpha+1)\,l}\left[\frac{16g}{l^2}\right]^{1/\alpha+1}, \qquad w_0 = \frac{PVl^2}{16C_\alpha}\left[\frac{(\alpha+1)\,l}{(\alpha+3)\,V}\right] \tag{7.30}$$

where $g(x) = x^2/4$, for $0 \leqslant x \leqslant l/2$, and $g(x) = l(l - x)/4$, for $l/2 \leqslant x < l$.

Later we shall prove that the deflection w, for the shape $S_0(x)$ determined above, occurs as a result of application of the load $q = P\delta(t - \xi)$ and that it satisfies the inequality $w \leqslant w_0$, for any x and ξ in the interval $(0, l)$.

We consider first the case $\alpha = 1$ and find, for the shape of S_0, the distribution of deflections due to the action of a point load, in accordance with Eqs. (7.29) and (7.30). To shorten our arguments, we can apply Betti's reciprocal theorem: for fixed values of x and ξ, if we apply a unit point load at point ξ and compute the deflection at point x, this is equal to the deflection computed at ξ if the unit load is applied at point x. Therefore, we have

$$
\begin{aligned}
w &= \omega_1(\xi, x) \qquad 0 < \xi \leqslant x \leqslant l/2 \\
w &= \omega_1(x, \xi) \qquad 0 < x \leqslant \xi \leqslant l/2 \\
w &= \omega_2(\xi, x) \qquad 0 < \xi \leqslant l/2 \leqslant x < l \\
w &= \omega_2(x, \xi) \qquad 0 < x \leqslant l/2 \leqslant \xi < l
\end{aligned}
\tag{7.31}
$$

$$\omega_1(\xi, x) \equiv \frac{Pl^2\xi}{8C_1V}\left[\frac{\xi x}{l} - \xi + \frac{x}{2} + \frac{x^2}{l} - 2x \ln\left(\frac{2x}{l}\right)\right]$$

$$\omega_2(\xi, x) \equiv \frac{Pl^2\xi}{8C_1V}\left[\xi\left(\frac{x}{l} - 1\right) + \frac{l + x}{2} - \frac{x^2}{l}\right]$$

Let us substitute into the inequality $w \leqslant w_0$ the expressions given in Eq. (7.30) for w and w_0, with $\alpha = 1$, and use Eq. (7.31). We also make use of the symmetry that is apparent in Eq. (7.31) to reduce the number of inequalities from four to two. We also introduce the dimensionless variables $x' = x/l$ and $\xi' = \xi/l$, with the primes omitted.

Our inequalities now become

$$
\begin{aligned}
&(1 - x)\xi^2 + \xi x(1/2 + x - 2\ln 2x) + 1/8 \geqslant 0 \\
&\qquad 0 < \xi \leqslant x < 1/2 \\
&(1 - x)\xi^2 + \xi(1/2(1 + x) - x^2) + 1/8 \geqslant 0 \\
&\qquad 0 < \xi \leqslant 1/2 \leqslant x < 1
\end{aligned}
\tag{7.32}
$$

Let us look at the first inequality in Eq. (7.32). The expression given in the left-hand side of this inequality is a quadratic expression in ξ, containing three terms, with a positive coefficient of ξ^2. The nonnegative property of this expression can be determined for $0 < \xi \leqslant x_2 \leqslant 1/2$ if its discriminant is nonpositive, i.e., if $2x^2(1/2 + x - 2\ln 2x)^2 - 1 + x \leqslant 0$. This inequality may be transformed into

$$
\begin{aligned}
&T_1(x) \leqslant T_2(x), \qquad T_1(x) \equiv \sqrt{2}\,(1/2 + x - 2\ln 2x) \\
&T_2(x) \equiv \sqrt{1 - x}/x
\end{aligned}
\tag{7.33}
$$

At the point $x = 1/2$, the function $T_1(x)$ and $T_2(x)$ are equal, i.e., $T_1(1/2) = T_2(1/2) = \sqrt{2}$. As $x \to 0$, we have $T_1 \to \infty$ and $T_2 \to \infty$. Hence, to prove Eq. (7.33), it suffices to show that the derivatives T_{1x} and T_{2x} satisfy the inequality

$T_{1x} \geqslant T_{2x}$, for $0 \leqslant x \leqslant 1/2$. Substituting into this inequality the expressions for derivatives $T_{1x} = \sqrt{2}[1 - (2/x)]$, $T_{2x} = (x - 2)/[2x^2/(1 - x)^{1/2}]$ and performing some elementary manipulations, we arrive at the inequality $8x^2(1 - x) \leqslant 1$ that must be satisfied for all x in the interval of definition. Hence, the proof of the first inequality in Eq. (7.32) is complete.

We shall now prove the second inequality in Eq. (7.32). As in the preceding case, the expression in the left-hand side of this inequality is quadratic in ξ, having three terms, and the coefficient of ξ^2 is positive. In the interval $1/2 \leqslant x < 1$ the discriminate of this polynomial is nonpositive, a can be shown rather easily. Consequently, the second inequality in Eq. (7.32) is true.

Hence, in the case $\alpha = 1$, the solution of the original optimization problem of Eqs. (7.26)-(7.28) is the function $S = S_0(x)$, which is given by Eq. (7.30).

Let us now consider the cases $\alpha = 2$ and $\alpha = 3$. We intend to find, by means of Eqs. (7.29) and (7.30), the deflection caused by application of a point load P to the beam at the point ξ. We carry out some substitutions, obtaining

$$
\begin{aligned}
w &= \omega_3(\xi, x), \quad 0 < \xi \leqslant x \leqslant l/2 \\
w &= \omega_3(x, \xi), \quad 0 < x \leqslant \xi \leqslant l/2 \\
w &= \omega_4(\xi, x), \quad 0 < \xi \leqslant l/2 \leqslant x < l \\
w &= \omega_4(x, \xi), \quad 0 < x \leqslant l/2 \leqslant \xi < l \\
\omega_3 &= D\xi\Big[\Big(\frac{x}{l} - 1\Big)\xi^{2/(\alpha+1)} + \frac{x^{(\alpha+3)/(\alpha+1)}}{l} + \frac{\alpha+3}{\alpha+1}x^{2/(\alpha+1)} \\
&\quad - \frac{x(\alpha^2+3\alpha+4)}{\alpha^2-1}\Big(\frac{l}{2}\Big)^{(1-\alpha)/(1+\alpha)}\Big] \\
D &= \frac{(\alpha+1)P}{2(\alpha+3)C_\alpha}\Big[\frac{(\alpha+1)l}{(\alpha+3)V}\Big]^{\alpha}\Big(\frac{l}{2}\Big)^{2\alpha/(\alpha+1)} \\
\omega_4(\xi, x) &= D\xi\Big[\Big(\frac{x}{l} - 1\Big)\xi^{2/(\alpha+1)} - x\frac{2(\alpha+2)}{(\alpha+1)l}\Big(\frac{l}{2}\Big)^{2/(\alpha+1)} \\
&\quad - \frac{1}{l}(l - x)^{(\alpha+3)/(\alpha+1)} + \frac{2(\alpha+2)}{(\alpha+1)}\Big(\frac{l}{2}\Big)^{2/(\alpha+1)}\Big]
\end{aligned}
\tag{7.34}
$$

We begin the proof that $w \leqslant w_0$ by checking the inequality $\omega_3(\xi, x) \leqslant w_0$ for $0 < \xi \leqslant x \leqslant l/2$. This check is carried out by making the following estimates:

$$\omega_3(\xi, x) \leqslant \omega_3(x, x) \leqslant \omega_3(l/2, l/2) \tag{7.35}$$

First, let us prove the first inequality in Eq. (7.35), i.e., $\omega_3(\xi, x) \leqslant \omega_3(x, x)$ for $0 < \xi \leqslant x \leqslant l/2$. We compute the derivative $\omega_{3\xi}(\xi, x) \geqslant 0$ in the domain covered by x and ξ. After some elementary algebra, we have the following estimate:

$$
\begin{aligned}
\omega_{3\xi} = D \Bigg[& \frac{\alpha+3}{\alpha+1}\left(\frac{x}{l}-1\right)\xi^{2/(\alpha+1)} + \frac{x^{(\alpha+3)/(\alpha+1)}}{l} \\
& + \frac{\alpha+3}{\alpha+1}\, x^{2/(\alpha+1)} - \frac{2x(\alpha^2+3\alpha+4)}{l(\alpha^2-1)}\left(\frac{l}{2}\right)^{2/(\alpha+1)} \Bigg] \\
\geqslant \frac{2xD}{l(\alpha+1)} \Bigg[& (\alpha+2)\, x^{2/(\alpha+1)} + \frac{\alpha+3}{\alpha-1}\, l x^{(1-\alpha)/(1+\alpha)} \\
& - \frac{\alpha^2+3\alpha+4}{\alpha-1}\left(\frac{l}{2}\right)^{2/(\alpha+1)} \Bigg]
\end{aligned}
\tag{7.36}
$$

It is easy to show (see Ref. 12 for detailed arguments) that the expression inside the square bracketed in Eq. (7.36) is positive on the interval $0 < x < l/2$ (for $\alpha = 2$ and 3). Hence, the proof of the first inequality in Eq. (7.35) is complete. The second inequality in Eq. (7.35) can be proved by a direct computation of the derivative $[\omega_3(x, x)]_x$ and by checking that this derivative is nonnegative on $0 < x \leqslant l/2$. In a similar manner, we can prove that $\omega_4(\xi, x) \leqslant w_0$ for $0 < \xi \leqslant l/2 \leqslant x < 1$. To complete the proof, we need the following estimates:

$$
\omega_4(\xi, x) \leqslant \omega_4(\xi, l/2) \leqslant \omega_4(l/2, l/2) = w_0 \tag{7.37}
$$

The validity of the first and second inequalities in Eq. (7.37) follows, respectively, from the nonnegative property of the derivative $\omega_{4x}(\xi, x)$ for $0 < \xi \leqslant l/2 \leqslant x < l$ and non-negativeness of the derivative of $\omega_4(\xi, l/2)$ with respect to ξ (see Ref. 12). The proof of the inequality $w \leqslant w_0$ is completed by confirmation of the inequalities of Eq. (7.37). Hence, the function $S = S_0(x)$ solves the original optimization problem of Eqs. (7.26)-(7.28), for all cases we have considered—for $\alpha = 2$, $\alpha = 3$, and for $\alpha = 1$.

7.4. Design of Plates for Certain Classes of Loads

We consider the problem studied in Ref. 16 concerning bending of an elastic plate supported along its boundary Γ in the x–y plane. The domain Ω bounded by the curve Γ is assumed to be convex. The load applied to the plate, which acts perpendicular to the x-y plane, is denoted by $q = q(x, y)$, $(x, y) \in \Omega$. The type of load is not determined *a priori*. We consider simultaneously all possible admissible loads that obey the constraints

$$q(x, y) \geqslant 0, \qquad \iint_{\Omega} q(x, y)\, dx\, dy \leqslant P \tag{7.38}$$

where P is a given positive constant.

The equations of equilibrium and the boundary conditions assigned on the curve Γ are expressed in terms of dimensionless variables,

$$\begin{aligned} &L(h)\, w \equiv (W_{w_{xx}})_{xx} + 2\,(W_{w_{xy}})_{xy} + (W_{w_{yy}})_{yy} = q \\ &w = 0, \qquad h^{\alpha}\left(\Delta w - \frac{(1-\nu)}{R}\frac{\partial w}{\partial n}\right) = 0, \qquad (x, y) \in \Gamma \\ &W = h^{\alpha}\,[(\Delta w)^2 - 2\,(1-\nu)\,(w_{xx}w_{yy} - w_{xy}^2)] \end{aligned} \tag{7.39}$$

where w and h denote, respectively, the deflection and the thickness of the plate, Δ is the Laplace operator, and ν is Poisson's ratio. Subscripts denote partial differentiation with respect to the indicated variables. The parameter α assumes the values 1 or 3, corresponding, as we have already indicated, to the cases of a trilayer plate and a uniform plate.

In optimization problems discussed below, the thickness $h(x, y)$ assumes the role of the control variable, which must satisfy the constraint

$$\iint_{\Omega} h(x, y)\, dx\, dy = V \tag{7.40}$$

where V is a given constant. The deflection $w(x, y)$, $(x, y) \in \Omega$, depends on the coordinates of the point (x, y) and has a functional dependence on h and q, i.e., $w(x, y, h, q)$. For a given distribution of thickness, the maximum deflection is determined by

$$J = J(h) = \max_q \max_{xy} w(x, y, h, q) \tag{7.41}$$

where $(x, y) \in \Omega$ and $q(x, y)$ satisfies the constraints of Eq. (7.38).

The optimization problem consists of finding a thickness function $h(x, y)$ satisfying the constraint of Eq. (7.40) and minimizing the maximum deflection, i.e.,

$$J_* = \min_h J(h) = \min_h \max_q \max_{xy} w(x, y, h, q) \tag{7.42}$$

The optimization problem of Eqs. (7.38)-(7.42) is a game-theoretic problem and the quality criterion is local.

The minimax problem formulated above may be considered as an optimization problem with incomplete information concerning the applied loads. We do not know the specific loads applied, only the set of admissible loads satisfying Eq. (7.38).

We investigate some properties of the problem given by Eqs. (7.38)-(7.42), which permit us to reduce the number of unknown variables. We shall prove the following assertion:

Theorem 1. For an arbitrary choice of the thickness function, the load causing the maximum deflection is a point load.

To prove this theorem, we consider two deflection functions w and w^c corresponding, respectively, to a distributed load $q(x, y)$ satisfying Eq. (7.38) and to an admissible point load of magnitude P applied at the point (x_1, y_1), where w attains its maximum. The values of w and w^c at the point (x_1, y_1) are denoted by w_1 and w_1^c, respectively. For the deflection functions we have the following estimate:

$$\iint_\Omega qw \, dx \, dy \leqslant (\max_{x,\, y} w) \iint_\Omega q \, dx \, dy \leqslant w_1 P \tag{7.43}$$

after making use of Eq. (7.38). Furthermore, using Eq. (7.43) and the principle of minimum total potential energy, we obtain the following inequalities:

$$\begin{aligned} \iint_\Omega W(h, w^c) \, dx \, dy - 2Pw_1^c \leqslant \iint_\Omega W(h, w) \, dx \, dy \\ - 2Pw_1 \leqslant \iint_\Omega \{W(h, w) - 2qw\} \, dx \, dy \end{aligned} \tag{7.44}$$

The energy equations

$$Pw_1^c = \iint_\Omega W(h, w^c) \, dx \, dy, \qquad \iint_\Omega qw \, dx \, dy = \iint_\Omega W(h, w) \, dx \, dy$$

express the equality between the work performed by the external forces and the strain energy of the elastic deformation. They permit us to simplify the inequalities of Eq. (7.44) to

$$Pw_1^c \geqslant 2Pw_1 - \iint_\Omega qw\,dx\,dy \geqslant \iint_\Omega qw\,dx\,dy \tag{7.45}$$

The inequalities of Eq. (7.45) imply that $w_1^c \geqslant w_1$. We have shown that for an arbitrary admissible load q [i.e., satisfying Eq. (7.38)], it is possible to specify a point of application of a concentrated load P such that the deflection of the beam at the point of application of this load is not smaller than the maximum deflection caused by the distributed load $q(x, y)$. Hence, the proof of Theorem 1 is complete.

Let us consider the expression of Eq. (7.41) for the functional $J = J(h)$ that specifies the maximum deflection of the plate. Let this maximum with respect to the position (x, y) as, in Eq. (7.41), be realized at the point $x = x_1$ and $y = y_1$ and a maximum with respect to q is realized for $q = P\delta(x - x_2, y - y_2)$, i.e., when the concentrated load is applied at a point (x_2, y_2).

Theorem 2. The maximum deflection $J = J(h)$ is attained at the point of application of the load, i.e., at $x_1 = x_2$ and $y_1 = y_2$.

Note that the maximum is taken over all $(x, y) \in \Omega$ and all admissible loads q. We offer a proof by contradiction. Assuming that the maximum is not attained at the point of application of the load $P(x_2, y_2)$, but at a different point (x_1, y_1). The deflection at the point (x_2, y_2) is w_2. We also consider the deflection w^a caused by the force P applied at the point (x_1, y_1). We denote by w_1^a the magnitude of w^a at the point (x_1, y_1). Using a well-known variational principle, we have

$$\begin{aligned}\iint_\Omega W(h, w^a)\,dx\,dy - 2Pw_1^a \leqslant \iint_\Omega W(h, w)\,dx\,dy \\ - 2Pw_1 \leqslant \iint_\Omega W(h, w)\,dx\,dy - 2Pw_2\end{aligned} \tag{7.46}$$

Applying the energy equations

$$Pw_2 = \iint_\Omega W(h, w)\,dx\,dy, \qquad Pw_1^a = \iint_\Omega W(h, w^a)\,dx\,dy$$

we transform the inequality of Eq. (7.46) into the form

$$Pw_1^a \geqslant 2Pw_1 - \iint_\Omega W(h, w)\,dx\,dy \geqslant \iint_\Omega W(h, w)\,dx\,dy \tag{7.47}$$

The right-hand side inequality in Eq. (7.47) implies that

$$Pw_1 - \iint_\Omega W(h, w)\,dx\,dy \geqslant 0$$

Using this estimate, together with the left-hand side inequality in Eq. (7.47), we conclude that $w_1^a \geqslant w_1$. This contradiction proves theorem 2.

These theorems permit us to consider only concentrated loads of magnitude P and only the deflections measured at the point of application of such loads.

For the sake of convenience, we also study the dual problem

$$\begin{aligned} P_* &= \max_h \min_{\xi,\eta} \frac{1}{w_0} \iint_\Omega W(h, w)\,dx\,dy \\ w(\xi, \eta, \xi, \eta, h) &= w_0 \end{aligned} \tag{7.48}$$

i.e., the problem of maximizing with respect to h the minimum value of the force causing the deflection w_0 at the point (ξ, η) of application of the load. The solutions of the original and dual problems may be obtained from each other by means of a simple computation. Presume, for example, that for a given value w_0, we have the solution to the dual problem of Eqs. (7.39), (7.40), and (7.48), i.e., we have the optimum distribution of thickness h, the deflection function w, and the magnitude of the force P_*. Then the optimum soultion h', w' of the original problem of Eqs. (7.38)–(7.42), for the given value of P in Eq. (7.38) may be obtained from the formulas

$$h' = h, \qquad w' = Pw/P_*$$

In the next part we shall give results related to the case $\alpha = 1$. We shall consider the problem of Eqs. (7.39), (7.40), and (7.48), assuming that the point (ξ, η) of application of the point load is fixed and that the operation $\min_{\xi\eta}$ in Eq. (7.48) can be omitted. In this case we have exactly the same problem that the author investigated in Ref. 13, so the results of this research can be applied directly. A necessary and sufficient condition for a maximum of the functional of Eq. (7.48) with respect to h, with the isoperimetric condition of Eq. (7.40), has the form

$$(w_{xx} + w_{yy})^2 - 2(1 - \nu)(w_{xx}w_{yy} - w_{xy}^2) = \lambda^2, \quad (x, y) \in \Omega \qquad (7.49)$$

where λ^2 is an unknown constant. By making use of the optimality condition of Eq. (7.49), we reduce the optimization problem to two boundary-value problems involving partial differential equations of second order. The optimum deflection function may be found by solving the boundary-value problem for Eq. (7.49), with the condition $(w)_\Gamma = 0$. The optimum thickness function can then be found by solving the equation of equilibrium, with the boundary condition $(h)_\Gamma = 0$, after substituting into that equation the previously found optimum deflection function. The boundary-value problem that is solved to find the optimum deflection function is given in terms of the variable $v = w/\lambda$ as

$$\begin{aligned} &(v_{xx} + v_{yy})^2 - 2(1 - \nu)(v_{xx}v_{yy} - v_{xy}^2) = 1, \quad (x, y) \in \Omega \\ &v = 0, \quad (x, y) \in \Gamma \end{aligned} \qquad (7.50)$$

If the solution of Eq. (7.50) is derived, then the value of the force P_0 can be found from

$$P_0 = w_0 V / v^2(\xi, \eta) \qquad (7.51)$$

Further, we should note that the solution of Eq. (7.50) does not depend on the location of the point (ξ, η) at which the load is applied and is fully determined by the shape of the region Ω. This permits us to solve Eq. (7.50) for a given shape of the region Ω. Having found a solution $v(x, y)$ for a specific case, we perform some simple computations to find the optimum deflection function and the value of the maximized load P_0, for an arbitrary location of the point (ξ, η) in Ω.

Let the maximum value of the function $v(x, y)$ be attained at the point (x_*, y_*); i.e., $\nabla v = iv_x + jv_y = 0$ at $x = x_*$ and $y = y_*$. Then it is clear from Eq. (7.51) that in the limit as $(\xi, \eta) \to (x_*, y_*)$ [in a suitably small neighborhood of (x_*, y_*)], the magnitude of P_0 decreases.

The properties of the function $v(x, y)$ leads to a conjecture that a solution $h^*(x, y)$ of an auxiliary problem with $\xi = x_*$ and $\eta = y_*$ is actually a solution of the problem of Eqs. (7.39), (7.40), and (7.48), and therefore of the original problem of Eqs. (7.38)-(7.42). To confirm this statement, we check that when

$h = h^*$, the deflection w_0 at an arbitrary point (ξ, η) is the result of applying a concentrated load $P(\xi, \eta) \geqslant P_*$; i.e., the point $\xi = x_*$, $\eta = y_*$ is the point at which the function $P(\xi, \eta)$ attains its minimum. Let us estimate the quantity $\Delta P = iP_x + jP_y$. We have

$$\nabla P = \frac{1}{w_0} \iint [L(h^*) w] (iw_\xi + jw_\eta)\, dx\, dy =$$
$$= -\frac{P}{w_0} (iw_x + jw_y)_{x=\xi,\, y=\eta}$$

When $\xi = x_*$ and $\eta = y_*$, this formula and the condition $\nabla v = 0$ imply $\nabla P = -P\lambda w_0^{-1} \nabla v = 0$. It can be shown that the quadratic form $\Lambda(\xi, \eta) = P_{\xi\xi} s^2 + 2P_{\xi\eta} st + P_{\eta\eta} t^2$ is positive definite. Therefore, the point (x_*, y_*) is the point at which the function $P(\xi, \eta)$ assumes a minimum and our conjecture was shown to be true. We can conclude that the solution given in Section 3.3, for circular and elliptic plates, is a solution of the game-theoretic problem of Eqs. (7.38)-(7.42) and it is optimum for a class of applied loads.

7.5. Optimization of Beams Subjected to Bending and Torsion: Multicriteria Optimization Problems

In preceding sections, we have considered problems of optimal design of beams and plates with incomplete information concerning the bending loads. As we stated in the Introduction, optimization problems with incomplete information are equivalent to certain multipurpose optimal design problems. Thus, the problems discussed in Sections 7.2-7.4 may also be regarded as problems of optimal beam design, in which we successively subject the beam to various types of load with the resultant load not exceeding some previously assigned magnitude. The program of loading the beam should include all admissible loads of the given type. An important property of problems in Sections 7.2-7.4 was the following: As we load the beam we permit only bending loads. Therefore, in our optimization problems we always consider the same equation of state (i.e., the beam bending equation). It is at least of equal interest, both theoretically and from the point of view of applications, to study optimum shapes when loads of different types are applied to the beam. In that case, we must consider different defining equations and properties of the structure.

We offer below some solutions to problems of this kind.[164] In this type of a problem we seek the optimum shape of an elastic rod that is consecutively subjected to bending and torsion.

We first wish to minimize the cross-sectional area of an elastic cylindrical bar, whose torsional rigidity (K) and bending rigidity (C) satisfy the inequalities $K \geqslant K_0$, $C \geqslant C_0$, where K_0 and C_0 are given constants. We assume that the bar is not subjected to simultaneous bending and torsion, but that it works either in bending or in torsion as the loads change. We shall offer a rigorous mathematical formulation for this problem. First let us call the basic relations concerning the pure bending or pure torsion of bars. Let a cylindrical bar be twisted by a couple M applied to both ends. The resultant angle of twist θ (per unit length of the beam) is proportional to M, i.e., $M = K\theta$, where K is the bar torsional rigidity. To compute the value of the torsional rigidity, we introduce the stress function $\varphi = \varphi(x, y)$, which solves the following boundary-value problem:

$$\varphi_{xx} + \varphi_{yy} = -2, \qquad (x, y) \in \Omega, \quad (\varphi)_\Gamma = 0 \tag{7.52}$$

Here Ω is a simply connected domain occupying the transverse cross section of the bar and Γ is the boundary of Ω. The torsional rigidity K of the bar is given, in terms of the stress function $\varphi(x, y)$, by the formula

$$K = 2G \iint_\Omega \varphi \, dx \, dy \tag{7.53}$$

where G denotes the shear modulus. We assign the constraint

$$K \geqslant K_0 \tag{7.54}$$

where K_0 is a positive constant.

Now, let us consider the case in which the rod is used as a beam and is subjected to bending. The most important mechanical property characterizing a beam is the bending rigidity against transverse loading. Assuming that bending takes place in the y–z plane (z is the axis coinciding with the axis of the beam) and denoting Young's modulus by E, we recall the general formula for bending rigidity

$$C = E \iint_\Omega y^2 \, dx \, dy \tag{7.55}$$

For beams with constant cross section, not only the static rigidity against transverse loading, but also the spectrum of natural frequencies in vibration depends on the magnitude of the rigidity modulus C. By increasing the bending rigidity of a beam (for a given constant mass) we decrease the size of its deflections caused by transverse loads and increase the frequencies of free vibrations. Therefore, constraints on the greatest permissible deflection and on the dynamic properties can be reduced to the following inequality

$$C \geqslant C_0 \tag{7.56}$$

where C_0 is a given positive constant. The optimization problem consists of finding the cross-sectional shape of the elastic rod that satisfies conditions of Eqs. (7.54)-(7.56) and minimizes the area of the cross section

$$S(\Gamma) = \iint_{\Omega} dx\, dy \to \min \tag{7.57}$$

To formulate this problem, we first analyze all possible types of solutions. To do this, let us first consider the problem of Eqs. (7.52)-(7.54), and (7.57); i.e., the problem of minimizing the area of transverse cross section, with only a single constraint concerning the torsional rigidity. It is easy to show that a dual problem is one of maximizing torsional rigidity for a given area S of rod cross section. In the original problem (as shown in Ref. 110), the optimum rod has a circular cross section, having the radius $r = [2K_0/(\pi G)]^{1/4}$. The circular cross section is also optimum for the dual problem.

The torsional and bending rigidities are related to each other by the formula $K = 2GC/E = C/(1 + \nu)$, where ν is Poisson's ratio. This functional dependence has a graph that is a straight line separating the plane with parameters K and C as coordinates into two regions I and II, as indicated in Fig. 7.3. The slope of this line varies between 45° and 33.7°, as ν varies between the limits $0 \leqslant \nu \leqslant 1/2$. To each rod with a circular cross section corresponds a point on

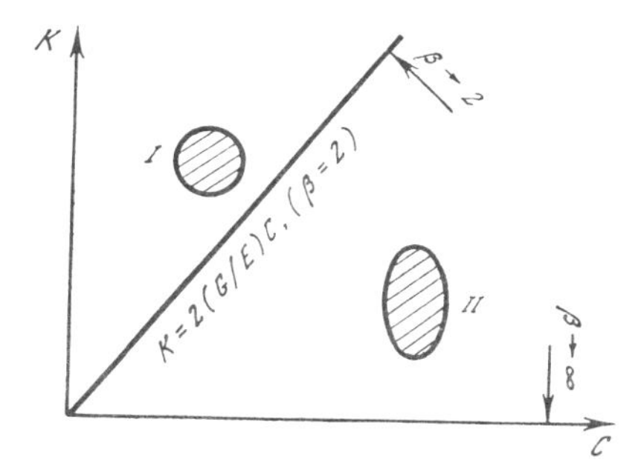

Figure 7.3

this line in the K–C plane. If constants K_0 and C_0 satisfy the inequality

$$K_0 \geqslant 2GC_0/E \tag{7.58}$$

i.e., if points corresponding to such a constant lie in region I (see Fig. 7.3) then the optimum rod has a circular cross section. This follows from the fact that if the inequality of Eq. (7.58) is true, then the torsional rigidity of a round rod [which satisfies the inequality of Eq. (7.54)] is greater than C_0, i.e., satisfies the inequality of Eq. (7.56).

If the values of the parameters K_0 and C_0 violate the inequality of Eq. (7.58); i.e., they lie in region II of Fig. 7.3, then both conditions of Eqs. (7.54) and (7.56) must be taken into account in finding the optimum shape of the cross section (which we assume to be convex) of a rod. In this case, we take into account the constraints of Eqs. (7.54) and (7.56) by the use of Lagrange multipliers. We note that no regions exist in the quadrant $K \geqslant 0, C \geqslant 0$ in which the optimum shape may be determined by the use of the single inequality of Eq. (7.56). The explanation is that we could let the cross-sectional area approach zero, without violating the convexity condition, while retaining the same value for the bending rigidity of the rod. Let us explain this phenomenon by offering an example of a rectangular cross section with width b and height h, for which $S = bh$ and $C = Ebh^3/12$. Consequently, we can retain a constant value of torsional rigidity $C = C_0$ as $h \to \infty$ while the cross-sectional area S tends to zero.

Let us determine an optimum shape for region II of Fig. 7.3. To obtain necessary conditions for optimality, it is convenient to transform the basic relations by eliminating from our considerations the differential equation of Eq. (7.52) (see Section 1.4 for a detailed discussion). The constraint of Eq. (7.54) can be written in the form

$$K = \min_{\varphi} \frac{G}{2} \iint_{\Omega} (\varphi_x^2 + \varphi_y^2 - 4\varphi)\, dx\, dy \geqslant K_0$$

where we seek the minimum with respect to φ, in the class of once continuously differentiable functions that satisfy the boundary condition $\varphi = 0$ on Γ. Let λ_1 and λ_2 denote constant multipliers. We construct a Lagrangian functional

$$\Pi = \iint_{\Omega} \left[1 + \lambda_1 \frac{G}{2} (\varphi_x^2 + \varphi_y^2 - 4\varphi) + \lambda_2 E y^2\right] dx\, dy$$

Applying Eq. (1.112), we derive the necessary condition for optimality of the contour Γ: $1 - {}^1\!/_2\, \lambda_1 G(\varphi_x^2 + \varphi_y^2) + \lambda_2 Ey^2 = 0$, which may be rewritten in the form

$$\varphi_x^2 + \varphi_y^2 = \mu_1 + \mu_2 y^2 \tag{7.59}$$

where $\mu_1 = 2/(\lambda_1 G)$ and $\mu_2 = 2\lambda_2 E/(\lambda_1 G)$. To find the constants μ_1 and μ_2, we set $K = K_0$ and $C = C_0$. Thus, the solution of our problem of shape optimization [that is determining the optimum shape of the transverse cross section and of the corresponding stress function $\varphi(x, y)$], may be found by use of Eqs. (7.52)-(7.56), and (7.59), where strict equality is assumed in Eqs. (7.54) and (7.56).

Starting with the necessary condition of optimality of Eq. (7.59), we look for the solution of our optimization problem, in the form

$$\begin{aligned} \Gamma:\ x^2 + ay^2 &= b \\ \varphi &= N\,(b - x^2 - ay^2) \end{aligned} \tag{7.60}$$

where a, b, and N are unknown constants, which must be determined together with the Lagrange multipliers μ_1 and μ_2 by the basic relations of our problem. It is easy to check that for Γ and φ in the form given by Eq. (7.60), the boundary conditions of Eq. (7.52) is automatically satisfied. Substituting the expression for φ into the optimality condition of Eq. (7.59), we obtain

$$x^2 + (a^2 - \mu_2/4N)\, y^2 = \mu_1/4N \tag{7.61}$$

Since Eqs. (7.60) and (7.61) describe the same curve, we obtain two relations between the unknown coefficients $a^2 - \mu_2/(4N^2) = a$ and $\mu_1/(4N^2) = b$. The remaining three equations that are necessary for determination of a, b, μ_1, μ_2, and N we obtain by substituting Eqs. (7.60) into Eq. (7.52) and into the constraints of Eqs. (7.54) and (7.56). [We use strict equality in Eqs. (7.54) and (7.56)]. Thus, we have the following system of algebraic equations:

$$\begin{aligned} a^2 - \mu_2/4N^2 &= a, \quad \mu_1/4N^2 = b, \quad 1 + a = 1/N \\ b^2 N/a^{1/2} &= K_0/\pi G, \quad b^2/a^{3/2} = 4C_0/\pi E \end{aligned}$$

Solving this system for the unknown quantities, we obtain

$$\begin{aligned} a &= 1/(\beta - 1), \quad b = (K_0\beta/\pi G)^{1/2}(1/\beta - 1)^{3/4}, \quad N = (\beta - 1)/\beta \\ \mu_1 &= 4\,(K_0/\pi G)^{1/2}(\beta - 1)^{5/4}/\beta^{3/2}, \quad \mu = 4\,(2 - \beta)/\beta^2 \end{aligned}$$

where $\beta = 4C_0 G/(EK_0)$. Equation (7.58) implies that $\beta \geqslant 2$ in the entire region II of Fig. 7.3. For this region, the shape of the cross section for a cylindrical rod of smallest weight and the corresponding stress function have the form

$$\Gamma : x^2 + \left(\frac{1}{\beta} - 1\right) y^2 = \left(\frac{K_0\beta}{\pi G}\right)^{1/2} (\beta^{-1} - 1)^{3/4}$$
$$\varphi = [(K_0\beta/\pi G)^{1/2} (\beta^{-1} - 1)^{3/4} - x^2 - (\beta^{-1} - 1)\, y^2]\, ((\beta - 1)/\beta) \tag{7.62}$$

Thus, the solution of Eq. (7.62) to the optimization problem is fully determined if the material constants E and G and the constants K_0 and C_0 satisfying the constraint $4GC_0/(EK_0) \geqslant 2$ are given. The cross-sectional area for the optimum rod is

$$S_{\text{opt}} = (\pi K_0/G)^{1/2} (\beta^2/(\beta - 1))^{1/4} \tag{7.63}$$

To estimate the effectiveness of the optimization process, we compare S_{opt} with the area of a rod having a circular cross section and having the same bending rigidity C_0 as the optimum rod. For a round rod, we have $S_0 = (4\pi C_0/E)^{1/2}$. The gain attained by optimizing is given by

$$(S_0 - S_{\text{opt}})/S_0 = 1 - (\beta - 1)^{-1/4}$$

It is clear from this formula that when $\beta = 2$ (the boundary of the region II), $S_0 = S_{\text{opt}}$ and a round rod is optimum. As β increases (with $\beta \geqslant 2$), the gain according to this formula also increases as the shape is optimized.

In addition to the above problem, we also consider two more basic problems that result in the optimization condition of Eq. (7.59). The first problem consists of maximizing the torsional rigidity of a cylindrical rod ($K \to \max$), with constraints on the area of the cross section and on the bending rigidity of the rod

$$C \geqslant C_0, \quad S \leqslant S_0 \tag{7.64}$$

In the second problem, we minimize the bending rigidity of the rod ($C \to$ max), with constraints on the area of the cross section and on the torsional rigidity

$$K \geqslant K_0, \quad S \geqslant S_0 \tag{7.65}$$

Let us describe the solutions to these problems, without going into details. In the first problem, the first quadrant of the C–S plane ($C \geqslant 0, S \geqslant 0$) is divided

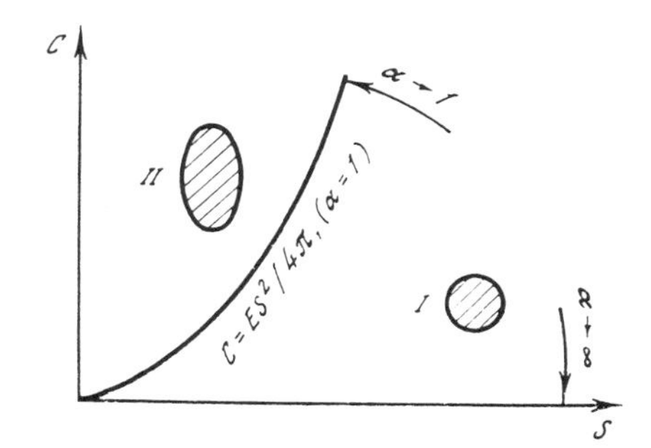

Figure 7.4

into two regions by the parabola $C = ES^2/(4\pi)$, as shown in Fig. 7.4. If the parameters C_0 and S_0 satisfy the inequality

$$C_0 \leqslant ES_0^2/4\pi \tag{7.66}$$

i.e., if they lie in region I of Fig. 7.4, then the cross-sectional area of the rod having the greatest torsional rigidity is a circular disk with radius $r = (\pi S_0)^{1/2}$. In region I, constraints on bending rigidity are automatically satisfied so they have no influence on the optimum shape of the rod.

If the inequality of Eq. (7.66) is violated [i.e., the point (C_0, S_0)] lies in region II of Fig. 7.4], then both constraints of Eq. (7.64) influence the optimum shape of the cross section. As before, the optimum cross section is an elliptic disc and Γ is given by Γ: $x^2 + \kappa y^2 = \sqrt{\kappa}\, S_0/\pi$, $\kappa = (ES_0/4\pi C_0)^2$, with $\varphi = [\sqrt{\kappa}\, S_0/(\pi - x^2 - \kappa y^2)]/(1 + \kappa)$. The torsional rigidity is equal to

$$K_{\text{opt}} = \sqrt{\varkappa}\, GS_0^2/\pi\,(1 + \varkappa)$$

Now let us describe the solution of the second problem. As in the preceding problem, in the case of Eq. (7.65), the first quadrant $K \geqslant 0, S \geqslant 0$ is divided into two regions by the parabola $K = GS^2/2\pi$, shown in Fig. 7.5.

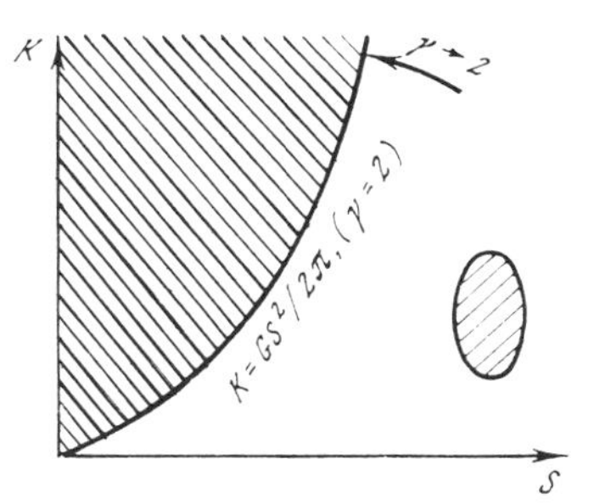

Figure 7.5

One aspect of this problem is completely different from the preceding ones. If the parameters K_0 and S_0 are such that $K_0 \geqslant GS_0^2/2\pi$; i.e., if the point (K_0, S_0) lies in the shaded region shown in Fig. 7.5, this problem has no solutions. We offer the following explanation: For an arbitrary cylindrical rod having the cross-sectional area $S \leqslant S_0$, the torsional rigidity satisfies the inequality $K \leqslant GS^2/2\pi$. In the unshaded region of Fig. 7.5, in which $K_0 \leqslant GS_0^2/2\pi$, the optimum shape of the boundary and the corresponding stress function φ satisfy $x^2 + \delta y^2 = \sqrt{\delta}\, S_0/\pi$, $\varphi = [\sqrt{\delta}\, S_0/(\pi - x^2 - \delta y^2)](1 + \delta)$, and the corresponding bending rigidity is given by

$$K_0 \geqslant GS_0^2/2\pi$$

We comment that the technique explained in this section, consisting of analyzing all possible types of solutions, may be directly generalized to multi-purpose design problems with a large number of constraints.

7.6. Design of a Circular Plate Having Minimum Weight with Constraints on Rigidity and Natural Frequencies of Vibrations

As a different example of a multi-purpose optimization problem, we shall consider (following Ref. 56) a circular plate of minimum weight, subjected to two different types of loading conditions and may be in two different states. In the first state, the plate performs free axially symmetric vibrations. In the second state, the plate is subjected to a constant bending load. In both cases the plate is freely supported along its boundary.

Constraints are imposed on the fundamental frequency and on the deflection at the center of the plate (i.e., on the rigidity). We introduce the basic equations for this problem. Let us denote by w_1 the amplitude function of plate deflection during free vibration, and by w_2 the deflection caused by a static load. For a given thickness function h, the functions $w_1(r)$ and $w_2(r)$ can be determined as solutions of the following system of equations (in terms of dimensionless variables):

$$L(h)\, w_1 = \omega^2 r h w_1 \tag{7.67}$$

$$L(h)\, w_2 = rq \tag{7.68}$$

with boundary conditions

$$\begin{gathered} w_i(1) = 0, \quad \left[h^3\left(w_{irr} + \frac{\nu}{r} w_{ir}\right)\right]_{r=1} = 0, \qquad w_{ir}(0) = 0 \\ [h^3(w_{irr} + w_{ir}/r)_r + (h^3)_r(w_{irr} + \nu w_{ir}/r)]_{r=0} = 0, \qquad i = 1, 2 \end{gathered} \tag{7.69}$$

where r denotes the distance from the center of the plate and $L(h)$ stands for the differential operator

$$L(h)\,w = \left\{ r\left[h^3\left(w_{rr} + \frac{1}{r}\, w_r \right)_r + (h^3)_r \left(w_{rr} + \frac{\nu}{r}\, w_r \right) \right] \right\}_r$$

The dimensionless deflection at the center of the bent plate is

$$a = \int_0^1 r w_2\, dr \Big/ \int_0^1 r\, dr = 2 \int_0^1 r w_2\, dr \tag{7.70}$$

The dimensionless volume of the plate is given by

$$V = \int_0^1 rh\, dr \tag{7.71}$$

Constraints on the fundamental frequency of vibration and on the deflection at the center of the plate are

$$\omega \geqslant \omega_0, \qquad a \leqslant a_0 \tag{7.72}$$

where ω_0 and a_0 are given dimensionless constants.

We formulate the following optimization problem: We need to find functions $h(r)$, $w_1(r)$, and $w_2(r)$ that satisfy differential equations of Eqs. (7.67) and (7.68), boundary conditions of Eq. (7.69), and constraints of Eq. (7.72), while the function $h(r)$ realizes the minimum value for the volume functional of Eq. (7.71).

We first derive necessary conditions for optimality. Since the boundary-value problem of Eqs. (7.67), (7.69), and (7.68), (7.69) are given by self-adjoint and positive definite operators, the fundamental frequency and the deflection at the center may be found from variational principles; i.e., from Rayleigh's principle and from the law of minimum potential energy, respectively (see Section 1.4),

$$\begin{aligned} \omega^2(h) &= \min_{w_1} \frac{(L(h)\,w_1,\, w_1)}{(rhw_1,\, w_1)} \\ a(h) &= -\frac{2}{q} \min_{w_2} [(L(h)\,w_2,\, w_2) - 2(rq,\, w_2)] \end{aligned} \tag{7.73}$$

where the inner product $(L(h)\, w_i, w_i)$ is defined by

$$(L(h)\, w_i, w_i) = \int_0^1 h^3 \left[w_{irr}^2 + \frac{2\nu}{r} w_{ir} w_{irr} + \left(\frac{w_{ir}}{r}\right)^2 \right] r\, dr$$

The minima in Eq. (7.73) are sought in certain classes of functions w_1 and w_2 that satisfy the boundary conditions $w_{1r}(0) = w_1(1) = w_{2r}(0) = w_2(1) = 0$.

The constraints of Eq. (7.72) can be restated in the form $\omega^2(h_0) - \omega_0^2 \geqslant 0$ and $a_0 - a(h) \geqslant 0$. Using Eq. (7.73) and allowing a multiplication by an arbitrary constant, the gradients of functionals given in Eq. (7.73) are

$$\begin{aligned}
\operatorname{grad}(\omega^2(h) - \omega_0^2) &= r \left\{ 3h^2 \left[w_{1rr}^2 + \frac{2\nu}{r} w_{1r} w_{1rr} + \left(\frac{w_{1r}}{r}\right)^2 \right] - \omega^2 w_1 \right\} \\
&= r\psi_1(h, w_1) \\
\operatorname{grad}(a_0 - a(h)) &= 3rh^2 \left[w_{2rr}^2 + \frac{2\nu}{r} w_{2rr} w_{2r} + \left(\frac{w_{2r}}{r}\right)^2 \right] = r\psi_2(h, w_2)
\end{aligned}$$

Since the magnitude of the gradient of the functional given by Eq. (7.71) (i.e., of the total volume) is equal to r, if we assume that $h(r) > 0$ for all $r(0, 1)$, the following necessary conditions for optimality of $h(r)$ are immediately derived:

$$\mu_1 \psi_1(h, w_1) + \mu_2 \psi_2(h, w_2) = 1 \tag{7.74}$$

$$\mu_1 (\omega^2(h) - \omega_0^2) = 0, \qquad \mu_1 \geqslant 0 \tag{7.75}$$

$$\mu_2 (a(h) - a_0) = 0, \qquad \mu_2 \geqslant 0 \tag{7.76}$$

where μ_1 and μ_2 denote Lagrange multipliers. Thus, we have Eqs. (7.67), (7.68), (7.72), (7.74)-(7.76), and (7.69) available for determining the unknown functions $h(r)$, $w_1(r)$, and $w_2(r)$, and the unknown constants μ_1 and μ_2.

We proceed to study the dependence of the optimum solution of our problem on the parameters q, ω_0, and a_0. First, let us examine the case in which the first constraint in Eq. (7.72) is a strict equality, i.e., when $\omega^2(h) = \omega_0^2$, with $a(h) < a_0$. We conclude from Eqs. (7.74)-(7.76) that in this case $\mu_1 > 0$ and $\mu_2 = 0$. We recall that the operator $L(h)$ is homogeneous. Using the properties of the functional of Eq. (7.71) and the first functional in Eq. (7.73), we can easily show that the optimum thickness function is given by

$$h_1 = \omega_0 h_1^0(r)$$

where $h_1^0(r)$ is a solution of Eqs. (7.67), (7.69), and (7.74)-(7.76), with $\mu_2 = 0$, $\mu_1 > 0$, $\omega_0 = 1$. We have an inequality that must be satisfied by the parameters q, a_0, and ω_0, if the function $h_1(r)$ solves the original optimization problem. Hence, we need to solve the boundary-value problem of Eqs. (7.68) and (7.69) for h_1 and compute the magnitude of the deflection in the center of the plate. Since the functional $a(h)$ in the inequality $a(h_1) \leqslant a_0$ is homogeneous, we have the following constraint:

$$q/a_0\omega_0^3 \leqslant \beta_1, \qquad \beta_1 \equiv q/a\,(h_1^0) \tag{7.77}$$

Let us consider the other possibility, namely that the second inequality in Eq. (7.72) becomes a strict equality, i.e., $a(h) = a_0$, while $\omega^2(h) > \omega_0$. For this case we have, as a consequence of Eqs. (7.74)-(7.76), $\mu_1 = 0$, $\mu_2 > 0$. The optimum solution is

$$h_2 = (q/a_0)^{1/3}\, h_2^0\,(r)$$

where h_2^0 is a solution of Eqs. (7.68), (7.69), and (7.72)-(7.74), with $\omega_0 = 0$ and $(q/a) = 1$. Let us solve the eigenvalue problem of Eqs. (7.67) and (7.69) for h_2 and find $\omega^2(h_2)$. Since $\omega^2(h_2)$ is a homogeneous functional and $\omega^2(h_2) \geqslant \omega_0^2$, we have the inequality

$$q/a_0\omega_0^3 \geqslant \beta_2, \qquad \beta_2 \equiv [\omega^2\,(h_2^0)]^{-3/2} \tag{7.78}$$

The function h_2 satisfying the above condition solves our optimization problem.

Let us examine the influence of the parameter $\beta = q/(a_0\,\omega_0^3)$. As we have indicated previously for $\beta \leqslant \beta_1$ (see Fig. 7.6, region I) the function $h_1 = \omega_0 h_1^0(r)$ solves our optimization problem. If $\beta \geqslant \beta_2$ (region II), the function $h_2 = (q/a)^{1/3}$ ·

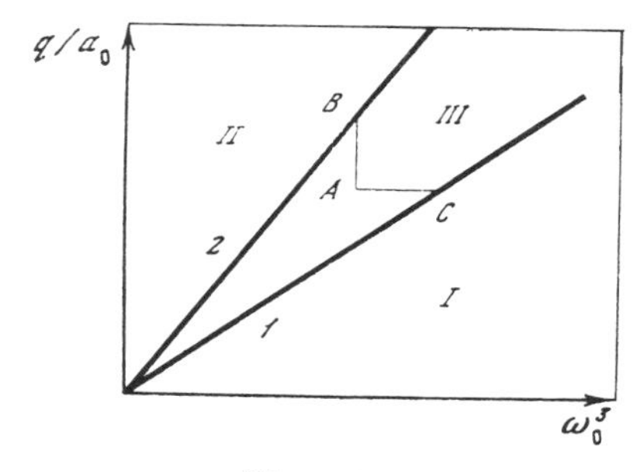

Figure 7.6

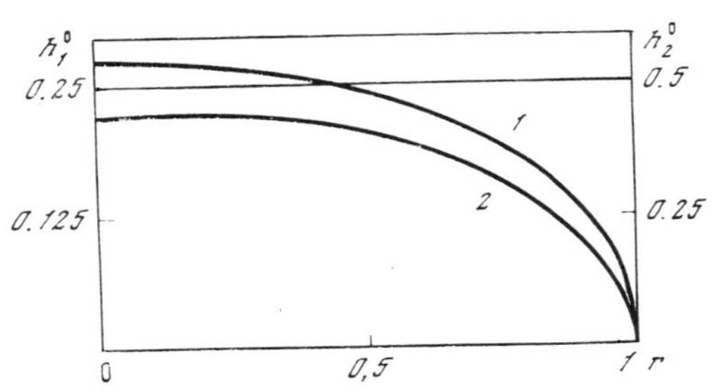

Figure 7.7

$h_2^0(r)$ is the optimum solution. Curve 1 in Fig. 7.6 corresponds to the value $\beta = \beta_1$ and curve 2 to $\beta = \beta_2$.

The dependence of $h_1^0(r)$ and $h_2^0(r)$ on r shown by curves 1 and 2, computed numerically for $\nu = 0.3$, is illustrated in Fig. 7.7. The corresponding volumes are $v_1^0 = 0.088$ and $v_2^0 = 0.130$.

Parameter values β that satisfy the two sided inequalities $\beta_1 < \beta < \beta_2$ (see Fig. 7.6, region III) correspond to case $(\mu_1 > 0, \mu_2 > 0)$ when both constraints of Eq. (7.72) are replaced by strict equalities, as can be easily deduced from Eq. (7.72).

Quasi-optimal solutions may now be calculated. Let the parameters of our problem, q, a_0, and ω_0 be such that β satisfies the two-sided inequality $\beta_1 \leqslant \beta \leqslant \beta_2$ (this is illustrated by point A in Fig. 7.6). To this point corresponds a solution $h(r)$ with the value of the cost functional V (i.e., the volume). Then to the point B, shown in Fig. 7.6, correspond the parameters $a_0' = a_0 \beta/\beta_2$ and $\omega_0' = \omega_0$. Consequently, the optimum solution is $h_2 = (\beta_2 q/\beta a_0)^{1/3} h_2^0(r)$, with volume

$$V_2 = \left(\frac{\beta_2 q}{\beta a_0}\right)^{1/3} V_2^0, \qquad V_2^0 = \int_0^1 h_2^0 r \, dr$$

The solution $h_2(r)$ is an admissible solution to our original problem if the relations $a_0' \leqslant a_0$ and $\omega_0' = \omega_0$ are satisfied. Therefore, either

$$V_2 \geqslant V$$
$$V \leqslant (\beta_2 q/\beta a_0)^{1/3} V_2^0$$

In complete analogy with the preceding arguments, if we take a point C shown in Fig. 7.6 that corresponds to an admissible solution $h_1(r)$, we derive an estimate

$$V \leqslant \omega_0 (\beta/\beta_1)^{1/3} V_1^0$$

We combine these inequalities as follows:

$$V \leqslant V_q = \min [\omega_0 (\beta/\beta_1)^{1/3} V_1^0, \ (\beta_2 q/\beta a_0)^{1/3} V_2^0] \tag{7.79}$$

An optimum solution obtained from $h_1(r)$ and $h_2(r)$, which corresponds to the minimum of the right-hand side in Eq. (7.79) is called a *quasi-optimal solution,* for the value V_q of the volume functional.

We also have the estimate

$$\max [\omega_0 V_1^0, \ (q/a_0)^{1/3} V_2^0] \leqslant V \tag{7.80}$$

because the function $h(r)$ that solves our problem with two constraints cannot assign to the volume functional a smaller value that the functions $\omega_0 h_1^*(r)$ and $(q/a_0)^{1/3} h_2^*(r)$, which actually solve the problem but satisfy only one constraint. It is easy to obtain the following inequality from Eqs. (7.79) and (7.80):

$$1 \leqslant V_q/V \leqslant [V_1^0/\beta_1^{1/3} V_2^0, \ V_2^0 \beta_2^{1/3}/V_1^0] \tag{7.81}$$

Thus, knowing the functions $h_1^0(r)$ and $h_2^0(r)$, and therefore the values of the constants β_1 and β_2 for an arbitrary point A in region III, we can construct the admissible solutions $h_1(r) = \omega_0 (\beta/\beta_1)^{1/3} h_1^0(r)$ and $h_2 = (q\beta_2/\beta a_0)^{1/3} h_2^0(r)$. We choose from these two admissible thickness functions the one that minimizes Eq. (7.79) as the quasi-optimal solution. The estimate of Eq. (7.81) is available for comparing the quasi-optimal solution with an optimum solution. As our computations indicate, $1 \leqslant V_q/V \leqslant 1.0028$.

7.7. Construction of Quasi-Optimal Solutions to the Multipurpose Design Problems

In the preceding section we studied a specific example of a quasi-optimal solution, using the homogeneous property of the cost functional. As was shown in Ref. 227, it is this property that permits us to construct admissible solutions and to obtain two-sided estimates on the cost functional, in much more general problems with many constraint conditions.

Let us now study an extremization problem with several constraints.

$$\begin{gathered} \min J(h), \quad h \in \mathscr{K} \\ J_i(h) \geqslant c_i > 0, \ i = 1, 2, \ldots, n \end{gathered} \tag{7.82}$$

where c_i are given constants.

We assume that J and J_i $(i = 1, 2, \ldots, n)$ is positive in $\overline{K}$ and homogeneous in h, with positive degrees of homogeneity β and β_i, respectively. We consider the problem of Eq. (7.82), but with only one ith constraint, where i is chosen arbitrarily. We denote the solution of this problem by h_i^*, while the solution of the problem of Eq. (7.82) that obeys all constraints is denoted by h^*.

We compute the values of $J_j(h_i^*)$, $j = 1, 2, \ldots, n$. It is clear that certain constraints of the problem of Eq. (7.82) may be violated. Let us choose a multiplier γ_i such that the function $\gamma_i h_i^*$ satisfies all constraints. It is easy to see that γ_i can be taken as

$$\gamma_i = \max_{j=1,2,\ldots,n} [c_j/J_j(h_i^*)]^{1/\beta_j} \tag{7.83}$$

In fact

$$J_j(\gamma_i h_i^*) = \gamma_i^{\beta_j} J_j(h_i^*) \geqslant [(c_j/J_j(h_i^*))^{1/\beta_j}]^{\beta_j} J_j(h_i^*) = c_j$$

Thus the function $\gamma_i h_i^*$ is admissible, since all constraints of the problem of Eq. (7.82) are satisfied. In general, however, this function is not optimum. Consequently, the estimate $J(h^*) \leqslant J(\gamma_i h_i^*)$ is true. On the other hand, $J(h_i^*) \leqslant J(h^*)$, because the function h_i^* that solves the problem of Eq. (7.82) with only one constraint may not assign to the cost functional a larger value than the function h^* that solves the same problem, but satisfies all n constraints. Combining these inequalities, we have $J(h_i^*) \leqslant J(h^*) \leqslant J(\gamma_i h_i^*)$. Since i is an arbitrary index, we have

$$\max_{i=1,2,\ldots,n} J(h_i^*) \leqslant J(h^*) \leqslant \min_{i=1,2,\ldots,n} J(\gamma_i h_i^*)$$

An admissible solution $h_q = \gamma_p h_p^*$ that assigns a minimum to the right-hand side of this relation will be called a quasi-optimal solution, where p is the index for which the minimum is realized. According to this definition

$$J(h_q) = J(\gamma_p h_p^*) = \min_{i=1,2,\ldots,n} J(\gamma_i h_i^*)$$

From the last two relations given above, we can derive the following estimate:

$$1 \leqslant \frac{J(h_q)}{J(h^*)} \leqslant \frac{\min\limits_{i=1,2,\ldots,n} J(\gamma_i h_i^*)}{\max\limits_{i=1,2,\ldots,n} J(h_i^*)} \tag{7.84}$$

Therefore, if we know the solution h_i^* of an extremization problem with one constraint $(i = 1, 2, \ldots, n)$ we can compute the constants γ_i according to Eq. (7.83), construct the quasi-optimal solution, and estimate how well it approximates the optimum solution with multiple constraints, with respect to the value of the cost functional.

It is important to find the dependence of the upper estimate in Eq. (7.84) on the parameters c_i $(i = 1, 2, \ldots, n)$ that appear in Eq. (7.82)

$$h_i^* = c_i^{1/\beta_i} h_1^0 \tag{7.85}$$

where, as before, h_i^* denotes the solution of the problem of Eq. (7.82), with only the ith constraint satisfied and $c_i = 1$. Using Eq. (7.85) and homogeneity of the cost functional, we represent the constants γ_i in the form

$$\gamma_i = c_i^{1/\beta_i} \max_{j=1,2,\ldots,n} c_j^{1/\beta_j} (J_j(h_i^0))^{-1/\beta_j} = c_i^{1/\beta_i} c_{k_i}^{1/\beta_{k_i}} (J_{k_i}(h_i^0))^{-1/\beta_{k_i}} \tag{7.86}$$

In this expression, the index k_i denotes the integer j for which the expression given inside the square brackets in Eq. (7.83) attains its maximum.

We transform Eq. (7.84) by substituting Eqs. (7.85) and (7.86) to obtain:

$$\begin{aligned} 1 \leqslant \frac{J(h_q)}{J(h^*)} &\leqslant \frac{\min\limits_{i=1,2,\ldots,n} J(\gamma_i h_i^*)}{\max\limits_{i=1,2,\ldots,n} J(h_i^*)} = \min_{\substack{i=1,2,\ldots,n\\ j=1,2,\ldots,n}} \frac{J(\gamma_i h_i^*)}{J(h_j^*)} \\ &= \min_{\substack{i=1,2,\ldots,n\\ j=1,2,\ldots,n}} \left[\frac{c_{k_i}^{\beta/\beta_{k_i}} (J_{k_i}(h_i^0))^{-\beta/\beta_{k_i}} J(h_i^0)}{c_j^{\beta/\beta_j} J(h_j^0)} \right] \\ &\leqslant \min_{i=1,2,\ldots,n} \frac{J(h_i^0) J_{k_i}^{-\beta/\beta_{k_i}}(h_i^0)}{J(h_{k_i}^0)} \end{aligned} \tag{7.87}$$

The last inequality is true since the minimum of a set of numbers cannot decrease as the number of elements of the set is decreased. The estimates above do not contain explicitly the parameters c_i, but they depend on these numbers because the vector $k = (k_1, k_2, \ldots, k_n)$ is determined by relations that these parameters must obey [see Eq. (7.86)].

We remark that as we vary the parameters c_i, the components of the vector k do not assume all possible values. Let us explain this remark. Whether a vector k with components $k_1, k_2, \ldots, k_n$ exists or not can be determined by using Eq. (7.86) to construct a system of linear inequalities in variables c_i^{1/β_i}

$$[c_{k_i}/J_{k_i}(h_i^0)]^{1/\beta_{k_i}} \geqslant [c_j/J_j(h_j^0)]^{1/\beta_j}$$
$$c_i^{1/\beta_i} > 0, \; i, j = 1, 2, \ldots, n; \; j \neq k_i \tag{7.88}$$

In the n-dimensional space spanned by c_i^{1/β_i} ($i = 1, 2, \ldots, n$), the inequalities of Eq. (7.88) define a region R_k. If this region is nonempty, then for an arbitrary vector $c = (c_1^{1/\beta_1}, \ldots, c_n^{1/\beta_n})$, $c \in R_k$, Eq. (7.87) offers us an estimate of closeness of the quasi-optimal solution to the optimum solution by comparing the values of the cost functionals. If we consider for the right-hand side of Eq. (7.87) a maximum with respect to all possible values of the vectors k, such that the corresponding region R_k is nonempty, we arrive at an absolute estimate of the closeness of the quasi-optimal solution in the quadrant $c_i^{1/\beta_i} > 0$, $i = 1, 2, \ldots, n$.

We remark that $h_i^* = c_i^{1/\beta_i} h_i^0$ are optimum solutions of the original problem of Eq. (7.82) if the parameters c_i satisfy the following system of inequalities:

$$c_j^{1/\beta_j} c_i^{-1/\beta_i} \leqslant J_j^{1/\beta_j}(h_i^*), \qquad j = 1, 2, \ldots, n; \qquad j \neq i \tag{7.89}$$

We can perform a direct check to confirm that for the parameters c_i satisfying the above inequalities, all constraints in Eq. (7.82) are satisfied, $\gamma_i = 1$, and the quasi-optimal solution coincides with the optimum solution, i.e., $h_q = h_i^* = h^*$. In this case, components k_i of the vector k are equal to i.

We observe that in an analysis of the region in which the specific optimum and quasi-optimal solutions are realized, we can introduce certain substitutions, for example $t_i = c_i^{1/\beta_i}/c_n^{1/\beta_n}$, $i = 1, 2, \ldots, n - 1$. Instead of considering the n-dimensional space of parameters c_i^{1/β_i}, $i = 1, 2, \ldots, n$, we transfer our problem to an $n - 1$ dimensional space of parameters t_i, $i = 1, 2, \ldots, n - 1$.

As an example, we offer a solution to the problem of minimizing the volume of a beam, with three different constraints: one on the fundamental natural frequency of transverse vibrations, the second on the minimum value of the axial load that causes loss of stability, and the third on the magnitude of the deflection in the center of the span when a point load is applied to that point (see Ref. 124). The free vibrations, the compression of the beam by axial loads, and the deflection caused by transverse loads are all considered separately; i.e., we assume that when axial loads are applied to the ends of the beam, causing compressive stress, there is no transverse load, and the beam does not perform a transverse vibration. If, on the other hand, we consider the free vibrations, then both types of static load are absent. The cross section of the beam is a rectangle with variable height and constant width b. The beam is simply supported at both ends.

The entire analysis is carried out in dimensionless variables. The functional to be minimized (the volume of the beam) is given by

$$V = \int_0^1 h(x)\,dx \tag{7.90}$$

Formulas for the fundamental frequency of free vibration, the force causing the loss of stability, and the deflection w at the center of the span are given by

$$\omega^2(h) = \min_{w_1} \frac{\int_0^1 h^3 w_{1xx}^2\,dx}{\int_0^1 h w_1^2\,dx}, \qquad p(h) = \min_{w_2} \frac{\int_0^1 w_{2x}^2\,dx}{\int_0^1 w_2^2 h^{-3}\,dx}$$

$$w(h) = \int_0^1 \frac{\psi(x)}{h^3(x)}\,dx, \qquad \psi(x) = \begin{cases} 3x^2/8, & 0 \leqslant x \leqslant 1/2 \\ 3(1-x)^2/8, & 1/2 \leqslant x \leqslant 1 \end{cases} \tag{7.91}$$

The minimum in Eq. (7.91) is sought in a certain class of functions w_1 and w_2 that satisfy the boundary conditions $w_1(0) = w_1(1) = w_2(0) = w_2(1) = 0$.

We assign constraints to the natural frequencies of vibrations, to the critical load causing the loss of stability, and to the deflection at the center of the beam in the form

$$\omega^2 \geqslant \omega_0^2, \; p \geqslant p_0, \; 1/w \geqslant 1/w_0 \tag{7.92}$$

where ω_0, p_0, and w_0 are given dimensionless constants.

We formulate the optimization problem as follows: we wish to find a function $h(x) > 0$, $x \in (0, 1)$, that assigns a minimum to the functional of Eq. (7.90) and satisfies the constraints of Eq. (7.92), in which the quantities ω^2, p, and w are defined by Eq. (7.91). The solution to the problem of minimizing the fundamental frequency of free transverse vibrations was given by F. Niordson in Ref. 207. Analytic solutions of the buckling problems for a column with various boundary conditions were studied in Refs. 171, 184, and 233. Solving the problem of optimization of deflection with only one constraint presents no difficulties.

For the problem formulated above, we derive necessary conditions of optimality as

$$\sum_{i=1}^{3} \mu_i \psi_i = 1, \qquad x \in (0, 1)$$
$$(\omega^2 - \omega_0^2)\,\mu_1 = 0, \quad \mu_1 \geqslant 0, \quad (p - p_0)\,\mu_2 = 0, \quad \mu_2 \geqslant 0 \tag{7.93}$$
$$(w - w_0)\,\mu_1 = 0, \;\; \mu_3 \geqslant 0$$

For given values of the parameters ω_0, p_0, and w_0 that occur in Eqs. (7.91)–(7.93), we can find the functions $h(x)$, $w_1(x)$, and $w_2(x)$, and the multipliers μ_1, μ_2, and μ_3, thus realizing an optimum solution.

Depending on relations between the parameters of the problem, it is possible to restate certain constraints. For example, $\omega = \omega_0$, $p > p_0$, and $w < w_0$ correspond, according to Eq. (7.93), to $\mu_1 > 0$ and $\mu_2 = \mu_3 = 0$. Therefore, in this case, solution of the optimization problem is reduced to a problem having only the first constraint given in Eq. (7.92).

It follows directly from Eqs. (7.90) and (7.91) that the functions $V(h)$, $\omega^2(h)$, $p(h)$, and $w^{-1}(h)$ are positive if $h(x) > 0$ and are homogeneous in h, with the degree of homogeneity $\beta = 1$, $\beta_1 = 2$, $\beta_2 = 3$, and $\beta_3 = 3$, respectively. Solutions of the optimization problem with only a single ith constraint are given by

$$h_1^* = \omega_0 h_1^0(x), \qquad h_2^* = p_0^{1/3} h_2^0(x), \qquad h_3^* = w_0^{-1/3} h_3^0(x) \tag{7.94}$$

The functions $h_1^0(x)$ and $h_2^0(x)$ may be computed using the gradient technique in the space of admissible functions, where at each step of the gradient procedure we solve an eigenvalue problem. The functions $h_1^0(x)$ and $h_2^0(x)$ are shown in the Fig. 7.8. Because of symmetry of these functions about the point $x = \frac{1}{2}$ the sketch illustrates only a half of the graph.

Solution of our optimization problem with only a single constraint, namely the third constraint given in Eq. (7.92), is not hard to derive analytically: $h_3^0(x) = 2^{-5/6}\sqrt{x}$, for $0 \leqslant x \leqslant 1/2$, and $h_3^0(x) = 2^{-5/6}(1 - x)^{1/2}$ for $1/2 \leqslant x \leqslant 1$. Substituting the values of h_1^0, h_2^0, and h_3^0, we can compute the matrix [see Eq.

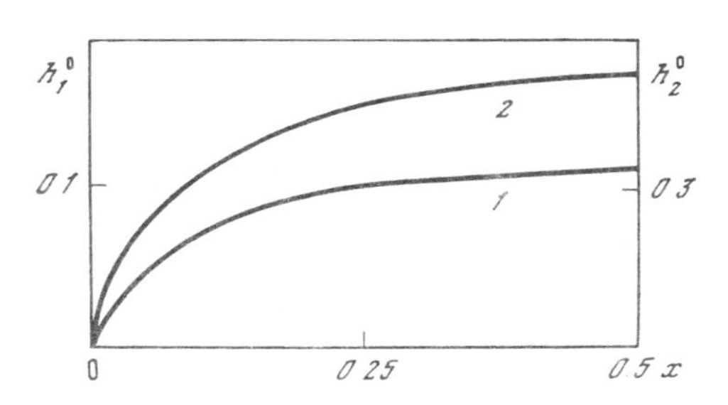

Figure 7.8

(7.86)]

$$\gamma_{ij} = c_j^{1/\beta_j} c_i^{-1/\beta_i} J_j^{-1/\beta_j}(h_i^0) \tag{7.95}$$

According to the inequality of Eq. (7.89) the solutions h_1^*, h_2^*, and h_3^* are realized, respectively, in regions I, II, and III, determined by parameter inequalities

$$c_i^{1/\beta_i} c_j^{\beta/\beta_j} \geqslant J_j^{-1/\beta_j}(h_i^0), \qquad j = 1, 2, \ldots, n; \quad j \neq i$$

For convenience we introduce parameters $t_1 = w_0^{-1/3}\omega_0^{-1}$, $t_2 = (w_0 p_0)^{-1/3}$, and $t_3 = \omega_0 p_0^{-1/3} = t_2/t_1$ and restate the inequalities defining the regions I, II, and III, respectively as

$$t_3 \geqslant a_1,\ t_1 \leqslant a_2; \quad t_3 \leqslant b_1,\ t_2 \leqslant b_2; \quad t_1 \geqslant d_1,\ t_2 \geqslant c_2$$

where $a_1 = 4.6043$, $a_2 = 0.3346$, $b_1 = 4.5993$, $b_2 = 1.5349$, $d_1 = 0.3513$, and $d_2 = 1.6329$. Figure 7.9 illustrates the regions I, II, and III. Thus, if the values of the parameters ω_0, p_0, and w_0 are such that the point (t_1, t_2) lies in region I, II, or III, the functions given by Eq. (7.94) solve the optimization problem.

Let us consider the case in which the parameter values are such that the corresponding point (t_1, t_2) does not belong to any of these regions. In that case we can use Eq. (7.83), compute the quantities γ_i, construct a quasi-optimal solution, and estimate [using Eq. (7.84)] how close this solution is to the optimum solution that minimizes the functional V.

We proceed to derive absolute (i.e., independent of the parameters ω_0, p_0, and w_0) estimates for closeness of a quasi-optimal solution to an optimum one. First, we construct the domains R_k according to Eq. (7.88). In our case, it turns out that among the sets of all possible vectors k, only the vectors (3, 3, 1),

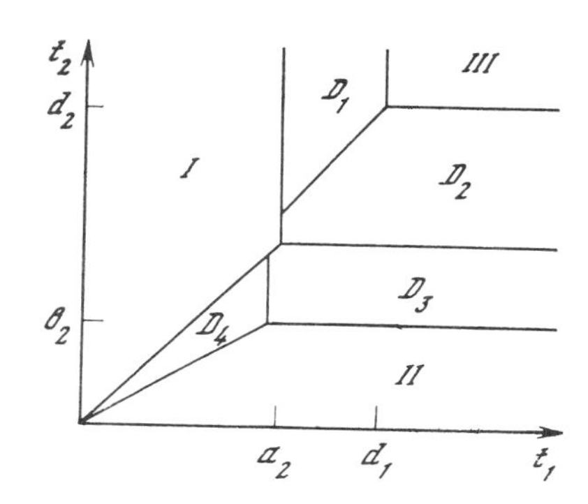

Figure 7.9

(3, 3, 2), (2, 1, 2), and (2, 3, 2) correspond to a nonempty set R_k, in agreement with the inequalities of Eq. (7.88) (vectors k whose components are $k_i = i$ are not considered). In the $(t_1 - t_2)$ plane, the inequalities of Eq. (7.88) define regions D_1, D_2, D_3, and D_4, as shown in Fig. 7.9:

$$D_1: \quad a_2 \leqslant t_1 \leqslant d_1, \quad t_2/t_1 \geqslant d_2/d_1$$

$$D_2: \quad t_1 \geqslant a_2, \quad t_2/t_1 \leqslant d_2/d_1, \quad a_1 a_2 \leqslant t_2 \leqslant d_2$$

$$D_3: \quad t_1 \geqslant b_2/b_1, \quad t_2/t_1 \leqslant a_1, \quad b_2 \leqslant t_2 \leqslant a_1 a_2$$

$$D_4: \quad t_1 \leqslant b_2/b_1, \quad b_1 \leqslant t_2/t_1 \leqslant a_1$$

In agreement with Eq. (7.87), in each of these regions we have the following estimates:

$$1 \leqslant \frac{V(h_q)}{V(h^*)} \leqslant \min\left\{(w(h_1^0))^{1/3} \frac{V(h_1^0)}{V(h_3^0)}, \quad (w(h_2^0))^{1/3} \frac{V(h_2^0)}{V(h_3^0)}, \right.$$

$$\left. \frac{V(h_3^0)}{\omega(h_3^0)\, V(h_1^0)} \right\} = (w(h_1^0))^{1/3} \frac{V(h_1^0)}{V(h_3^0)} = 1.021, \qquad (t_1, t_2) \in D_1$$

$$1 \leqslant \frac{V(h_q)}{V(h^*)} \leqslant \min\left\{(w(h_1^0))^{1/3} \frac{V(h_1^0)}{V(h_3^0)}, \quad (w(h_2^0))^{1/3} \frac{V(h_2^0)}{V(h_3^0)}, \right.$$

$$\left. \frac{V(h_3^0)}{p^{1/3}(h_3^0)\, V(h_2^0)} \right\} = \frac{V(h_1^0)}{V(h_3^0)} (w(h_1^0))^{1/3} = 1.021, \qquad (t_1, t_2) \in D_2$$

$$1 \leqslant \frac{V(h_q)}{V(h^*)} \leqslant \min\left\{\frac{V(h_1^0)}{p^{1/3}(h_1^0)\, V(h_2^0)}, \quad \frac{V(h_2^0)\, w^{1/3}(h_3^0)}{V(h_3^0)}, \right.$$

$$\left. \frac{V(h_3^0)}{p^{1/3}(h_3^0)\, V(h_2^0)} \right\} = \frac{V(h_1^0)}{p^{1/3}(h_1^0)\, V(h_2^0)} = 1.00064, \qquad (t_1, t_2) \in D_3$$

$$1 \leqslant \frac{V(h_q)}{V(h^*)} \leqslant \min\left\{\frac{V(h_1^0)}{p^{1/3}(h_1^0)\, V(h_2^0)}, \quad \frac{V(h_2^0)}{\omega(h_2^0)\, V(h_1^0)}, \right.$$

$$\left. \frac{V(h_3^0)}{p^{1/3}(h_3^0)\, V(h_2^0)} \right\} = \frac{V(h_2^0)}{\omega(h_2^0)\, V(h_1^0)} = 1.00046, \qquad (t_1, t_2) \in D_4$$

It is clear from these estimates that for arbitrary non-negative values of the parameters ω_0, p_0, and w_0, the quasi-optimal solutions of this problem assign cost functional values which do not exceed the optimum value of that functional by more than 2.1%.

Therefore, by making use of the homogeneous properties of cost functionals we are able to construct quasi-optimal solutions and to obtain estimates for closeness between the quasi-optimal and optimum solutions, based on the values of the cost functional. The advantages of quasi-optimal solutions consist in being able to solve n much simpler optimization problems, each having only one constraint, and being able to estimate the optimum solution with n constraints. This technique of constructing quasi-optimal solutions may be applied to optimum designs of structures with multiple constraints, for mechanical and structural system such as beams, rods, arches, plates, and shells, whose equations of state are homogeneous differential equations with homogeneous boundary conditions. The constraints may be imposed on natural frequencies of vibrations, critical buckling loads, compliance, and maximum value of deflection or stress components.

Bibliography

1. Abgarjan, K. A., *On the Theory of Beams of Minimal Weight, in the Monograph Strength Computations*, Mashgiz, Moscow, 8th edition, 1962, pp. 136–151.
2. Annin, B. D., *Modern Models for Plastic Bodies*, N. G. U. Novossibirsk, 1975.
3. Annin, B. D., "Optimal Design of Elastic, Anisotropic, Inhomogeneous Bodies," *Third National Congress on Theoretical and Applied Mechanics*, Varna, Bulgaria, 1977, pp. 275–280.
4. Armand, J. L., *Applications of the Theory of Optimal Control for Systems with Distributed Parameters to the Problems of Optimal Design*, Mir, Moscow, 1977.
5. Armand, J. L., Lur'e, K. A., and Cherkaev, A. V., "On the Solution of Problems of Optimizing the Eigenvalues Arising in Designs of Elastic Structures," *Izv. Akad. Nauk SSSR, MTT*, No. 5, 1978, pp. 159–162.
6. Arutjunjan, N. H., and Abramjan, B. L., *Torsion of Elastic Bodies*, Fizmatgiz, Moscow, 1963.
7. Ashkenazi, E. K., *Anisotropy in Materials Used for Machine Elements*, Mashinostroenie, L'vov, 1969.
8. Banichuk, N. V., "Computation of Loading for an Elasto-Plastic Body," *Izv. Akad. Nauk SSSR, MTT*, No. 1, 1969, pp. 128–135.
9. Banichuk, N. V., "A Game–Theoretic Approach to the Problem of Optimization of Elastic Bodies," *PMM*, Vol. 37, No. 6, 1973, pp. 1098–1108.
10. Banichuk, N. V., "Some Problems in Optimal Design of Beams for Certain Classes of Loads," *Izv. Akad. Nauk SSSR, MTT*, No. 5, 1973, pp. 102–110.
11. Banichuk, N. V., "Optimization of Stability for a Beam with an Elastic Reinforcement," *Izv. Akad. Nauk SSSR, MTT*, No. 4, 1974, pp. 150–154.
12. Banichuk, N. V., "Optimal Design in One-Dimensional Bending Problems for Stationary and Moving Loads," *Izv. Akad. Nauk SSSR, MTT*, No. 5, 1974, pp. 113–123.
13. Banichuk, N. V., "Optimal Shapes of Elastic Plates in Bending Problems," *Izv. Akad. Nauk SSSR, MTT*, No. 5, 1975, pp. 180–188.
14. Banichuk, N. V., "A Variational Problem with Unknown Boundary Concerning the Optimal Shapes of Elastic Bodies," *PMM*, Vol. 39, No. 6, 1975, pp. 1082–1092.
15. Banichuk, N. V., "Determining the Optimal Shape of Elastic Curved Beams," *Izv. Akad. Nauk SSSR, MTT*, No. 6, 1975, pp. 124–133.
16. Banichuk, N. V., "A Certain Game–Theoretic Problem of Optimizing an Elastic Body," *Dokl. Akad. Nauk SSSR*, Vol. 226, No. 3, 1976, pp. 497–499.
17. Banichuk, N. V., "Certain Problems in Optimization of Elastic Structures," *Second National Congress on Theoretical and Applied Mechanics*, Varna, 1973; publication of Bulgarian Academy of Sciences, Sofia, Vol. 2, 1976, pp. 619–627.

18. Banichuk, N. V., "On a Certain Two-Dimensional Optimization Problem in the Theory of Torsion of Elastic Beams," *Izv. Akad. Nauk SSSR, MTT,* No. 5, 1976, pp. 45–52.
19. Banichuk, N. V., "The Problem of Optimizing the Shape of a Hole in a Plate Subjected to Bending," *Izv. Akad. Nauk SSSR, MTT,* No. 3, 1977, pp. 81–88.
20. Banichuk, N. V., "Optimality Conditions in the Problem of Determining the Shapes of Holes Inside Elastic Bodies," *PMM,* Vol. 41, No. 5, 1977, pp. 920–925.
21. Banichuk, N. V., "Optimization of the Shape and of the Distribution of Moduli of Elastic Bodies," *Proceedings of the 14th Yugoslav Congress on Theoretical and Applied Mechanics*, Portorozh, Vol. C, 1978, pp. 319–326.
22. Banichuk, N. V., "Optimal Anisotropy of Torsioned Beams," *Izv. Akad. Nauk SSSR, MTT,* No. 4, 1978, pp. 73–79.
23. Banichuk, N. V., "A Certain Extremal Problem for a System with Distributed Parameters and Determining the Optimal Properties of an Elastic Medium," *Dokl. Akad. Nauk SSSR*, Vol. 242, No. 5, 1978, pp. 1042–1045.
24. Banichuk, N. V., "Optimization of Anisotropic Properties of Deformable Media in the Framework of Two-Dimensional Theory of Elasticity," *Izv. Akad. Nauk SSSR, MTT,* No. 1, 1979, pp. 71–77.
25. Banichuk, N. V., Kartvelishvili, V. M., and Mironov, A. A., "Methods of Successive Optimization for a Numerical Solution of Problems of Optimal Design of Structures," *Numerical Techniques of Nonlinear Programming, Proceedings of the Second All-Union Seminar*, Kharkov, 1976, pp. 54–59.
26. Banichuk, N. V., Kartvelishvili, V. M., and Mironov, A. A., "Numerical Solution of a Two-Dimensional Problem of Optimization of Elastic Plates," *Izv. Akad. Nauk SSSR, MTT,* No. 1, 1977, pp. 68–78.
27. Banichuk, N. V., Kartvelishvili, V. M., and Mironov, A. A., "Optimization Problems with Local Quality Criteria in the Theory of Bending of Plates," *Izv. Akad. Nauk SSSR, MTT,* No. 1, 1978, pp. 124–131.
28. Banichuk, N. V., Kartvelishvili, V. M., and Mironov, A. A., "A Numerical Technique for Solving Two-Dimensional Optimization Problem in the Theory of Elasticity," Proceedings of the Fifth All-Union Conference on Numerical Techniques for Solving Problems in the Theory of Elasticity and Plasticity, Part 2, Novosibirsk, *All-Union SO, Akad. Nauk SSSR*, 1978, pp. 3–14.
29. Banichuk, N. V., Kartvelishvili, V. M., and Chernous'ko, F. L., "On Finite Difference—Quadrature Approximations to Convex Integral Functionals," *Dokl. Akad. Nauk SSSR*, Vol. 231, No. 2, 1976, pp. 269–272.
30. Banichuk, N. V., "Minimization of the Weight of a Wing with Constraints on the Velocity of Divergence," Scientific Notes, *TsAGI,* Vol. 9, No. 5, 1978.
31. Banichuk, N. V., and Mironov, A. A., "Optimization of the Vibration Frequency for a Plate Submerged in an Ideal Liquid," *PMM,* Vol. 39, No. 5, 1975, pp. 889–899.
32. Banichuk, N. V., and Mironov, A. A., "Optimal Design of Plates in the Framework of Dynamic Problems of Hydroelasticity," *Proceedings of the Tenth All-Union Conference on the Theory of Shells and Plates*, Tbilisi, Metsniereba, 1975, pp. 35–44.
33. Banichuk, N. V., and Mironov, A. A., "Problems in Optimization of Vibrating Plates Submerged in an Ideal Fluid," *PMM,* Vol. 4, No. 3, 1976, pp. 520–527.
34. Banichuk, N. V., and Mironov, A. A., "A Scheme of a Jet Flow for Investigating the Equilibrium Shapes of Elastic Plates Submerged in a Stream of a Fluid and Some Optimization Problems," *PMM,* Vol. 43, No. 1, 1979, pp. 83–90.
35. Birkhoff, G., *Hydrodynamics*, Foreign Translations Series, Moscow, 1963.
36. Birjuk, V. I., Lipin, E. K., and Frolov, V. M., *Techniques of Designing for Aircraft Structures*, Mashinostroenie, Moscow, 1977.
37. Birjuk, V. I., and Moiseenko, V. P., "An Application of the Discrete–Continuous Version of the Maximum Principle to Problems of Optimal Structural Design," Scientific Notes, *TsAGI,* Vol. 4, No. 4, 1973.

38. Bisplinghoff, P. L., Ashley, H., and Halfmann, R. L., *Aeroelasticity*, Addison-Wesley, Reading, Massachusetts, 1955.
39. Bliss, G., *Lectures on the Calculus of Variations*, University of Chicago Press, Chicago, 1961.
40. Bolotin, V. V., "Basic Equations of the Theory of Reinforced Media," *Polym. Mech.*, Vol. 1, No. 2, 1965, pp. 27–37.
41. Bolotin, V. V., "A Two-Dimensional Theory of Elasticity for Parts Made from Reinforced Materials," in *Strength Computations*, Mashinostroenie, Moscow, No. 12, 1966, pp. 3–31.
42. Bryson, A., and Ho, Y. C., *Applied Optimal Control*, Halsted Press, 1979; Russian translation: Mir, Moscow, 1972.
43. Bryzgalin, G. I., "On Rational Design of Flat Anisotropic Bodies with a Weak Binder," *Izv. Akad. Nauk SSSR, MTT*, No. 4, 1969, pp. 123–131.
44. Butkovskii, A. G., *Optimal Control Theory for Systems with Distributed Parameters*, Nauka, Moscow, 1965.
45. Vakulenko, L. D., and Mazalov, V. N., *Optimal Design of Structures, Library Guide to the Motherland and Foreign Literature for the Period 1948–1974*, Novosibirsk, Institute of Hydrodynamics, *S. O. Akad. Nauk SSSR*, 1975, parts I and II.
46. Vigderhaus, S. B., "An Integral Equation for the Inverse Problem of the Two-Dimensional Theory of Elasticity," *PMM*, Vol. 40, No. 3, 1976, pp. 566–569.
47. Vigderhaus, S. B., "A Special Case of the Inverse Problem of the Two-Dimensional Theory of Elasticity," *PMM*, Vol. 41, No. 5, 1977, pp. 902–908.
48. Vladimirov, V. S., *Equations of Mathematical Physics*, Dekker, 1971; recent Russian edition: Nauka, Moscow, 1976.
49. Galin, L. A., *Contact Problems in the Theory of Elasticity*, Gostekhizdat, Moscow, 1953.
50. Gel'fand, I. M., and Fomin, S. V., *Calculus of Variations*, Prentice-Hall, 1963; original Russian edition: Fizmatgiz, Moscow, 1961.
51. Golubev, I. S., *Analytical Methods for Designing Wing Structures*, Mashinostroenie, 1970.
52. Gol'dshtein, Iu. B., and Solomeshch, M. A., "Optimal Design of Beams for Dynamic Loading," *Structural Mechanics and Computation of Reinforcement*, No. 4, 1968.
53. Grigorovich, V. K., Sobolev, N. C., and Fridman, Ia. B., "On the Most Suitable Direction of Fibers in Fabrics Made from Anisotropic Materials," *Dokl. Akad. Nauk SSSR*, Vol. 86, No. 4, 1952, pp. 703–706.
54. Grinev, V. B., and Filippov, A. P., "Optimal Design of Structures with Prescribed Natural Frequencies," *Prikl. Mekh.*, Vol. 7, No. 10, 1971, pp. 19–25.
55. Gura, N. M., "A Shell of Maximum Rigidity Subjected to Torsion," *Izv. Akad. Nauk SSSR, MTT*, No. 1, 1979.
56. Gura, N. M., and Seiranjan, A. P., "The Optimal Circular Plate Satisfying Constraints on Rigidity and Fundamental Frequency of Vibrations," *Izv. Akad. Nauk SSSR, MTT*, No. 1, 1977, pp. 138–145.
57. Gunther, N. M., *A Course in Calculus of Variations*, OGIZ, GITL, Moscow and Lenin-grad, 1941.
58. Dubovitskii, A. B., and Miljutin, A. A., "Extremal Problems Subject to Constraint *Zh. Vychisl. Mat. Mat. Fiz.*, Vol. 5, No. 3, 1965, pp. 395–453.
59. Germain, P., *Mechanics of Continuous Media*, Mir, Moscow, 1965.
60. Sommefeld, A., *Mechanics of Deformable Media*, Foreign Literature Transla Moscow, 1954.
61. Ivanov, G. M., and Kosmodemjanskii, A. S., "Inverse Bending Problems for Th isotropic Plates." *Izv. Akad. Nauk SSSR, MTT*, No. 5, 1974, pp. 53–56.
62. Iegi, E. M., "An Optimal Structure and Its Design," *Proceedings of the Tallin F nical Institute*, No. 257, 1967, pp. 63–85.

63. Ishlinskii, A. Iu., *A Beam Having Cross Sections of Equal Strength*, Scientific Notes, Moscow State University, No. 39, 1940, pp. 87–90.
64. Kartvelishvili, V. M., "Numerical Solution of Two Contact Problems for Elastic Plates," *Izv. Akad. Nauk, USSR, MTT*, No. 6, 1974, pp. 68–72.
65. Keldysh, M. V., "Conformal Map of a Multiconnected Region into a Canonical Region," *Usp. Mat. Nauk*, No. 6, 1939.
66. Kiselev, V. A., *Rational Forms of Trusses and Suspension Systems*, Gosstroiizdat, Moscow, 1953.
67. Collatz, L., *Eigenvalue Problems*, Nauka, Moscow, 1968; translated from the German.
68. Collatz, L., *Functional Analysis and Computational Mathematics*, Mir, Moscow, 1969; translated from the German.
69. Komkov, V., *Optimal Control Theory for the Damping of Vibrations of Simple Elastic Systems*, Lecture Notes in Mathematics, Vol. 253, Springer-Verlag, 1972; Russian translation: Mir, Moscow, 1975 (Foreword by T. Selezov).
70. Kornishin, M. S., and Alexandrov, M. A., "Numerical Computation of Flexible Plates and Shallow Shells of Minimal Weight with Mixed Boundary Conditions," *Proceedings of the Fifth All-Union Conference on Numerical Solution Techniques in the Theory of Elasticity and Plasticity, Part 2*, Novosibirsk, *V.Ts. SO. Akad. Nauk SSSR*, 1978, pp. 76–81.
71. Krasovskii, N. N., *Motion Control Theory*, Nauka, Moscow, 1968.
72. Krasovskii, N. N., *Theory of Optimal Controlled Systems, Mechanics in USSR During the Last 50 Years*, Nauka, Moscow, Vol. 1, 1968, pp. 179–244.
73. Krein, M. G., "On Certain Maximal and Minimal Problems for Eigenvalues and on Ljapunov's Zones of Stability," *PMM*, Vol. 15, No. 3, 1951, pp. 323–348.
74. Courant, R., *Partial Differential Equations*, Mir, Moscow, 1964.
75. Courant, R., and Hilbert, D., *Methods of Mathematical Physics*, Wiley-Interscience, Vol. 1, 1953; Vol. 2, 1962.
76. Courant, R., *Dirichlet's Principle, Conformal Maps and Minimal Surfaces*, Foreign Literature Translation, Moscow, 1953.
Kurzhanskii, A. B., *Control and Observation with Uncertainty*, Nauka, Moscow, 1977.
...rshin, L. M., "On the Problem of Determining the Cross Section of a Beam Having ...ximal Torsional Rigidity," *Dokl. Akad. Nauk SSSR*, Vol. 223, No. 3, 1975, pp. 585–

...n, L. M., and Onoprjenko, P. N., "Determining the Shape of a Doubly-Connected ...ction for Beams of Maximal Torsional Rigidity," *PMM*, Vol. 40, No. 6, 1976, ... 1084.
... M. A., and Ljusternik, L. A., *A Course in Calculus of Variations*, Gostek... ...ow and Leningrad, 1950.
... A., and Shabat, B. V., *Hydrodynamic Problems and Their Mathemat...* ...uka, Moscow, 1973.
... *Variational Techniques for Solution of Problems in the Theory of Elas-* ...t, Moscow, 1943.
... ...cations of Pontryagin's Maximality Principle to Problems of Strength, ...s," ...ns in Thin-Walled Structures," in *Mechanics*, Mir, Moscow, No.

...ons of Pontryagin's Maximality Principle in Designing of Cylin- ...ions, ...asto-Plastic Material," in *Advances in Mechanics of Deform-* ...w, 1975, pp. 340–349.
...n An- ...*ic Plates*, 2nd Edition, Gordon, 1968; original Russian edi- ...957.
...olytech- ... *Anisotropic and Inhomogeneous Beams*, Nauka, Mos-

...*Systems Described by Partial Differential Equations*, ... French).

89. LITVINOV, V. G., "Certain Problems in Optimization of Plates and Shells," *Prikl. Mekh.*, Vol. 8, No. 11, 1972, pp. 33–42.
90. LITVINOV, V. G., "Certain Inverse Problems in Bending of Plates," *PMM*, Vol. 40, No. 4, 1976, pp. 682–691.
91. LOMAKIN, V. A., *Theory of Elasticity of Inhomogeneous Bodies*, Moscow State University Publications, 1976.
92. LUR'E, A. I., *On Small Deformations of Thin-Curved Beams*, Research Works of Leningrad Politechnical Institute Named after M. I. Kalinin, No. 3, 1951.
93. LUR'E, A. I., *Application of the Maximum Principle to Some Simplest Problems of Mechanics*, Research Works of Leningrad Politechnical Institute, Leningrad and Moscow, Mashinostroenie, No. 252, 1965, pp. 34–46.
94. LUR'E, A. I., *Theory of Elasticity*, Nauka, Moscow, 1970.
95. LUR'E, K. A., *Optimal Control in Problems of Mathematical Physics*, Nauka, Moscow, 1975.
96. LUR'E, K. A., and CHERKAEV, A. V., "Application of Prager's Theorem to the Problem of Optimal Design of Thin Plates," *Izv. Akad. Nauk SSSR, MTT*, No. 6, 1976, pp. 157–159.
97. MALMEISTER, A. K., TAMUZH, V. P., and TETERS, G. A., *Strength of Rigid Polymer Materials*, Zinatne, Riga, 1967.
98. MARCHENKO, V. M., *Temperature Fields and Stresses in Structures of Flying Machines*, Mashinostroenie, Moscow, 1965.
99. MIKHLIN, S. G., *Variational Methods in Mathematical Physics*, Nauka, Moscow, 1970.
100. MOISEEV, N. N., *Numerical Techniques in the Theory of Optimal Systems*, Nauka, Moscow, 1971.
101. MUSKHELISHVILI, N. I., *Some Basic Problems of the Mathematical Theory of Elasticity*, Nauka, Moscow, 1954.
102. MUSHTARI, KH. A., "On the Bending Theory for Rectangular Plates of Variable Cross Section," *Inzh. Zh.*, Vol. 4, No. 1, 1964, pp. 45–49.
103. MUSHTARI, KH. A., "Theory of Bending for Plates Having Minimal Weight Which are Made from a Composite Material," *Prikl. Mekh.*, Vol. 3, No. 4, 1967, pp. 1–7.
104. NAIMARK, M. A., *Linear Differential Operators*, Nauka, Moscow, 1969.
105. NEMIROVSKII, JU. B., "Rational Design of Reinforced Structures From the Point of View of Strength and Stability," All-Union Interuniversity Collection *Applied Problems of Strength and Plasticity*, Gorkii, Issue No. 6, 1977, pp. 70–80.
106. NIKOLAI, E. L., *Collected Works in Mechanics*, Gostekhizdat, Moscow, 1955.
107. NIKOLAI, E. L., *Lagrange's Problem on the Best Shape of a Column*, Notes of the St. Petersburg Politechnical Institute, Vol. 8, 1907.
108. OBRAZTSOV, I. F., and VASIL'EV, V. V., *Certain Problems in Computation and Design of Optimal Structures Made from Oriented Glasslike Plastics*, Research Notes of the Moscow Aviation Institute, No. 180, 1971, pp. 201–206.
109. OBRAZTSOV, I. F., VASIL'EV, V. V., and BUNAKOV, V. A., *Optimal Reinforcement of Shells of Revolution Made from Composite Materials*, Mashinostroenie, Moscow, 1977.
110. POLYA, G., and SZEGÖ, G., *Isoperimetric Inequalities in Mathematical Physics*, Fizmatgiz, Moscow, 1962; published in English by Princeton University Press, 1951.
111. PONTRYAGIN, L. S., BOLTJANSKII, V. G., GAMKRELIDZE, R. V., and MISHCHENKO, E. F., *Mathematical Theory of Optimal Processes*, 3rd edition, Nauka, Moscow, 1969; English translation, Wiley-Interscience, New York, 1962.
112. PRAGER, W., *Introduction to Structural Optimization*, Springer-Verlag, 1974.
113. RABINOVICH, I. M., *On the Theory of Statically Indeterminate Trusses*, Transzheldor-Izdat, Moscow, 1933.
114. RABINOVICH, I. M., "Systems of Beams of Minimal Weight," in *Proceedings of the Second All-Union Conference on Theoretical and Applied Mechanics: Expository Lectures*, Nauka, Moscow, 3rd edition, 1966, pp. 265–275.
115. RADTSIG, IU. A., *Statically Indeterminate Trusses of Least Weight*, Kazan University, Kazan, 1969.

116. RZHANITSYN, A. R., *Computation of Reinforcements Taking into Account Plastic Properties of the Materials*, Stroivoenmorizdat, 1949.
117. RZHANITSYN, A. R., *Stability of the Equilibrium of Elastic Systems*, Gostekhizdat, Moscow, 1955.
118. SAVIN, G. N., *Stress Distribution in the Vicinity of Holes*, Nauk Dumka, Kiev, 1968.
119. SAVIN, G. N., *Stress Concentration in the Vicinity of Holes*, Gostekhteoretizdat, Moscow, Lenningrad, 1951.
120. SAMSONOV, A. M., "The Optimal Location of a Thin Rib for an Elastic Plate," *Izv. Akad. Nauk, USSR, MTT*, No. 1, 1978, pp. 132–138.
121. SEIRANJAN, A. P., "Elastic Plates and Beams of Least Weight in the Presence of Several Types of Bending Loads," *Izv. Akad. Nauk SSSR, MTT*, No. 5, 1973, pp. 95–101.
122. SEIRANJAN, A. P., "Optimal Design of Beams with Constraints on the Bending Moment," *Izv. Akad. Nauk Arm. SSR, Mechanics*, No. 6, 1975, pp. 24–33.
123. SEIRANJAN, A. P., "Optimal Design of Beams with Constraints on the Natural Frequency of Vibration and on the Magnitude of the Force Causing the Loss of Stability," *Izv. Akad. Nauk SSSR, MTT*, No. 1, 1976, pp. 147–152.
124. SEIRANJAN, A. P., "Quasi-Optimal Solutions for the Problem of Optimal Design with Various Constraints," *Prikl. Mekh.*, No. 6, 1977, pp. 18–26.
125. SEIRANJAN, A. P., "Study of the Extremum in the Optimality Problem for Vibrations of a Circular Plate," *Izv. Akad. Nauk SSSR, MTT*, No. 6, 1978, pp. 113–118.
126. SAINT VENANT, B., *Memoir on the Torsion of Prismatic Beams, Memoir on the Bending of Prismatic Beams*, Russian translation: Fizmatgiz, Moscow, 1961.
127. SMIRNOV, V. I., *A Course of Higher Mathematics*, Vol. 4, Part 1, Pergamon Press, New York, 1974 (translated from Russian).
128. TIMOSHENKO, S. P., *Computation of Elastic Trusses*, Gostekhizdat, Moscow, 1933.
129. TIMOSHENKO, S. P., *The History of the Strength of Materials*, Gostekhizdat, Moscow, 1957.
130. TIMOSHENKO, S. P., and WOINOWSKI-KRIEGER, S., *The Theory of Plates and Shells*, 2nd edition, McGraw-Hill, 1959.
131. TIMOSHENKO, S. P., and GOODIER, D., *Theory of Elasticity*, McGraw-Hill, New York, 1954.
132. TIMOSHENKO, S. P., *Stability and Vibrations of Structural Elements*, Nauka, Moscow, 1975.
133. TROITSKII, V. A., "Optimization of Freely Vibrating Elastic Beams," *Izv. Akad. Nauk SSSR, MTT*, No. 3, 1976, pp. 145–152.
134. TROITSKII, V. A., *Optimal Vibration Phenomena in Mechanical Systems*, Mashinostroenie, Leningrad, 1976.
135. UKRAINTSEV, G. V., and FROLOV, V. M., "A Method of Strength Optimization with Respect to Rigidity for a Wing with Variable Distribution of the Thickness of the Cross Section," *Uchen. Zap. Ts. AGI.*, Vol. 2, No. 4, 1972.
136. UMANSKII, A. A., *Structural Mechanics of the Aeroplane*, Oborongiz, Moscow, 1961.
137. FEODOS'EV, V. I., *Selected Problems in Strength of Materials*, Nauka, Moscow, 1973.
138. FILIN, A. P., and GUREVICH, A. I., *Applications of the Calculus of Variations to the Finding of a Rational Shape for a Structure*, research notes of the Leningrad Institute of Railway and Transport Engineering, Edition No. 190, 1962, pp. 161–187.
139. FILIN, A. P., SOLOMESHCH, M. A., and GOL'DSHTEIN, JU. B., "Classical Calculus of Variations and the Problem of Optimizing Systems of Elastic Beams," in *Research on the Theory of Reinforcement*, Stroiizdat, Moscow, Vol. 19, 1972, pp. 156–163.
140. HILL, R., *Mathematical Theory of Plasticity*, Gostekhteoretizdat, Moscow, 1956 (translated from English).
141. HILL, R., *Elastic Properties of Composite Media; Some Theoretical Principles—Mechanics*, Translation of Periodicals and Foreign Collections of Articles, No. 5, 1964, pp. 127–143.
142. CHENTSOV, N. G., "Columns of Least Weight," Research Articles, *TsAGI*, No. 265, 1936, pp. 1–48.
143. CHEREPANOV, G. P., "Certain Problems in the Theory of Elasticity and Plasticity with

Unknown Boundaries," in *Applications of the Theory of Functions to Mechanics of Continua*, Nauka, Moscow, Vol. 1, 1965.

144. CHEREPANOV, G. P., "Inverse Problems in the Two-Dimensional Theory of Elasticity," *Prikl. Matem. i Mekh.*, Vol. 38, No. 6, 1974, pp. 963–979.
145. CHERNOUS'KO, F. L., "Technique of Local Perturbations for Numerical Solution of Variational Problems," *Zh. Vychisl. Mat. Mat. Fiz.*, Vol. 5, No. 4, 1965, pp. 749–754.
146. CHERNOUS'KO, F. L., "Certain Problems of Optimal Control With a Small Parameter," *Prilk. Mat. Mekh.*, Vol. 32, No. 1, 1968, pp. 15–26.
147. CHERNOUS'KO, F. L., "Certain Optimal Shapes of Bifurcating Beams," *Izv. Akad. Nuak SSSR, MTT*, No. 3. 1979.
148. CHERNOUS'KO, F. L., and BANICHUK, N. V., *Variational Problems of Mechanics and Control Theory*, Nauka, Moscow, 1973.
149. CHERNOUS'KO, F. L., and KOLMANOVSKII, V. B., "Computational and Approximate Techniques in Optimal Control," in *Advances in Science and Technology, Mathematical Analysis*, VINITI, Moscow, Vol. 14, 1977.
150. CHERNOUS'KO, F. L., and MELIKJAN, A. A., *Game-Theoretic Problems of Control and Search*, Nauka, Moscow, 1978.
151. CHIRAS, A. A., *Optimization Theory in the Limit Analysis of a Deformable Solid Body*, Mintis, Vilnius, 1971.
152. SHATROVSKII, L. I., "A Computational Technique for Solving Problems of Optimal Control," *Zh. Vychisl. Mat. Mat. Fiz.*, Vol. 2, No. 3, 1962, pp. 488–491.
153. SHERMAN, D. I., "On the Stress Distribution in Partitions, An Elastic Heavy Medium Which is Weakened by Elliptic Holes," *Izv. Akad. Nauk SSSR, OTN*, No. 7, 1952, pp. 992–1010.
154. SHERMAN, D. I., "Concerning a Technique of Solution of Certain Problems of Bending and Torsion in the Two-Dimensional Theory of Elasticity for Multiply-Connected Domains," *Prikl. Mekh.*, Vol. 3, No. 4, 1957, pp. 363–377.
155. SHERMAN, D. I., "A Certain Viewpoint Concerning Boundary Value Problems in the Theory of Functions and Two Dimensional Problems in the Theory of Elasticity," in *Mechanics of Continuous Media and Our Country's Problems in Analysis*, Nauka, Moscow, 1972, pp. 635–665.
156. SHERMERGOR, T. D., *Theory of Elasticity of Micro-Inhomogeneous Media*, Nauka, Moscow, 1977.
157. ENEEV, T. M., "Applications of the Gradient Technique in Problems of Optimal Control Theory," *Cosmic Res.*, Vol. 4, No. 5, 1966.
158. ARMAND, J.-L., "Minimum-Mass Design of a Platelike Structure for Specified Fundamental Frequency," *AIAA J*, 1971, Vol. 9, No. 9, pp. 1739–1745.
159. ARMAND, J.-L., "Applications of Optimal Control Theory to Structural Optimization: Analytical and Numerical Approach," in *Proceedings of IUTAM Symposium on Optimization in Structural Design, Warsaw, 1973*, Springer-Verlag, Berlin, 1975, pp. 15–39.
160. ASHLEY, H., and MCINTOSH, S. C., JR., "Applications of Aeroclastic Constraints in Structural Optimization," in *Proceedings of the 12th International Congress of Applied Mechanics, Stanford University, 1968*, Springer-Verlag, Berlin, 1969, pp. 100–113.
161. BANICHUK, N. V., "Game Problems in the Theory of Optimal Design," in *Proceedings of IUTAM Symposium on Optimization in Structural Design, Warsaw, 1973*, Springer-Verlag, Berlin, 1975, pp. 111–121.
162. BANICHUK, N. V., "Optimization in Elastic Bars in Torsion," *Int. J. Solids Struct.*, 1976, Vol. 12, No. 4, pp. 275–286.
163. BANICHUK, N. V., "Minimax Approach to the Structural Optimization Problems," *J. Optimiz. Theory and Appl.*, 1976, Vol. 20, No. 1, pp. 111–127.
164. BANICHUK, N. V., and KARIHALOO, B. L., "Minimum-Weight Design of Multipurpose Cylindrical Bars," *Int. J. Solids Struct.*, 1976, Vol. 12, No. 4, pp. 267–273.
165. BANICHUK, N. V., and KARIHALOO, B. L., "On the Solution of Optimization Problems with Singularities," *Int. J. Solids Struct.*, 1977, Vol. 13 No. 8, pp. 725–733.

166. BARNETT, R. L., "Survey of Optimum Structural Design," *Exp. Mech.*, 1966, Vol. 6, No. 12, pp. 19–26.
167. BARNETT, R. L., "Minimum-Weight Design of Beams for Deflection," *Proc. ASCE J. Engng. Mech. Div.*, 1961, Vol. 87, No. 1, pp. 75–109.
168. BARNETT, R. L., Minimum Deflection Design of a Uniformly Accelerating Cantilever Beam," *J. Appl. Mech. Trans. ASME*, 1963, Vol. 30, No. 3, pp. 466–467.
169. BLASIUS, H., "Minimum Deflection Supports and Bars of Maximum Buckling Resistance with Specified Materials Usage," *Z. Math. Phys.*, Vol. 62, 1914, pp. 182–197.
170. BUDIANSKY, B., FRAUENTHAL, J. C., and HUTCHINSON, J. W., "On Optimal Arches, *J. Appl. Mech. Trans. ASME*, 1969, Vol. 36, No. 4, pp. 239–240.
171. CLAUSEN, T., "About the Shape of Architectural Columns," *Bull. Phys. Math. Acad. St. Peterbourg, T. 9*, 1851, pp. 279–294.
172. FRAUENTHAL, J. C., "Constrained Optimal Design of Circular Plates Against Buckling," *J. Struct. Mech.*, Vol. 1, 1972, pp. 159–186.
173. GALILEI, G., "Mathematical Discourses and Demonstrations," (Illustrations), Leiden, 1638.
174. GURVITCH, E. L., "On Isoperimetric Problems for Domains with Partly Known Boundaries," *J. Optimiz. Theory and Appl.*, 1976, Vol. 20, No. 1, pp. 65–79.
175. HASHIN, Z., "Theory of Mechanical Behaviour of Heterogeneous Media," *Appl. Mech. Rev.*, 1964, Vol. 17, No. 1.
176. HASHIN, Z., "On Elastic Behaviour of Fibre-Reinforced Materials of Arbitrary Transverse Phase Geometry," *J. Mech. and Phys. Solids*, 1965, Vol. 13, No. 3.
177. HASHIN, Z., and ROSEN, B. W., "The Elastic Moduli of Fiber-Reinforced Materials," *J. Appl. Mech.*, 1964, Vol. 31, No. 2.
178. HEGEMIER, G. A., and TANG, H. G., "A Variational Principle, the Finite Element Method, and Optimal Structural Design for Given Deflection," in *Proceedings of IUTAM Symposium on Optimization in Structural Design, Warsaw, 1973*, Springer-Verlag, Berlin, 1975, pp. 464–483.
179. HOFF, N. J., "Approximate Analysis of the Reduction in Torsional Rigidity and of the Torsional Buckling of Solid Wings under Thermal Stresses," *J. Aeronaut. Sci.*, 1956, Vol. 23, No. 6, pp. 603–604.
180. HUANG, N. C., "Optimal Design of Elastic Beams for Minimum–Maximum Deflection, *J. Appl. Mech.*, 1971, No. 4.
181. HUTCHINSON, J. W., and NIORDSON, R. I., "Designing Vibrating Membranes," in *Mechanics of Continuous Media and Problems of Analysis in this Country*, Nauka, Moscow, 1972, pp. 581–590.
182. KARIHALOO, B. L., and NIORDSON, F. I., "Optimum Design of Vibrating Cantilevers," *J. Optimiz. Theory and Appl.*, 1973, Vol. 11, No. 6, pp. 638–654.
183. KARIHALOO, B. L., and NIORDSON, F. I., "Optimum Design of Circular Shaft in Forward Precession," in *Proceedings of IUTAM Symposium on Optimization in Structural Design, Warsaw, 1973*, Springer-Verlag, Berlin, 1975, pp. 142–151.
184. KELLER, J. B., "The Shape of the Strongest Column," *Arch. Rational Mech. and Anal.*, 1960, Vol. 5, No. 4, pp. 275–285.
185. KELLER, J. B., and NIORDSON, F. I., "The Tallest Column," *J. Math. and Mech.*, 1966. Vol. 16, No. 5, pp. 433–446.
186. KLOSOWICZ, B., "On the Optimal Nonhomogeneity of a Torqued Bar," *Bull. Acad. Polon. Sci. Ser. Sci. Techn.*, Vol. 18, No. 8, 1970, pp. 611–615.
187. KLOSOWICZ, B., and LUR'E, K. A., "On the Optimal Nonhomogeneity of Torsional Elastic Bar," *Arch. Mech., Warszawa, 1971*, Vol. 24, No. 2, pp. 239–249.
188. LAGRANGE, J. L., "On the Shape of the Columns," *Miscellanea Taurinensia, 1770–1773*, p. 5.
189. LEPIK, U., "Application of Pontryagin's Maximum Principle for Minimum-Weight Design of Rigid-Plastic Circular Plates," *Int. J. Solids and Struct.*, 1973, Vol. 9, pp. 615–624.

190. Lepik, U., "Minimum-Weight Design of Circular Plates with Limited Thickness," *Int. J. Nonlinear Mech.*, 1972, Vol. 7, No. 4, pp. 353–360.
191. Lepik, U., "Optimal Design of Beams with Minimum Compliance," *Int. J. Nonlinear Mech.*, 1978, Vol. 13, pp. 33–42.
192. Majerczyk-Gomulkowa, J., and Mioduchowski, A., "Optimal Plastic Inhomogeneity for a Twisted Rod Concerning the Limiting Load Capacity," *Rozpr. Inz.*, 1969.
193. Mansfield, E. H., "Optimum Tapers of Eccentrically Loaded Ties," *J. Roy. Aeronaut. Soc.*, 1967, Vol. 71, No. 681, pp. 647–650.
194. Martin, J. B., "Optimal Design of Elastic Structures for Multi-Purpose Loading," *J. Optimiz. Theory and Appl.*, 1970, Vol. 6, No. 1, pp. 22–40.
195. Masur, E. F., "Optimum Stiffness and Strength of Elastic Structures," *ASCE J. Engr. Mech. Div.*, 1970, Vol. 96, No. 5, pp. 621–640.
196. Masur, E. F., "Optimality in the Presence of Discreteness and Discontinuity," in *Proceedings of IUTAM Symposium on Optimization in Structural Design, Warsaw, 1973*, Springer-Verlag, Berlin, 1975, pp. 441–453.
197. Maxwell, C., "Scientific Paper," *Cambridge University Press*, 1980, Vol. 2, pp. 175–177.
198. McIntosh, S. C., and Eastep, F. E., "Design of Minimum-Mass Structures with Specified Stiffness Properties," *AIAA Journal*, 1968, Vol. 6, pp. 962–964.
199. Michell, A. G. M., "The Limits of Economy of Material in Framestructures," *Phil. Mag.*, 1904, Vol. 8, No. 47.
200. Michell, A. J. M., and Melbourne, M. C. S., "The Limits of Economy of Material," *Phil. Mag. Ser. 6*, 1904, Vol. 8, pp. 589–597.
201. Miele, A., "Drag Minimization as the Extremization of Products of Powers of Integrals," in *Problems of Hydrodynamics and Mechanics of Continuous Media*, Nauka, Moscow, 1969.
202. Mroz, Z. "Optimal Design of Elastic Structures Subjected to Dynamic Harmonically-Varying Loads," *ZAMM*, 1970, Vol. 50, No. 5, pp. 303–309.
203. Mroz, Z., and Taylor, J. E., "Pre-Stress for Maximum Strength," *Int. J. Solids and Struct.*, 1973, Vol. 9, No. 12, pp. 1535–1541.
204. Mroz, Z., and Shamiev, F. G., "On Optimal Design of Reinforced Annular Slabs," *Arch. Inz. Lad.*, 1970, Vol. 16, No. 4, pp. 575–584.
205. Neuber, H., "The Flat Bar Under Extensional Stress, with Optimum Cross Section Variation," *Forsch. Ingenieurwes.*, Vol. 35, 1969, pp. 29–30.
206. Neuber, H., "To the Optimization of the Stress," in *Mechanics of Continuous Media and Related Problems Analyses*, Nauka, Moscow, 1972, pp. 375–380.
207. Niordson, F. I., "On the Optimal Design of a Vibrating Beam," *Quart. Appl. Math.*, 1965, Vol. 23, No. 1, pp. 47–53.
208. Niordson, F. I., and Pedersen, P., "A Review of Optimal Structural Design," *Proc. 13th Internat. Congr. Theoret. and Appl. Mech. Moscow University, 1972*, Springer-Verlag, Berlin, 1973, pp. 264–278.
209. Olhoff, N., "Optimal Design of Vibrating Circular Plates," *Int. J. Solids and Struct.*, 1970, Vol. 6, No. 1, pp. 139–156.
210. Olhoff, N., "Optimal Design of Vibrating Rectangular Plates," *Int. J. Solids and Struct.*, 1974, Vol. 10, No. 1, pp. 93–109.
211. Olhoff, N., "On Singularities, Local Optima and Formation of Stiffeners in Optimal Design of Plates," in *Proceedings of IUTAM Symposium on Optimization in Structural Design, Warsaw 1973*, Springer-Verlag, Berlin, 1975, pp. 82–103.
212. Olhoff, N., and Rusmussen, S. H., "On Single and Bimodal Optimum Buckling Loads of Clamped Columns," *Int. J. Solids and Struct*, 1977, Vol. 13, pp. 605–614.
213. Onat, E. T., Shumann, W., and Shield, R. T., "Design of Circular Plates for Minimum Weight," *ZAMP*, 1957, Vol. 8, No. 6, pp. 485–499.
214. Pedersen, P., "Optimal Joint Positions for Space Trusses," *J. Struct. Div. Proc. Amer. Soc. Civ. Eng.*, 1973, Vol. 99, No. ST12, pp. 2459–2476.

215. Pierson, B. L., "A Survey of Optimal Structural Design Under Dynamic Constraints," *Int. J. Numer. Meth. Eng.*, 1972, Vol. 4, pp. 491–499.
216. Pierson, B. L., "An Optimal Control Approach to Minimum-Weight Vibrating Beam Design," *J. Struct. Mech.*, 1977, Vol. 5, pp. 147–148.
217. Polya, G., "Is the Largest Strain Located on the Surface?" *Z. Angew. Math. Mech.*, Vol. 10, No. 4, 1930, pp. 353–360.
218. Polya, G., "Torsional Rigidity, Principal Frequency, Electrostatic Capacity and Symmetrization," *Quart. Appl. Math.*, 1948, Vol. 6, No. 3.
219. Prager, W., "Optimality Criteria in Structural Design," *Proc. Nat. Acad. Sci. USA*, 1968, Vol. 61, No. 3, pp. 794–796.
220. Prager, W., "Optimization of Structural Design," *J. Optimiz. Theory and Appl.*, 1970, Vol. 6, No. 1, pp. 1–21.
221. Prager, W., "Optimal Thermoelastic Design for Given Deflection," *Int. J. Mech. Sci.*, 1970, Vol. 12, pp. 705–709.
222. Prager, W., and Shield, R. T., "Optimal Design of Multipurpose Structures," *Int. J. Solids and Struct.*, 1968, Vol. 4, No. 4, pp. 469–475.
223. Prager, W., and Taylor, J. E., "Problems of Optimal Structural Design," *J. Appl. Mech. Trans. ASME*, 1968, Vol. 35, No. 1, pp. 102–106.
224. Prandtl, L., "Examples of Application of a Hencky's Law to Plastic Equilibrium," *Z. Angew. Math. Mech.*, Vol. 3, No. 6, 1923.
225. Save, M. A., "Some Aspects of Minimum-Weight Design," in *Engineering Plasticity* (J. Heyman, F. A. Leckie, Ed.), Cambridge University Press, 1968, pp. 611–626.
226. Save, M., and Prager, W., "Minimum-Weight Design of Beam Subjected to Fixed and Moving Loads," *J. Mech. and Phys. Solids*, 1963, Vol. 11, No. 4, pp. 255–267.
227. Seyranian, A. P., "Homogeneous Functionals and Structural Optimization Problems," *Int. J. Solids and Struct.*, 1979, Vol. 15, No. 4.
228. Sheu, C. Y., "Elastic Minimum-Weight Design for Specified Fundamental Frequency," *Int. J. Solids and Struct.*, 1968, Vol. 4, No. 10, pp. 953–958.
229. Sheu, C. Y., and Prager, W., "Recent Developments in Optimal Structural Design," *Appl. Mech. Rev.*, 1968, Vol. 21, No. 10, pp. 985–992.
230. Shield, R. T., "Optimum Design Methods for Multiple Loading," *ZAMP*, 1963, Vol. 14, pp. 38–45.
231. Shield, R. T., "Optimum Design of Structures through Variational Principles," *Lect. Notes Phys.*, 1973, Bd. 21, No. 1.
232. Shield, R. T., and Prager, W., "Optimal Structural Design for Given Deflection," *ZAMP*, 1970, Vol. 21, No. 2.
233. Tadjbaksh, I., and Keller, J. B., "Strongest Columns and Isoperimetric Inequalities for Eigenvalues," *J. Appl. Mech.*, 1962, Vol. 29, No. 1, pp. 159–164.
234. Taylor, J. E., "Minimum Mass Bar for Axial Vibration at Specified Natural Frequency," *AIAA Journal*, 1967, Vol. 5, No. 10, pp. 1911–1913.
235. Taylor, J. E., "The Stronges Column, an Energy Approach," *J. Appl. Mech. Trans. ASME*, 1967, Vol. 34, No. 2, pp. 486–487.
236. Taylor, J. E., "Optimum Design of a Vibrating Bar with Specified Minimum Cross Section," *AIAA Journal*, 1968, Vol. 6, pp. 1379–1381.
237. Turner, M. J., "Design of Minimum Mass Structures with Specified Natural Frequencies," *AIAA Journal*, 1966, Vol. 5, No. 3, pp. 406–412.
238. Tvergaard, V., "On the Optimum Shape of a Fillet in a Flat Bar with Restrictions," in *Proceedings of IUTAM Symposium on Optimization in Structural Design, Warsaw 1973*, Springer-Verlag, Berlin, 1975, pp. 181–195.
239. Valentine, F. A., "The Problem of Lagrange with Differential Inequalities as Added Side Conditions," *The Calculus of Variations, 1933–1937*, University of Chicago Press, Chicago, 1937, pp. 433–447.

240. Wasiutynski, Z., "On the Equivalence of Design Principles: Minimum Potential-Constant Volume and Minimum Volume—Constant Potential," *Bull. Acad. Pol. Sci. Ser. Sci. Tech.*, 1966, Vol. 14, No. 9, pp. 883–885.
241. Wasiutynski, Z. "On the Criterion of Minimum Deformability Design of Elastic Structures; Effect of Own Weight of Material," *Bull. Acad. Pol. Sci. Ser. Sci. Tech.*, 1966, Vol. 14, No. 9, pp. 875–878.
242. Wasiutynski, Z., "On the Congruency of the Forming According to the Minimum Potential Energy with that According to the Equal Strength," *Bull. Acad. Pol. Sci. Ser. Sci. Tech.* 1960, Vol. 8, No. 6, pp. 259–268.
243. Wasiutynski, Z., and Brandt, A. "The Present State of Knowledge in the Field Optimum Design of Structures," *Appl. Mech. Revs*, 1963, Vol. 16, No. 5, pp. 341–350.
244. Weisshaar, T. A., "Optimization of Simple Structures with Highermode Frequency Constraints," *AIAA Journal*, 1972, Vol. 10, pp. 691–693.
245. Wheeler, L., "On the Role of Constant-Stress Surfaces in the Problem of Minimizing Elastic Stress Concentration," *Int. J. Solids and Struct.*, 1976, Vol. 12, No. 11, pp. 779–789.
246. Wu, C. H., "The Strongest Circular Arch-A Perturbation Solution," *J. Appl. Mech. Trans. ASME*, 1968, Vol. 35, No. 3, pp. 476–480.

Addendum to Bibliography

A recent review of distributed parameter structural optimization literature by E. J. Haug[342] focuses its attention on the following aspects of structural optimization:

1. Buckling of columns and beams;
2. Vibration of bars, beams, and plates;
3. Deflection and compliance of structures;
4. Dynamic response;
5. Shape of two- and three-dimensional elements;
6. Multipurpose structures and special problems.

We quote below articles given in this literature review that were not included in the bibliography given by the author. We also append some other relevant articles.

247. Adali, S., "Optimal Shape And Nonhomogeneity of a Nonuniformly Compressed Column," *Int. J. Solids Struct.*, Vol. 15, 1979, pp. 935–949.
248. Ainola, L. I., "On the Inverse Problem of Natural Vibrations of Elastic Shells," *PMM*, Vol. 35, No. 2, 1971, pp. 358–364.
249. Alblas, J. B., "Optimal Strength of a Compound Column," *Int. J. Solids Struct.*, Vol. 13, 1977, pp. 307–320.
250. Andreev, L. V., Mossakovsii, V. I., and Obodan, N. I., "On Optimal Thickness of a Cylindrical Shell Loaded by External Pressure," *PMM*, Vol. 36, No. 4, 1972, pp. 677–685.
251. Aristov, M. V., and Troitskii, V. A., "Elastic Circular Plate of Minimum Weight," *MTT*, No. 3, 1975, pp. 153–156.

252. Armand, J. L., "Numerical Solutions in Optimization of Structural Elements," Paper presented at *First International Conference on Computational Methods in Nonlinear Mechanics*, Austin, Texas, 1974.
253. Armand, J. L., and Lodier, B., "Optimal Design of Bending Elements," *Int. J. Num. Methods Eng.*, Vol. 13, 1978, pp. 373–384.
254. Armand, J. L., and Vitte, W. J., *Foundations of Aeroelastic Optimization and Some Applications to Continuous Systems*, Report No. SUDAAR-390, Department of Aeronautics and Astronautics, Stanford University, 1970.
255. Arora, J. S., and Haug, E. J., "Optimum Structural Design with Dynamic Constraints," *ASCE J. Struct. Div.*, Vol. 103, No. ST10, 1977, pp. 2071–2074.
256. Banichuk, N. V., "Optimality Conditions and Analytical Methods of Shape Optimization," *Optimization of Distributed Parameter Structures* (E. J. Haug and J. Cea, Ed.), Sijthoff & Noordhoff, Alphen aan den Rijn, Netherlands, 1981, pp. 973–1004.
257. Banichuk, N. V., "Design of Plates for Minimum Deflection and Stress," *Optimization of Distributed Parameter Structures* (E. J. Haug and J. Cea, Ed.), Sijthoff & Noordhoff, Alphen aan den Rijn, Netherlands, 1981, pp. 333–361.
258. Barnes, E. R., "The Shape of the Strongest Column and Some Related Extremal Eigenvalue Problems," *Q. Appl. Math.*, Vol. 35, 1977, pp. 393–409.
259. Batterman, S. C., and Pavicic, N., "Optimum Design of Fiber-Reinforced Shell of Revolution," *Optimization in Structural Design*, Springer-Verlag, New York, 1975.
260. Benedict, R. L., "Optimal Design for Elastic Bodies in Contact," *Optimization of Distributed Parameter Structures* (E. J. Haug and J. Cea, Ed.), Sijthoff & Noordhoff, Alphen aan den Rijn, Netherlands, 1981, pp. 1553–1599.
261. Bert, C. W., "Optimal Design of A Composite Material Plate to Maximize Its Fundamental Frequency," *J. Sound Vib.*, Vol. 50, 1977, pp. 229–237.
262. Bhargava, S., and Duffin, R. J., "Dual Extermum Principles Relating to Optimum Beam Design," *Arch. Rat. Mech. Anal.*, Vol. 50, 1973, pp. 314–330.
263. Bhatti, M. A., *Optimal Design of Localized Nonlinear Systems with Dual Performance Criteria Under Earthquake Excitations*, Report No. EERC 79-15, University of California, Berkeley, July, 1979.
264. Bhatti, M. A., Essebo, T., Nye, W., Pister, K. S., Polak, E., Sangiovanni-Vincentelli, X., and Titz, A. "A Software System for Optimization-Based Interactive Computer-Aided Design," *Optimization of Distributed Parameter Structures* (E. J. Haug and J. Cea, Ed.), Sijthoff & Noordhoff, Alphen aan den Rijn, Netherlands, 1981, pp. 602–620.
265. Bhatti, M. A., Pister, K. S., and Polak, E., *Optimal Design of an Earthquake Isolation System*, Report No. EERC 78-22, University of California, Berkeley, October, 1978.
266. Bhatti, M. A., and Pister, K. S., "Application of Optimal Design to Structures Subject to Earthquake Loading," *Optimization of Distributed Parameter Structures*, (E. J. Haug and J. Cea, Ed.), Sijthoff & Noordhoff, Alphen aan den Rijn, Netherlands, 1981, pp. 620–649.
267. Bhatti, M. A., Polak, E., and Pister, K. S., *OPTDYN—A General Purpose Optimization Program for Problems with or without Dynamic Constraints*, Report No. EERC 79-16, University of California Berkeley, July, 1979.
268. Bhavikatti, S. S., and Ramakrishnan, C. V., "Optimum Design of Fillets in Flat and Round Tensions Bars," *ASME Paper, 77-DET-45*, 1977.
269. Bjorkman, G. S., and Richards, R., "Harmonic Holes—An Inverse Problem in Elasticity," *J. Appl. Mech.*, Vol. 43, 1976, pp. 414–418.
270. Blachut, J., and Gajewski, A., "A Unified Approach to Optimal Design of Columns," *Solid Mech. Arch.*, December 1980.
271. Blachut, J., and Gajewski, A., "On Unimodal and Bimodal Optimal Design of Funicular Arches," *Int. J. Solids Struct.*, to appear, 1981.
272. Brach, R. M., "On the Extremal Fundamental Frequencies of Vibrating Beams," *Int. J. Solid Struct.*, Vol. 4, 1968, pp. 667–674.

273. Brach, R. M., "Minimum Dynamic Response for a Class of Simply Supported Beam Shapes," *Int. J. Mech. Sci.*, Vol. 10, 1968, pp. 429–439.
274. Brach, R. M., "Optimum Design of Beams for Sudden Loading," *J. Eng. Mech. Div., ASCE*, Vol. 96, No. EM6, 1968, pp. 1395–1407.
275. Brach, R. M., "Optimized Design: Characteristic Vibration Shapes and Resonators," *J. Acoust. Soc. Am.*, Vol. 53, No. 1, 1973, pp. 113–119.
276. Budiansky, B., Frauenthal, J. C. and Hutchinson, J. W., "On Optimal Arches," *J. Appl. Mech.*, Vol. 36, 1969, pp. 880–882.
277. Cantu, E., and Cinquini, C., "Iterative Solutions for Problems of Optimal Elastic Design," *Comput. Meth. Appl. Mech. Eng.*, Vol. 20, 1979, pp. 257–266.
278. Cardou, A., and Warner, W. H., "Minimum-Mass Design of Sandwich Structures with Frequency and Section Constraints," *J. Optimiz. Theory Appl.*, Vol. 14, No. 6, 1974, pp. 633–647.
279. Carmichael D., "Singular Optimal Control Problems In the Design of Vibrating Structures," *J. Sound Vib.*, Vol. 53, 1977, pp. 245–253.
280. Carmichael, D., "Optimal Control in the Design of Material Continua," *Arch. Mech.*, Vol. 30, 1978, pp. 743–755.
281. Carmichael, D. G., "Computation of Pareto Optima In Structural Design," *Int. J. Num. Meth. Eng.*, Vol. 15, 1980, pp. 925–952.
282. Carmichael, D., and Goh, B. S., "Optimal Vibrating Plates and a Distributed Parameter Singular Control Problem," *Int. J. Control*, Vol. 26, 1977, pp. 19–31.
283. Cassis, J. H., and Schmit, L. A., "Optimum Design with Dynamic Constraints," *J. Struct. Div., ASCE*, Vol. 102, No. ST10, 1976, pp. 2053–2071.
284. Cea, J., *A Numerical Method for Search for an Optimum Domain*, IMAN de Universite de Nice, 1976.
285. Cea, J., *Optimization—Theory and Algorithms*, Tata Institute, Bombay, 1978.
286. Cea, J., "Problems of Shape Optimal Design," *Optimization of Distributed Parameter Structures* (E. J. Haug and J. Cea, Ed.), Sijthoff & Noordhoff, Alphen aan den Rijn, Netherlands, 1981, pp. 1005–1048.
287. Cea, J., "Numerical Methods for Shape Optimal Design," *Optimization of Distributed Parameter Structures* (E. J. Haug and J. Cea, Ed.), Sijthoff & Noordhoff, Alphen aan den Rijn, Netherlands, 1981, pp. 1049–1087.
288. Cea, J., Gioan, A., and Michel, J., *Some Results on Identification of Domains*, Calcolo, 1973.
289. Cheng, F. Y., and Botkin, M. E., "Nonlinear Optimum Design of Dynamic Damped Frames," *J. Struct. Div., ASCE*, Vol. 102, No. ST3, 1976, pp. 609–628.
290. Cheng, F. Y., and Srifuengfung, D., "Earthquake Structural Design Based on Optimality Criterion," *Sixth World Conference on Earthquake Engineering*, Vol. 5, Earthquake-Resistant Design, 1977.
291. Cheng, F. Y., and Srifuengfung, D., "Optimal Structural Design for Simultaneous Multicomponent Static and Dynamic Input," *Int. J. Num. Meth. Eng.*, Vol. 13, 1978, pp. 353–371.
292. Cheng, K.-T., and Olhoff, N., *An Investigation Concerning Optimal Design of Solid Elastic Plates*, DCAMM-Report No. 174, The Danish Center for Applied Mathematics and Mechanics, 1980.
293. Cherkaev, A. V., "On the Question of Formulating the Problem of Optimal Design of Freely Oscillating Structures," *PMM*, Vol. 42, No. 1, 1978, pp. 194–197.
294. Chern, J. M., "Optimal Thermoelastic Design for Given Deformation," *J. Appl. Mech.*, Vol. 38, No. 2, 1971, pp. 538–540.
295. Chern, J. M., "Optimal Structural Design for Given Deflection in Presence of Body Forces," *Int. J. Solids Struct.*, Vol. 7, 1971, pp. 363–382.
296. Chern, J. M., "Optimal Design of Beams for Alternative Loads and Constraints on Generalized Compliance and Stiffness," *Int. J. Mech. Sci.*, Vol. 13, No. 8, 1971, pp. 661–674.

297. Chern, J. M., and Martin, J. B., "The Multipurpose Optimal Design of Elastic Structure with Piecewise Uniform Cross Section," *Z. Angew. Math. Phys.*, Vol. 22, No. 5, 1971, pp. 834–855.
298. Chern, J. M., and Prager, W., "Optimal Design for Prescribed Compliance Under Alternative Loads," *J. Optimiz. Theory Appl.*, Vol. 5, 1970, pp. 424–431.
299. Chern, J. M., and Prager, W., "Optimal Design of Rotating Disk for Given Radial Displacement of Edge," *J. Optimiz. Theory Appl.*, Vol. 6, No. 2, 1970, pp. 161–170.
300. Choi, K. K., and Haug, E. J., "Optimization of Structures with Repeated Eigenvalues," *Optimization of Distributed Parameter Structures* (E. J. Haug and J. Cea, Ed.), Sijthoff & Nordhoff, Alphen aan den Rijn, Netherlands, 1981, pp. 219–277.
301. Chun, Y. W., and Haug, E. J., "Two-Dimensional Shape Optimal Design," *Int. J. Num. Meth. Eng.*, Vol. 13, 1978, pp. 311–336.
302. Chun, Y. W., and Haug, E. J., "Shape Optimal Design of an Elastic Body of Revolution," Preprint No. 3526. *ASCE Annual Meeting*, Boston, April 1979.
303. Cinquini, C., "Optimal Elastic Design for Prescribed Maximum Deflection," *J. Struct. Mech.*, Vol. 7, No. 1, 1979, pp. 21–34.
304. Conry, T. F., and Seireg, A., "A Mathematical Programming Method for Design of Elastic Bodies in Contact," *J. Appl. Mech.*, Vol. 48, 1971, pp. 387–392.
305. Dafalias, Y. F., and Dupuis, G., "Minimum-Weight Design of Continuous Beams under Displacement and Stress Constraints," *J. Optimiz. Theory Appl.*, Vol. 9, No. 2, 1971, pp. 137–154.
306. Danilov, V. Ja., and Fedorchenko, I. S., "On a Problem of Optimal Damping of Vibrations of an Elastic Beam," in *Some Problems in Mechanics of Continuous Media*, Institute of Mathematics, Ukrainiam Academy of Sciences, Kiev, 1978, pp. 152–163.
307. Dems, K., "Multiparameter Shape Optimization of Elastic Bars In Torsion," *Int. J. Num. Meth. Eng.*, Vol. 15, 1980, pp. 1517–1539.
308. Dems, K., and Mróz, Z., "Multiparameter Structural Shape Optimization by the Finite Element Method," *Int. J. Num. Meth. Eng.*, Vol. 13, 1978, pp. 247–263.
309. DeSilva, B. M. E., "Optimal Vibrational Modes of a Disk," *J. Sound Vib.*, Vol. 21, No. 1, 1972, pp. 19–34.
310. DeSilva, B. M. E., "Optimal Control Concepts in the Design of Turbine Disks and Blades," *Shock and Vibration Digest*, Vol. 7, 1975, pp. 63–76.
311. DeSilva, B. M. E., "Application of Pontryagin's Principle to a Minimum-Weight Design Problem," *J. Basic Eng.*, Vol. 92, 1979, pp. 245–250.
312. Dinkoff, B., Levine, M., and Luus, R., "Optimum Linear Tapering in the Design of Columns," *J. Appl. Mech.*, Vol. 46, 1979, pp. 956–958.
313. Distefano, N., and Todeschini, R., "Invariant Imbedding and Optimum Beam Design with Displacement Constraints," *Int. J. Solids Struct.*, Vol. 8, No. 8, 1972, pp. 1073–1088.
314. Dixon, L. C. W., "Pontryagin's Maximum Principle Applied to the Profile of a Beam," *J. R. Aeronaut. Soc.*, Vol. 71, 1967, pp. 513–515.
315. Dupuis, G., "Optimal Design of Statically Determinate Beams Subject to Displacement and Stress Constraints," *AIAA J.*, Vol. 9, No. 5, 1971.
316. Dupuis, G., "An Iterative Approach to Structural Optimization," *Int. J. Num. Meth. Eng.*, Vol. 4, 1972, pp. 331–336.
317. Durelli, A. J., and Rajaiah, K., "Optimum Hole Shapes in Finite Plates Under Uniaxial Load," *J. Appl. Mech.*, Vol. 46, 1979, pp. 691–695.
318. Durelli, A. J., Rajaiah, K., Hovanesian, J. D., and Hung, Y. Y., "General Method to Directly Design Stress-Wise Optimum Two-Dimensional Structures," *Mech. Res. Commun.*, Vol. 6, 1979, pp. 159–165.
319. Egorov, A. I., and Fomenko, A. V., "Optimal Stabilization of Elastic Systems," in *Dynamics of Controlled Systems*, Nauka, Novosibinsk, 1979, pp. 112–121.
320. Elwany, M. H. S., and Barr, A. D. S., "Some Optimization Problems In Torsional Vibration," *J. Sound Vib.*, Vol. 57, 1978, pp. 1–33.

321. Elwany, M. H. S., and Barr, A. D. S., "Minimum-Weight Design of Beams in Torsional Vibration with Several Frequency Constraints," *J. Sound Vib.*, Vol. 62, 1979, pp. 411–425.
322. Erbatur, F., and Mengi, Y., "Optimal Design of Plates Under the Influence of Dead Weight and Surface Loading," *J. Struct. Mech.*, Vol. 5, 1977, pp. 345–356.
323. Erbatur, F., and Mengi, Y., "On the Optimal Design of Plates for a Given Deflection," *J. Optimiz. Theory Appl.*, Vol. 21, 1977, pp. 103–110.
324. Farshad, M., "Optimum Shape of Continuous Columns," *Int. J. Mech. Sci.*, Vol. 16, No. 8, 1974, pp. 597–602.
325. Farshad, M., and Tadjbakhsh, I., "Optimum Shape of Columns with General Conservative End Loading," *J. Optimiz. Theory Appl.*, Vol. 11, No. 4, 1973, pp. 413–420.
326. Feng, T. T., Arora, J. S., and Haug, E. J., "Optimal Structural Design under Dynamic Loads," *Int. J. Num. Meth. Eng.*, Vol. 11, 1977, pp. 39–52.
327. Foley, M. H., "A Minimum Mass Square Plate with Fixed Fundamental Frequency of Free Vibration," *AIAA J.*, Vol. 16, 1978, pp. 1001–1004.
328. Foley, M., and Citron, S. J., "A Simple Technique for the Minimum Mass Design of Continuous Structural Members," *J. Appl. Mech.*, Vol. 44, 1977, pp. 285–290.
329. Fox, R. L., and Kapoor, M. P., "Structural Optimization in the Dynamics Response Regime: A Computational Approach," *AIAA J.*, Vol. 8, No. 10, 1970, pp. 1798–1804.
330. Francavilla, A., Ramakrishnan, C. V., and Zienkiewicz, O. C., "Optimization of Shape to Minimize Stress Concentration," *J. Strain Anal.*, Vol. 10, 1975, pp. 63–70.
331. Frauenthal, J. C., "Constrainted Optimal Design of Columns Against Buckling," *J. Struct. Mech.*, Vol. 1, 1972, pp. 79–89.
332. Frauenthal, J. C., "Constrainted Optimal Design of Circular Plates Against Buckling," *J. Struct. Mech.*, Vol. 1, No. 2, 1972, pp. 159–186.
333. Fuchs, M. B., and Brull, M. A., "A New Strain Energy Theorem and Its Use in the Optimum Design of Continuous Beams," *Comput. Struct.*, Vol. 10, 1979, pp. 647–657.
334. Gajewski, A., and Zyczkowski, M., "Optimal Design of Elastic Columns Subjected to the General Conservative Behavior of Loading," *Z. Angew. Math. Phys.*, Vol. 21, No. 52, 1971, pp. 806–818.
335. Grinev, V. B., and Garal, Y. A., "Optimization of the Parameters of Rotating Rods," *Sov. Appl. Mech.*, Vol. 13, No. 9, 1975, pp. 389–393.
336. Gupta, V. K., and Murthy, P. N., "Optimal Design of Uniform Nonhomogeneous Vibrating Beams," *J. Sound Vib.*, Vol. 59, 1978, pp. 521–531.
337. Gutowski, R., "Stability of Vibrations of an Elastic Strut with Account Taken of the Fluid Flow Around it," *Czech. Acad. Sciences, Inst. of Thermodynamics*, Proceedings of the XIth Conference on Dynamics of Machines, Prague, 1977, pp. 123–128.
338. Gutowski, R., "Sensitivity of the Solutions of Linear Vibration Equation for a Beam over an Elastic Base with Respect to Coefficient Changes," *Teor. Prim. Meh.*, Vol. 4, 1978, pp. 33–38.
339. Haug, E. J., *Minimum-Weight Design of Beams with Inequality Constraints on Stress and Deflection*, Ph.D. Thesis, Kansas State University, 1966.
340. Haug, E. J., "Two Methods of Optimal Structural Design," *Developments in Mechanics*, Vol. 5 (Weiss, H. J., Young, D. F., Riley, W. F., and Rogge, T. R., Ed.), Iowa State University Press, 1969, pp. 847–860.
341. Haug, E. J., "Optimization of Distributed Parameter Structures with Repeated Eigenvalues," *New Approaches to Nonlinear Problems in Dynamics* (P. J. Holmes, Ed.), SIAM, Philadelphia, 1980, pp. 511–520.
342. Haug, E. J., "A Review of Distributed Parameter Structural Optimization Literature," *Optimization of Distributed Parameter Structures* (E. J. Haug and J. Cea, Ed.), Sijthoff & Noordhoff, Alphen aan den Rijn, Netherlands, 1981, pp. 3–68.
343. Haug, E. J., "A Gradient Projection Method for Structural Optimization," *Optimization of Distributed Parameter Structures* (E. J. Haug and J. Cea, Ed.), Sijthoff & Noordhoff, Alphen aan den Rijn, Netherlands, 1981. pp. 446–473.

344. HAUG, E. J., and ARORA, J. S., "Design Sensitivity Analysis of Elastic Mechanical Systems," *Comput. Meth. Appl. Mech. Eng.*, Vol. 15, 1978, pp. 35–62.
345. HAUG, E. J., and ARORA, J. S., *Applied Optimal Design*, Wiley-Interscience, New York, 1979.
346. HAUG, E. J., and ARORA, J. S., "Distributed Parameter Structural Optimization for Dynamic Response," *Optimization of Distributed Parameter Structures* (E. J. Haug and J. Cea, Ed.), Sijthoff & Noordhoff, Alphen aan den Rijn, Netherlands, 1981, pp. 474–515.
347. HAUG, E. J., ARORA, J. S., and FENG, T. T., "Sensitivity Analysis and Optimization of Structures for Dynamic Response," *ASME J. Mech. Design*, Vol. 100, 1978, pp. 311–318.
348. HAUG, E. J., ARORA, J. S., and MATSUI, K., "A Steepest-Descent Method for Optimization of Mechanical Systems," *J. Optimiz. Theory Appl.*, Vol. 19, No. 3, 1976, pp. 401–424.
349. HAUG, E. J., and FENG, T. T., "Optimization of Distributed Parameter Structures under Dynamic Loads," *Control and Dynamic Systems* (C. T. Leondes, Ed.), Vol. 13, 1977, pp. 207–246.
350. HAUG, E. J., and FENG, T. T., "Optimal Design of Dynamically Loaded Continuous Structures," *Int. J. Num. Meth. Eng.*, Vol. 12, 1978, pp. 299–307.
351. HAUG, E. J., and KIRMSER, P. G., "Minimum-Weight Design of Beams with Inequality Constraints on Stress and Deflection," *J. Appl. Mech.*, Vol. 34, 1967, pp. 999–1007.
352. HAUG, E. J., and KOMKOV, V., "Sensitivity Analysis in Distributed-Parameter Mechanical Systems Optimization," *J. Optimiz. Theory Appl.*, Vol. 23, No. 3, 1977, pp. 445–464.
353. HAUG, E. J., and KWAK, B. M., "Contact Stress Minimization by Contour Design, *Int. J. Num. Meth. Eng.*, Vol. 12, 1978, pp. 917–930.
354. HAUG, E. J., PAN, K. C., and STREETER, T. D., "A Computational Method for Optimal Structural Design I: Piecewise Uniform Structures," *Int. J. Num. Meth. Eng.*, Vol. 5, 1972, pp. 171–184.
355. HAUG, E. J., PAN, K. C., and STREETER, T. D., "A Computational Method for Optimal Structural Design II: Continuous Problems," *Int. J. Num. Meth. Eng.*, Vol. 9, 1975, pp. 649–667.
356. HAUG, E. J., and ROUSSELET, B., "Design Sensitivity Analysis in Structural Mechanics I: Static Response Variations," *J. Struct. Mech.*, Vol. 8, No. 1, 1980, pp. 17–41.
357. HAUG, E. J., and ROUSSELET, B., "Design Sensitivity Analysis in Structural Mechanics II: Eigenvalues Variations," *J. Struct. Mech.*, Vol. 8, 1980, to appear.
358. HEGEMIER, G. A., and PRAGER, W., "On Michell Trusses" *Int. J. Mech. Sci.*, Vol. 11, 1969, p. 209.
359. HEGEMIER, G. A., and TANG, H. T., "A Variational Principle, The Finite Element Method, and Optimal Structural Design for Given Deflection," *Optimization in Structural Design* (A. Sawczuk and Z. Mroz, Ed.), Springer-Verlag, New York, 1975, pp. 464–483.
360. HENRY, A. S., *The Analytic Design of Torsion Members*, Ph.D. Thesis, University of Iowa, 1971.
361. HERSCH, J., and Payne, L. E., "Extremal Principles and Isoperimetric Inequalities for Some Mixed Problems of Stekloff's Type" *Z. Angew. Math. Phys.*, Vol. 19, 1968, pp. 802–817.
362. HILL, R. D., and ROZVANY, G. I. N., "Optimal Beam Layouts: The Free Edge Paradox," *J. Appl. Mech.*, Vol. 44, 1977, pp. 696–700.
363. HIRANO, Y. "Optimum Design of Laminated Plates Under Axial Compression," *AIAA J.*, Vol. 17, No. 9, 1979, pp. 1017–1019.
364. HORNBUCKLE, J. C., and BOYKIN, W. H., "Equivalence of a Constrained Minimum Weight and Maximum Column Buckling Load Problem with Solution," *J. Appl. Mech.*, Vol. 45, 1978, pp. 159–164.
365. HSIAO, M. H., HAUG, E. J., and ARORA, J. S., "A State Space Method for Optimal Design of Vibration Isolators," *ASME J. Mech. Design*, Vol. 101, 1979, pp. 309–314.
366. HU, K. K., and KIRMSER, P. G., "A Numerical Solution of A Nonlinear Differential-Integral Equation for the Optimal Shape of the Tallest Column," *Int. J. Eng. Sci.*, Vol. 18, 1980, pp. 333–339.

367. Huang, N. C., "Optimal Design of Elastic Structures for Maximum Stiffness, *Int. J. Solids Struct.*, Vol. 4, 1968, pp. 689–700.
368. Huang, N. C., "On the Principle of Stationary Mutual Complementary Energy and its Application to Structural Design," *Z. Angew. Math. Phys.*, Vol. 22, 1971, pp. 608–620.
369. Huang, N. C., "Optimal Design of Elastic Beams for Minimum–Maximum Deflection," *J. Appl. Mech.*, Vol. 38, 1971, pp. 1078–1081.
370. Huang, N. C., "Effect Shear Deformation on Optimal Design of Elastic Beams," *Int. J. Solids Struct.*, Vol. 7, 1971, pp. 321–326.
371. Huang, N. C., "Minimum Weight Design of Vibrating Elastic Structures with Dynamic Deflection Constraint," *J. Appl. Mech.*, Vol. 43, 1976, pp. 171–180.
372. Huang, N. C., and Sheu, C. Y., "Optimal Design of an Elastic Column of Thin-Walled Cross Section," *J. Appl. Mech.*, Vol. 35, No. 2, 1968, pp. 285–288.
373. Huang, N. C., and Sheu, C. Y., "Optimal Design of Elastic Circular Sandwich Beams for Minimum Compliance," *J. Appl. Mech.*, Vol. 37, 1970, p. 569.
374. Huang, N. C., and Tang, H. T., "Minimum-Weight Design of Elastic Sandwich Beams with Deflection Constraints," *J. Optimiz. Theory Appl.*, Vol. 4, No. 4, 1969, pp. 277–298.
375. Icerman, L. J., "Optimal Structural Design for Given Dynamic Deflection," *Int. J. Solids Struct.*, Vol. 5, 1969, pp. 473–490.
376. Ioffe, A. D., and Tihomirov, V. M., *Theory of Extremal Problems*, North-Holland, Amsterdam, 1979.
377. Jacquot, R. G., "Optimal Dynamic Vibration Absorbers for General Beam Systems," *J. Sound Vib.*, Vol. 60, 1978, pp 535–542.
378. Kamat, M. P., and Simitses, G. J., "Optimal Beam Frequencies by the Finite Element Displacement Method," *Int. J. Solids Struct.*, Vol. 9, 1973, pp. 415–429.
379. Kamat, M. P., "Effect of Shear Deformations and Rotary Inertia on Optimum Beam Frequencies," *Int. J. Num. Meth. Eng.*, Vol. 9, 1975, pp. 51–62.
380. Karihaloo, B. L., "Optimal Design of Multipurpose Tie Column of Solid Construction," *Int. J. Solids Struct.*, Vol. 15, 1979, pp. 103–109.
381. Karihaloo, B. L., "Optimal Design of Multipurpose Tie-Beams" *J. Optimiz. Theory Appl.*, Vol. 27, No. 3, 1979, pp. 427–438.
382. Karihaloo, B. L., "Optimal Design of Multipurpose Structures," *J. Optimiz. Theory Appl.*, Vol. 27, No. 3, 1979, pp. 449–461.
383. Karihaloo, B. L., and Parbery, R. D., "Optimal Design of Multipurpose Beam-Columns," *J. Optimiz. Theory Appl.*, Vol. 27, No. 3, 1979, pp. 439–448.
384. Karihaloo, B. L., and Wood, G. L., "Optimal Design of Multipurpose Sandwich Tie-Columns," *ASCE J. Eng. Mech. Div.*, Vol. 105, No. EM3, 1979, pp. 465–469.
385. Kato, B., Nakamara, Y., and Anraku, H., "Optimum Earthquake Design of Shear Buildings," *ASCE J. Eng. Mech. Div.*, Vol. 98, No. EM4, 1972, pp. 892–909.
386. Knightly, George H., and Sather, D., "Regularity and Symmetry Properties of Solutions of the John Shell Equations for a Spherical Shell," Special Session on Elastic Vibrations and Stability, American Mathematical Society, Pittsburgh, Pennsylvania, May 15–16, 1981, to appear in *Contemporary Mathematics Series*, American Mathematics Society (Vadim Kimkov, Ed.).
387. Komkov, V., "Classification of Boundary Conditions in Optimal Control Theory of Beams and Plates," *Proc. International Conference on Control Theory for Systems with Distributed Parameters*, I.F.A.C., Banff, Canada, No. 5.5, 1971, pp. 1–11.
388. Komkov, V., *Optimal Control Theory for the Damping of Elastic-Vibration of Simple Elastic Systems*, Springer-Verlag, Berlin 1972.
389. Komkov, V., "A Dual Form of Noether's Theorem with Applications to Continuum Mechanics," *J. Math. Anal. Appl.*, Vol. 75, No. 1, 1980, pp. 251–269.
390. Komkov, V., "A Dynamic Theory of Column Buckling (Abstract)," *17th Midwestern Mechanics Conference*, Ann Arbor, Michigan, May 6–8, 1981.
391. Komkov, V., "Application of Invariant Variational Principles to the Optimal Design of a Column," *Z. Angew. Math. Mech.*, Vol. 61, 1981, pp. 75–80.

392. Komkov, V., "An Optimal Design Problem, A Nonexistence Theorem," *Arch. Mech.*, Vol. 33, 1981, pp. 147–151.
393. Komkov, V., and Coleman, N., "Optimality of Design and Sensitivity Analysis of Beam Theory," *Int. J. Control*, Vol. 18, No. 4, 1973, pp. 731–740.
394. Komkov, V., and Coleman, N. P., "An Analytic Approach to Some Problems of Optimal Design of Beams and Plates," *Arch. Mech.*, Vol. 27, No. 4, 1975, pp. 565–575.
395. Komkov, V., and Haug, E. J., "On the Optimum Shape of Columns," *Optimization of Distributed Parameter Structures* (E. J. Haug and J. Cea, Ed.), Sijthoff & Noordhoff, Alphen aan den Rijn, Netherlands, 1981, pp. 399–425.
396. Komkov, V., and Haug, E. J., "Effects of Nonlinear Terms on the Buckling of Elastic Bodies," Special Session on Elastic Vibrations and Stability, American Mathematical Society, Pittsburgh, Pennsylvania, May 15–16, 1981, to appear in *Contemporary Mathematics Series*, American Mathematics Society (Vadim Komkov, Ed.).
397. Kristensen, E. S., and Madsen, N. F., "On the Optimum Shape of Fillets in Plates Subjected to Multiple In-Plane Loading Cases," *Int. J. Num. Meth. Eng.*, Vol. 10, 1976, pp. 1007–1019.
398. Krivenkov, Iu. P., "Sufficient Conditions of Optimality in Linear Problems of the Mathematical Theory of Optimal Processes with Phase Constraints," *PMM*, Vol. 42, 1978, pp. 623–632.
399. Kruzelecki, J., "Optimization of Shells Under Combined Loadings via the Concept of Uniform Stability," *Optimization of Distributed Parameter Structures* (E. J. Haug and J. Cea, Ed.), Sijthoff & Noordhoff, Alphen aan den Rijn, Netherlands, 1981, pp. 929–950.
400. Kunar, R. R., and Chan, A. S. L., "A Method for the Configurational Optimization of Structures," *Comput. Meth. Appl. Mech. Eng.*, Vol. 7, 1976, pp. 331–350.
401. Kunoo, K., and Yang, T. Y., "Minimum-Weight Design of a Cylindrical Shell with Multiple Stiffener Sizes," *AIAA J.*, Vol. 16, 1978, pp. 35–40.
402. Knightly, G. H., and Sather, D., "Regularity and Symmetry Properties of Solutions of the John Shell Equations for a Spherical Shell," Special Session on Elastic Vibrations and Stability, American Mathematical Society, Pittsburgh, Pennsylvania, May 15–16, 1981, to appear in *Contemporary Mathematics Series*, American Mathematics Society (Vadim Komkov, Ed.).
403. Levy, H. J., and Wolf, B. M., "Fully Stressed Dynamically Loaded Structures," *ASME Paper 740WA/DE-19, ASME*, New York, 1974.
404. Lions, J. L., *Optimal Control of Systems Governed by Partial Differential Equations*, Springer-Verlag, New York, 1971.
405. Lions, J. L., *Some Aspects of the Optimal Control of Distributed Parameter Systems*, SIAM Series in Regional Conferences in Applied Mathematics, 1972.
406. Litvinov, V. G., "A Problem Encountered in Bending of Plates of Variable Thickness," *Prikl. Mekh.*, Vol. 11, No. 5, 1975.
407. Litvinov, V. G., "A Problem Encountered in Bending of Plates of Variable Thickness," *Izv. Acad. Nauk SSSR, MTT*, Vol. 15, No. 2, 1980, pp. 174–180.
408. Lukasiewicz, S. A., "Optimum Design in Junction and Contact Problems," *Colloquium No. 110, Contact Problems and Load Transfer in Mechanical Assemblages*, Linkoping, Sweden, 1978.
409. Lur'e, K. A., "Some Problems of Optimal Bending of Plates," *Izv. Akad. Nauk SSSR, Mekh. Tverd. Tela*, Vol. 14, No. 6, 1979, pp. 86–93.
410. Maday, C. J., "A Class of Minimum Weight Shafts," *ASME J. Eng. Industry*, Vol. 96, No. 1, 1974, pp. 166–170.
411. Manevich, A. I., and Kaganov, M. Y., "Stability and Weight Optimization of Reinforced Spherical Shells Under External Pressure," *Prikl. Mekh.*, Vol. 9, No. 1, 1973, pp. 20–26.
412. Martin, J. B., "The Optimal Design of Beams and Frames with Compliance," *Int. J. Solids Struct.*, Vol. 7, 1971, pp. 63–81.
413. Masur, E. F., "Optimum Stiffness and Strength of Elastic Structures," *ASCE J. Eng. Mech. Div.*, Vol. 95, No. EM5, 1970, pp. 621–640.

414. Masur, E. F., "Optimal Structural Design for a Discrete Set of Available Structural Members," *Comput. Meth. Appl. Mech. Eng.*, Vol. 3, 1974, pp. 195–207.
415. Masur, E. F., "Optimal Placement of Available Sections in Structural Eigenvalue Problems," *J. Optimiz. Theory Appl.*, Vol. 15, 1975, pp. 69–84.
416. Masur, E. F., "Optimal Design of Symmetric Structures Against Postbuckling Collapse," *Int. J. Solids Struct.*, Vol. 14, 1978, pp. 319–326.
417. Masur, E. F., "Singular Problems of Optimal Design," *Optimization of Distributed Parameter Structures* (E. J. Haug and J. Cea, Ed.), Sijthoff & Noordhoff, Alphen aan den Rijn, Netherlands, 1981, pp. 200–218.
418. Masur, E. F., "Optimal Structural Design with Multiple Eigenvalue Constraint," *Abstract, 17th Midwestern Mechanics Conference*, Ann Arbor, Michigan, May 6–8, 1981.
419. Masur, E. F., and Mróz, Z., "On Nonstationary Optimality Conditions in Structural Design," *Int. J. Solids Struct.*, Vol. 15, 1979, pp. 503–512.
420. Masur, E. F., and Mróz, Z., "Singular Solutions in Structural Optimization Problems," *Proc. IUTAM Symposium on Variational Methods in the Mechanics of Solids* (S. Nemat-Nasser, Ed.), Pergamon Press, New York, 1980, pp. 344–348.
421. Mayne, R. W., "Optimization Techniques for Shock and Vibration Isolator Development," *Shock and Vibration Digest*, Vol. 11, No. 10, 1979, pp. 25–33.
422. McCart, B. R., Haug, E. J., and Streeter, T. D., "Optimal Design of Structure, with Constraints on Natural Frequency," *AIAA J.*, Vol. 8, No. 6, 1970, pp. 1012–1019.
423. McGlothin, G. E., "Optimal Control of Distributed Parameter Systems with Penalties on Special Derivatives of the State," *Int. J. Control*, Vol. 24, No. 4, 1976, pp. 145–166.
424. McIntosh, S. C., "Structural Optimization Via Optimal Control Techniques," *Proceedings, ASME Structural Optimization Symposium*, ASME, AMDT, New York, 1974.
425. McNamara, R. J., "Turned Mass Dampers for Buildings," *J. Struct. Div., ASCE*, Vol. 103, No. ST9, 1977, pp. 1785–1798.
426. Miele, A., Mangiavacchi, A., Mohanty, B. P., and Wu, A. K., "Numerical Determination of Minimum Mass Structures with Specified Natural Frequencies," *Int. J. Num. Meth. Eng.*, Vol. 13, No. 2, 1978, pp. 203–228.
427. Miersemann, E., "Eigenvalue Problems for Variational Inequalities," Special Session on Elastic Vibrations and Stability, American Mathematical Society, Pittsburgh, Pennsylvania, May 15–16, 1981, to appear in *Contemporary Mathematics Series*, American Mathematics Society (Vadim Komkov, Ed.).
428. Mikhlin, S. G., "Stability of Solutions to One-Sided Variational Problems; Application to the Theory of Plasticity," in *Numerical Techniques and Problems of Systematic Computation*, (LOMI seminar) Nauka, Moscow, Vol. 102, 1980, pp. 68–101.
429. Mroz, Z., "Optimal Design of Structures of Composite Materials," *Int. J. Solids Struct.*, Vol. 6, 1970, pp. 859–879.
430. Mróz, Z., "Multiparameter Optimal Design of Plates and Shells," *J. Struct. Mech.*, Vol. 1, No. 3, 1973, pp. 371–392.
431. Mróz, Z., "On Optimal Force Action and Reaction on Structures," in *Structural Control* (H. H. F. Leipholz, Ed.), North-Holland, Amsterdam, 1980, pp. 523–544.
432. Mróz, Z., and Garstecki, A., "Optimal Design of Structures with Unspecified Loading Distribution," *J. Optimiz. Theory Appl.*, Vol. 20, 1976, pp. 359–380.
433. Mróz, A., and Rozvany, G. I. N., "Optimal Design of Structures with Variable Support Conditions," *J. Optimiz. Theory Appl.*, Vol. 15, 1975, pp. 85–101.
434. Neuber, H., "Longitudinally Stressed Straight Columns with Optimal Distribution of Cross-Sections," *Forsch. Ingenieurwes.*, Vol. 35, 1969, pp. 29–30.
435. Neuber, H., "Optimization of Tensile Stress Concentration," in *Continuum Mechanics and Related Problems of Analysis*, Nauka, Moscow, 1972, pp. 375–380.
436. Oda, J., "On A Technique to Obtain an Optimum Strength Shape by the Finite Element Method," *Bull. JSME*, Vol. 20, 1977, pp. 160–167.
437. Olhoff, N., "On Singularities, Local Optima, and Formation of Stiffeners in Optimal

Design of Plates," in *Optimization in Structural Design* (Sawczuk and Mróz, Ed.), Springer-Verlag, New York, 1975, pp. 82–103.
438. OLHOFF, N., "Optimization of Vibrating Beams with Respect to Higher Order Natural Frequencies," *J. Struct. Mech.*, Vol. 4, 1976, pp. 87–122.
439. OLHOFF, N., "A Survey of the Optimal Design of Vibrating Structural Elements, Part I: Theory," and "Part II: Applications," *Shock and Vibration Digest*, Vol. 8, No. 8, 1976, pp. 3–10; and No. 9, 1976, pp. 3–10.
440. OLHOFF, N., "Maximizing Higher Order Eigenfrequencies of Beams with Constraints on the Design Geometry," *J. Struct. Mech.*, Vol. 5, 1977, pp. 107–134.
441. OLHOFF, N., "Optimization of Transversely Vibrating Beams and Rotating Shafts," in *Optimization of Distributed Parameter Structures* (E. J. Haug and J. Cea, Ed.), Sijthoff & Noordhoff, Alphen aan den Rijn, Netherlands, 1981, pp. 177–199.
442. OLHOFF, N., "Optimization of Columns Against Buckling," in *Optimization of Distributed Parameter Structures* (E. J. Haug and J. Cea, Ed.), Sijthoff & Noordhoff, Alphen aan den Rijn, Netherlands, 1981, pp. 152–176.
443. OLHOFF, N., and CHENG, K. T., "Optimal Design of Solid Elastic Plates," in *Optimization of Distributed Parameter Structures* (E. J. Haug and J. Cea, Ed.), Sijthoff & Noordhoff, Alphen aan den Rijn, Netherlands, 1981, pp. 278–303.
444. OLHOFF, N., LUR'E, K. A., CHERKAEV, A. V., and FEDOROV, A. V., *Sliding Regimes and Anisotropy in Optimal Design of Vibrating Axisymmetric Plates*, DCAMM-Rept., The Danish Center for Applied Mathematics and Mechanics, 1980.
445. OLHOFF, N., and NIORDSON, F. I., "Some Problems Concerning Singularities of Optimal Beams and Columns," *Z. Angew. Math. Mech.*, Vol. 59, 1979, pp. T16–T26.
446. OLHOFF, N., and TAYLOR, J. E., "Designing Continuous Columns for Minimum Total Cost of Material and Interior Supports," *J. Struct. Mech.*, Vol. 6, 1978, pp. 367–382.
447. OLHOFF, N., and TAYLOR, J. E., "On Optimal Structural Remodeling," *J. Optimiz. Theory Appl.*, Vol. 27, No. 5, 1979, pp. 571–581.
448. PAPPAS, M., and ALLENTUCH, A., "Automated Optimal Design of Frame Reinforced, Submersible, Circular, Cylindrical Shells," *J. Ship Res.*, Vol. 17, 1973, pp. 208–216.
449. PAPPAS, M., and ALLENTUCH, A., "Pressure Hull Optimization Using General Instability Equation Admitting More Than One Longitudinal Buckling Half-Wave," *J. Ship Res.*, Vol. 19, 1975, pp. 18–22.
450. PARBERY, R. D., and KARIHALOO, B. L., "Minimum-Weight Design of Hollow Cylinders for Given Lower Bounds on Torsional and Flexural Rigidities," *Int. J. Solids Struct.*, Vol. 13, 1977, pp. 1271–1280.
451. PARBERY, R. D., and KARIHALOO, B. L., "Minimum-Weight Design of Thin-Walled Cylinders Subject to Flexural and Torsional Stiffness Constraints," *J. Appl. Mech.*, Vol. 47, 1980, pp. 106–110.
452. PAYNE, L. E., "Stabilization of Ill-Posed Cauchy Problems in Nonlinear Elasticity," Special Session on Elastic Vibrations and Stability, American Mathematical Society, Pittsburgh, Pennsylvania, May 15–16, 1981, to appear in *Contemporary Mathematics Series*, American Mathematics Society (Vadim Komkov, Ed.).
453. PETERSEN, N. R., "Design of Large Scale Turned Mass Dampers," Preprint 3578, *ASCE National Meeting*, Boston, April, 1979.
454. PISTER, K. S., "Optimal Design of Structures under Dynamic Loading," in *Optimization of Distributed Parameter Structures* (E. J. Haug and J. Cea, Ed.), Sijthoff & Noordhoff, Alphen aan den Rijn, Netherlands, 1981. pp. 569–585.
455. PLAUT, R. H., "On Minimizing the Response of Structures to Dynamic Loading," *Z. Angew. Math. Phys.*, Vol. 21, 1970, pp. 1004–1010.
456. PLAUT, R. H., "Optimal Structural Design for Given Deflection under Periodic Loading," *Q. Appl. Math.*, Vol. 29, 1971, pp. 315–318.
457. PLAUT, R. H., "Approximate Solutions to Some Static and Dynamic Optimal Structural Design Problems," *Q. Appl. Math.*, Vol. 31, 1973, pp. 535–539.

458. Plaut, R. H., "Vibrations and Stability of Shallow Elastic Arches," Special Session on Elastic Vibrations and Stability, American Mathematical Society Pittsburgh, Pennsylvania, May 15–16, 1981, to appear in *Contemporary Mathematics Series*, American Mathematics Society (Vadim Komkov, Ed.).
459. Pochtman, Y. M., "Optimization of Structures with Constraints on Dynamic and Frequency Characteristics," *Dokl. Akad. Nauk SSSR* (in Russian), Vol. 203, No. 2, 1972, pp. 307–308; English Translation NASA TT F-14, 540, NASA, 1972.
460. Polak, E., "Algorithms for Optimal Design," in *Optimization of Distributed Parameter Structures* (E. J. Haug and J. Cea, Ed.), Sijthoff & Noordhoff, Alphen aan den Rijn, Netherlands, 1981, pp. 586–601.
461. Popelar, C. H., "Optimal Design of Beams Against Buckling: A Potential Energy Approach," *J. Struct. Mech.*, Vol. 4, 1976, pp. 181–196.
462. Popelar, C. H., "Optimal Design of Structures Against Buckling: A Complementary Energy Approach," *J. Struct. Mech.*, Vol. 5, 1977, pp. 45–66.
463. Prager, W., "Optimal Design of Statically Determinate Beams for Given Deflection," *J. Mech. Sci.*, Vol. 13, 1971, p. 893.
464. Prager, W., "Conditions for Structural Optimality," *Comput. Struct.*, Vol. 2, 1972, pp. 833–840.
465. Prager, S., and Prager W., "A Note on Optimal Design of Columns," *Int. J. Mech. Sci.*, Vol. 21, 1979, pp. 249–251.
466. Prager, W., and Rozvany, G. I. N., "Optimal Layout of Grillages," *J. Struct. Mech.*, Vol. 5, 1977, pp. 1–18.
467. Ramakrishnan, C. V., and Francavilla, A., "Structural Shape Optimization Using Penalty Functions," *J. Struct. Mech.*, Vol. 3, No. 4, 1975, pp. 403–432.
468. Rammerstorfer, R. G., "Increase of the First Natural Frequency and Buckling Load of Plates by Optimal Field of Initial Stresses," *Acta Mech.*, Vol. 27, 1977, pp. 217–238.
469. Rangacharyulu, M. A. V., and Done, G. T. S., "A Survey of Structural Optimization Under Dynamic Constraints," *Shock and Vibration Digest*, Vol. 11, No. 12, 1979, pp. 15–25.
470. Rao, S. S., and Singh, K., "Optimum Design of Laminates with Natural Frequency Constraints," *J. Sound Vib.*, Vol. 67, 1979, pp. 101–112.
471. Rao, S. S., "Structural Optimization Under Shock and Vibration Environment," *Shock and Vibration Digest*, Vol. 11, No. 2, 1979, pp. 3–12.
472. Ray, D., *Sensitivity Analysis for Hysteretic Dynamic Systems: Application to Earthquake Engineering*, EERC 74-5, Earthquake Engineering Research Center, University of California, Berkeley, April, 1974.
473. Ray, D., Pister, K. S., and Chopra, A. K., *Optimum Design of Earthquake-Resistant Shear Buildings*, EERC 74-3, Earthquake Engineering Research Center, University of California, Berkeley, January 1974.
474. Ray, W. H., "Some Recent Applications of Distributed Parameter Systems Theory—A Survey," *Automatica*, Vol. 14, 1978, pp. 281–287.
475. Ray, D., Pister, K. S., and Polak, E., "Sensitivity Analysis for Hysteretic Dynamic Systems: Theory and Applications," *Comput. Meth. Appl. Mech. Eng.*, Vol. 14, 1978, pp. 179–208.
476. Reiss, R., "Optimal Compliance Criterion for Axisymmetric Solid Plates," *Int. J. Solids Struct.*, Vol. 12, 1976, pp. 319–329.
477. Rikards, R. B., "Convexity of Some Classes of Optimization Problems for Multilayer Shells under Conditions of Stability and Vibration," *MTT*, Vol. 15, 1980, pp. 145–154.
478. Robinson, A. C., "A Survey of Optimal Control of Distributed Parameter Systems," *Automatica*, Vol. 7, No. 3, 1971, pp. 371–388.
479. Rockenback, P. C., *Minimum-Mass Response-Constrained Design of Vibrating Sandwich Beams*, Report R-604, Coordinated Science Laboratory, University of Illinois, Urbana, 1973.

480. ROMANO, G., "On the Energy Criterion for the Stability of Continuous Elastic Structures," *Mecanico*, Vol. 3, No. 10, 1975, pp. 198–202.
481. ROORDA, J., and REIS, A. J., "Nonlinear Interactive Buckling: Sensitivity and Optimality," *J. Struct. Mech.*, Vol. 5, 1977, pp. 207–232.
482. ROSSOW, M. P., and TAYLOR, J. E., "A Finite Element Method for the Optimal Design of Variable Thickness Sheets," *AIAA J.*, Vol. 11, 1973, pp. 1566–1569.
483. ROUSSELET, B., "Implementation of Some Methods of Shape Optimal Design," in *Optimization of Distributed Parameter Structures* (E. J. Haug and J. Cea, Ed.), Sijthoff & Noordhoff, Alphen aan den Rijn, Netherlands, 1981, pp. 1195–1220.
484. ROUSSELET, B., "Dependence of Eigenvalues with Respect to Shape," in *Optimization of Distributed Parameter Structures* (E. J. Haug and J. Cea, Ed.), Sijthoff Noordhoff, Alphen aan den Rijn, Netherlands, 1981, pp. 1221–1249.
485. ROUSSELET, B., "Singular Dependence of Repeated Eigenvalues," in *Optimization of Distributed Parameter Structures* (E. J. Haug and J. Cea, Ed.), Sijthoff & Noordhoff, Alphen aan den Rijn, Netherlands, 1981, pp. 1443–1456.
486. ROUSSELET, B., and HAUG, E. J., "Design Sensitivity Analysis of Shape Variation," in *Optimization of Distributed Parameter Structures* (E. J. Haug and J. Cea, Ed.), Sijthoff & Noordhoff, Alphen aan den Rijn, Netherlands, 1981, pp. 1397–1442.
487. ROZVANY, G. I. N., "Grillages of Maximum Strength and Maximum Stiffness," *Int. J. Mech. Sci.*, Vol. 15, 1972, pp. 651–666.
488. ROZVANY, G. I. N., "Analytical Treatment of Some Extended Problems in Structural Optimization, Part I and II," *J. Struct. Mech.*, Vol. 3, 1974–75, pp. 359–402.
489. ROZVANY, G. I. N., "A New Class of Structural Optimization Problems: Optimal Archgirds," *Comput. Meth. Appl. Mech. Eng.*, Vol. 19, 1979, pp. 127–150.
490. ROZVANY, G. I. N., "Optimality Criteria for Grids, Shells and Arches," in *Optimization of Distributed Parameter Structures* (E. J. Haug and J. Cea, Ed.), Sijthoff & Noordhoff, Alphen aan den Rijn, Netherlands, 1981, pp. 112–151.
491. ROZVANY, G. I. N., and MRÓZ, Z., "Column Design: Optimization of Support Conditions and Segmentation," *J. Struct. Mech.*, Vol. 5, 1977, pp. 279–290.
492. SAVE, M., "A General Criterion for Optimal Structural Design," *J. Optimiz. Theory Appl.*, Vol. 15, 1975, pp. 119–129.
493. SAWCZUK, A., and MRÓZ, Z., *Optimization in Structural Design*, Springer-Verlag, New York, 1975.
494. SCHNACK, E., "An Optimization Procedure for Stress Concentrations by the Finite Element Technique," *Int. J. Num. Meth. Eng.*, Vol. 14, 1979, pp. 115–124.
495. SHERMAN, Z., "Weight Minimization of Axisymmetric Clamped Plates Subject to Constraints," *Int. J. Solids Struct.*, Vol. 9, 1973, pp. 279–290.
496. SHERMAN, Z., and WANG, P. C., "Volume Minimization of Thin Plates Subject to Constraints," *ASCE J. Eng. Mech. Div.*, Vol. 97, No. EM3, 1971, pp. 741–754.
497. SHEU, C. Y., "Elastic Minimum-Weight Design for Specified Fundamental Frequency," *Int. J. Solids Struct.*, Vol. 4, 1968, pp. 953–958.
498. SHEU, C. Y., and PRAGER, W., "Minimum-Weight Design with Piecewise Constant Specific Stiffness," *J. Optimiz. Theory Appl.*, Vol. 2, 1968, pp. 179–189.
499. SIMITSES, G. J., "Optimal versus the Stiffened Circular Plate," *AIAA J.*, Vol. 11, No. 10, 1973, pp. 1409–1412.
500. SIMITSES, G. J., and KOTRAS, T., "The Optimal Euler–Bernoulli Cantilever," *ASCE J. Eng. Mech. Div.*, Vol. 101, No EM6, 1975, pp. 922–929.
501. SIMITSES, G. J., and SHEINMAN, I., "Optimization of Geometrically Imperfect Stiffened Cylindrical Shells Under Axial Compression," *Comput. Struct.*, Vol. 19, 1978, pp. 337–381.
502. SIPPEL, D. L., and WARNER, W. H., "Minimum-Mass Design of Multielement Structures under a Frequency Constraint," *AIAA J.*, Vol. II, No. 4, 1973, pp. 483–489.
503. SMIRNOV, A. B., and TROITSKII, V. A., "Optimization of Natural Vibrational Frequencies of Curvilinear Thin Elastic Rods," *MTT*, Vol. 14, 1979, pp. 162–168.

504. SOLODOVNIKOV, V. N., "Optimization of Elastic Shells of Revolution," *PMM*, Vol. 42, No. 3, 1978, pp. 511–520.
505. STADLER, W., "Natural Structural Shapes of Shallow Arches," *J. Appl. Mech.*, Vol. 44, 1977, pp. 137–164.
506. STADLER, W., "Uniform Shallow Arches of Minimum Weight and Minimum Maximum Deflection," *J. Optimiz. Theory Appl.*, Vol. 23, 1977, pp. 137–164.
507. STADLER, W., "Natural Structural Shapes ((The Static Case)," *Q. J. Mech. Appl. Math.*, Vol. 31, 1978, pp. 169–217.
508. SZELAG, D., and MRÓZ, Z., "Optimal Design of Elastic Beams with Unspecified Support Conditions," *Z. Angew. Math. Phys.*, Vol. 58, 1978, pp. 501–510.
509. SZELAG, D., and MRÓZ, Z., "Optimal Design of Vibrating Beams with Unspecified Support Reactions," *Comput. Meth. Appl. Mech. Eng.*, Vol. 19, 1979, pp. 333–349.
510. TADJBAKHSH, I., and FARSHAD, M., "On Conservatively Loaded Funicular Arches and Their Optimal Design," in *Optimization in Structural Design* (Sawczuk and Mróz, Ed.), Springer-Verlag, New York, 1975, pp. 215–228.
511. THOMPSON, J. M. T., "Optimization as a Generator of Structural Instability," *Int. J. Mech. Sci.*, Vol. 14, No. 9, 1972, pp. 627–630.
512. THOMPSON, J. M. T., and HUNT, G. W., "Dangers of Structural Optimization," *Eng. Optimiz.*, Vol. 1, 1974, pp. 99–110.
513. THOMPSON, J. M. T., and LEWIS, G. M., "On the Optimum Design of Thin-Walled Compression Members," *J. Mech. Phys. Solids*, Vol. 20, 1972, pp. 101–109.
514. THOMPSON, J. M. T., and SUPPLE, W. D., "Erosion of Optimum Design by Compound Branching Phenomena," *J. Mech. Phys. Solids*, Vol. 21, No. 3, 1972, pp. 135–144.
515. VAVRICK, D. J., and WARNER, W. H., "Duality Among Optimal Design Problems for Torsional Vibration," *J. Struct. Mech.*, Vol. 6, No. 2, 1978, pp. 233–246.
516. VELTE, W., and VILLAGGIO, P., "Are the Optimum Problems in Structural Design Well Posed?" *Arch. Rat. Mech. Anal.*, 1981.
517. VENKAYYA, V. B., "Structural Optimization: A Review and Some Recommendations," *Int. J. Num. Meth. Eng.*, Vol. 13, No. 2, 1978, pp. 203–228.
518. VITIELLO, E., and PISTER, K. S., *Applications of Reliability-Based Global Cost Optimization to Design of Earthquake-Resistant Structures*, Report No. EERC 74-10, University of California, Berkeley, August 1974.
519. VORONTSOV, M. A., and KORIABIN, A. V., "Active Stabilization of Systems with Distributed Parameters," Vestnik Mosk. Univ. Series Mat. and Mekh, *Mekhamika*, No. 4, 1978, pp. 93–100.
520. WALKER, N. D., and PISTER, K. S., *Study of a Method of Feasible Directions for Optimal Elastic Design of Framed Structures Subject to Earthquake Loading*, EERC 75-39 Earthquake Engineering Research Center, University of California, Berkeley, December 1975.
521. WARNER, W. H., and VAVRICK, D. J., "Optimal Design in Axial Motion for Several Frequency Constraints," *J. Optimiz. Theory Appl.*, Vol. 15, No. 1, 1975, pp. 159–166.
522. WEINBERGER, H. F., and SERRIN, J. B., "Optimal Shapes for Brittle Beams under Torsion," in *Complex Analysis and its Applications;* a collection of articles honoring the 70th birthday of I. N. Vekua, Nauka, Moscow, 1978, pp. 88–91.
523. WEISSHAAR, T. A., *An Application of Control Theory Methods to the Optimization of Structures Having Dynamic or Aeroelastic Constraints*, Report No. SUDAAR 412, Department Aeronautics and Astronautics, Stanford University, 1970.
524. WEISSHAAR, T. A., and PLAUT, R. H., "Structural Optimization under Nonconservative Loading," in *Optimization of Distributed Parameter Structures* (E. J. Haug and J. Cea, Ed.), Sijthoff & Noordhoff, Alphen aan den Rijn, Netherlands, 1981, pp. 843–864.
525. WIESNER, K. B., "Turned Mass Dampers to Reduce Building Wing Motion," Preprint 3510, *ASCE National Meeting*, Boston, April 1979.
526. YOUSSEF, N. A. N., and POPPLEWELL, N., "A Theory of the Greatest Maximum Response of Linear Structures," *J. Sound Vib.*, Vol. 56, 1978, pp. 21–33.

527. Zienkiewicz, O. C., and Campbell, J. S., "Shape Optimization and Sequential Linear Programming," *Optimum Structural Design* (R. H. Gallagher and O. C. Zienkiewicz, Ed.), John Wiley, New York, 1973, pp. 109–126.
528. Zolezzi, T., "Necessary Conditions for Optimal Control of Elliptic or Parabolic Problems," *J. Control*, Vol. 10, No. 4, 1972, pp. 594–607.
529. Zolesio, J. P., "The Material Derivative (Or Speed) Method for Shape Optimization," *Optimization of Distributed Parameter Structures* (E. J. Haug and J. Cea, Ed.), Sijthoff & Noordhoff, Alphen aan den Rijn, Netherlands, 1981, pp. 1089–1151.
530. Zolesio, J. P., "Domain Variational Formulation of Free Boundary Problems," in *Optimization of Distributed Parameter Structures* (E. J. Haug and J. Cea, Ed.), Sijthoff & Noordhoff, Alphen aan den Rijn, Netherlands, 1981, pp. 1152–1194.
531. Życzkowski, M., and Krużelecki, J., "Optimal Design of Shells with Respect to Their Stability," in *Optimization in Structural Design* (Sawczuk and Mroz, Ed.), Springer-Verlag, New York, 1975, pp. 229–247.

Index